AF349974

Precision Nutrition

Precision Nutrition

Editors

Andreu Palou
Barbara Reynés

MDPI • Basel • Beijing • Wuhan • Barcelona • Belgrade • Manchester • Tokyo • Cluj • Tianjin

Editors
Andreu Palou
Universitat de les Illes Balears
Palma
Spain

Barbara Reynés
Universitat de les Illes Balears
Palma
Spain

Editorial Office
MDPI
St. Alban-Anlage 66
4052 Basel, Switzerland

This is a reprint of articles from the Special Issue published online in the open access journal *Nutrients* (ISSN 2072-6643) (available at: https://www.mdpi.com/journal/nutrients/special_issues/ precision_nutrition_update).

For citation purposes, cite each article independently as indicated on the article page online and as indicated below:

LastName, A.A.; LastName, B.B.; LastName, C.C. Article Title. *Journal Name* **Year**, *Volume Number*, Page Range.

ISBN 978-3-0365-7484-4 (Hbk)
ISBN 978-3-0365-7485-1 (PDF)

Contents

Preface to "Precision Nutrition" . . . . **vii**

Marina Camblor Murube, Elena Borregon-Rivilla, Gonzalo Colmenarejo,
Elena Aguilar-Aguilar, J Alfredo Martínez, Ana Ramírez De Molina, et al.
Polymorphism of CLOCK Gene rs3749474 as a Modulator of the Circadian Evening
Carbohydrate Intake Impact on Nutritional Status in an Adult Sample
Reprinted from: *Nutrients* **2020**, *12*, 1142, doi:10.3390/nu12041142 . . . . **1**

Yu-Jin Kwon, Jung Oh Kim, Jae-Min Park, Ja-Eun Choi, Da-Hyun Park, Youhyun Song, et al.
Identification of Genetic Factors Underlying the Association between Sodium Intake Habits and
Hypertension Risk
Reprinted from: *Nutrients* **2020**, *12*, 2580, doi:10.3390/nu12092580 . . . . **15**

Sebastià Galmés, Andreu Palou and Francisca Serra
Increased Risk of High Body Fat and Altered Lipid Metabolism Associated to Suboptimal
Consumption of Vitamin A Is Modulated by Genetic Variants rs5888 (*SCARB1*), rs1800629
(*UCP1*) and rs659366 (*UCP2*)
Reprinted from: *Nutrients* **2020**, *12*, 2588, doi:10.3390/nu12092588 . . . . **29**

Sandra Bayer, Vincent Winkler, Hans Hauner and Christina Holzapfel
Associations between Genotype–Diet Interactions and Weight Loss—A Systematic Review
Reprinted from: *Nutrients* **2020**, *12*, 2891, doi:10.3390/nu12092891 . . . . **49**

Przemyslaw Czajkowski, Edyta Adamska-Patruno, Witold Bauer, Joanna Fiedorczuk,
Urszula Krasowska, Monika Moroz, et al.
The Impact of *FTO* Genetic Variants on Obesity and Its Metabolic Consequences is Dependent
on Daily Macronutrient Intake
Reprinted from: *Nutrients* **2020**, *12*, 3255, doi:10.3390/nu12113255 . . . . **77**

Donghyun Jee, ShaoKai Huang, Suna Kang and Sunmin Park
Polygenetic-Risk Scores for A Glaucoma Risk Interact with Blood Pressure, Glucose Control,
and Carbohydrate Intake
Reprinted from: *Nutrients* **2020**, *12*, 3282, doi:10.3390/nu12113282 . . . . **103**

Francisco J. Rupérez, Gabriel Á. Martos-Moreno, David Chamoso-Sánchez, Coral Barbas
and Jesús Argente
Insulin Resistance in Obese Children: What Can Metabolomics and Adipokine
Modelling Contribute?
Reprinted from: *Nutrients* **2020**, *12*, 3310, doi:10.3390/nu12113310 . . . . **117**

María Luisa Maycotte-Cervantes, Adriana Aguilar-Galarza, Miriam Aracely Anaya-Loyola,
Ma. de Lourdes Anzures-Cortes, Lorenza Haddad-Talancón, Akram Sharim Méndez-Rangel,
et al.
Influence of Single Nucleotide Polymorphisms of ELOVL on Biomarkers of Metabolic
Alterations in the Mexican Population
Reprinted from: *Nutrients* **2020**, *12*, 3389, doi:10.3390/nu12113389 . . . . **137**

Julio Alvarez-Pitti, Ana de Blas and Empar Lurbe
Innovations in Infant Feeding: Future Challenges and Opportunities in Obesity and
Cardiometabolic Disease
Reprinted from: *Nutrients* **2020**, *12*, 3508, doi:10.3390/nu12113508 . . . . **155**

Miguel Seral-Cortes, Sergio Sabroso-Lasa, Pilar De Miguel-Etayo, Marcela Gonzalez-Gross, Eva Gesteiro, Cristina Molina-Hidalgo, et al.
Interaction Effect of the Mediterranean Diet and an Obesity Genetic Risk Score on Adiposity and Metabolic Syndrome in Adolescents: The HELENA Study
Reprinted from: *Nutrients* **2020**, *12*, 3841, doi:10.3390/nu12123841 . . . . **177**

Preface to "Precision Nutrition"

Obesity is a multifactorial and complex disease. The causes of obesity are really variated, diet, physical activity and, ultimately, lifestyle play a fundamental role. The relationship between these factors and obesity development are modulated by large number of varied individual parameters, such as the genome, epigenome, microbiome, metabolome, transcriptome and resulting interactomes. Emerging scientific knowledge has manifested the importance of considering most of these factors and individual parameters in developing dynamic and accurate nutritional recommendations. In this context, the term Precision Nutrition is key, a new approach to integrate internal and external environmental factors. This Special Issue aims to provide a thorough overview of research progress on Precision Nutrition, exploring new factors that could modulate the energy balance, adiposity and variety of obesity related alterations, for instance diabetes, cardiovascular disease, etc.

Precision nutrition is an emerging area of research due to its key role to maintain health status. In this way Alvarez-Pitti et al. review the recent research on Precision Nutrition and its role on prevention and treatment of obesity during pediatric years. In this article, the authors revise the nutrition-based interventions, included in the two levels of personalization proposed by Ordovas et al. in 2018. The first level of personalization is considering the behavioral and phenotype level, including eating habits and physical activity and lifestyle, during pregnancy, lactation period, childhood and adolescence. The second level of personalization includes the biological parameters, including the different response to food and/or nutrients that are conditioned by biochemical parameters, genetic, microbiota and metabolomics. All in all, reveal as tailored nutrition represents a promising approach to prevent and manage obesity.

Regarding the biological parameters, there are controversial data about the interaction of single nucleotide polymorphisms (SNPs) and macronutrients consumption on the extent of body weight loss. Bayer et al, review 27 articles based on the interaction between some SNPs and macronutrients on obesity treatment. Regarding the evaluated data the authors consider that there is not a clear information to support that genotype-diet interaction is a main determinant of obesity treatment success. On the other hand, Seral-Cortes et al. show as genotype modulate the relation between Mediterranean Diet adherence and the adiposity and metabolic syndrome in adolescents. Likewise, in addition to the effect of genotype on the relation between diet and metabolic parameters, Czajkowski et al. reveal as daily macronutrient intake may modulate the impact of some SNPs of FTO on obesity and obesity-related metabolic alterations. It is special interesting to develop efficient dietary strategies to prevent or treat obesity, controlling the percentage of each macronutrient for each genotype. Beyond the influence of the diet macronutrients proportion on the genotype-obesity relationship, Camblor et al. reveal as some polymorphism determine the effect of carbohydrates proportion on body mass index, depending the time of its consumption. Jee et al. show as macronutrient proportion also modulate the relationship between genotype and glaucoma. Elevated carbohydrate consumption increases the risk of glaucoma by three-fold in adults with a high polygenic risk score. Additionally, beyond the environment influence Maycotte-Cervantes et al. evidences that clinical markers of chronic non-communicable diseases in Mexican population are influenced by specific SNPs of ELOVL genes, in a sex dependent manner.

Beyond the macronutrient proportion other nutrients or food compounds could modulate the relationship between some metabolic alteration and genotype. Kwon et al. indicate as the genetic susceptibility to hypertension differ according to sodium intake. This study reveals that it is possible to make more individualized nutritional recommendations for cardiovascular disease prevention.

Vitamin A consumption has a relevant impact on metabolism and energy balance. However, Galmés et al. give away as the Vitamin A effects are modulated by the influence of some genetic variants, such as SCARB1, UCP2 and UCP1. Moreover, this study shows that the combination of genetic information and an ex vivo intervention in peripheral blood cells, to evaluate the gen-nutrient-phenotype interaction, seems to be an efficient strategy for the Precision Nutrition.

Further the studies evaluating the importance of genetics on Precision Nutrition, Ruperez et al. comment in their review as metabolomics can provide valuable information which can be applied to the pursuance of Precision Nutrition.

All in all, the articles of this special issue point out the key role of Precision Nutrition on the prevention and treatment of obesity and related alterations.

Acknowledgment: We would like to thank all the Authors and Reviewers for their contributions to this collection of articles.

Andreu Palou and Barbara Reynés
Editors

 *nutrients*

Article

Polymorphism of CLOCK Gene rs3749474 as a Modulator of the Circadian Evening Carbohydrate Intake Impact on Nutritional Status in an Adult Sample

Marina Camblor Murube [1], Elena Borregon-Rivilla [2], Gonzalo Colmenarejo [3], Elena Aguilar-Aguilar [2], J. Alfredo Martínez [2,4,5], Ana Ramírez De Molina [6], Guillermo Reglero [7,8] and Viviana Loria-Kohen [2,*]

[1] Departamento de Farmacología, Facultad de Medicina, Universidad Complutense de Madrid, 28040 Madrid, Spain; mcambl01@ucm.es
[2] Nutrition and Clinical Trials Unit, GENYAL Platform IMDEA-Food Institute, CEI UAM + CSIC, 28049 Madrid, Spain; elena.borregon@imdea.org (E.B.-R.); elena.aguilar@imdea.org (E.A.-A.); jalfredo.martinez@imdea.org (J.A.M.)
[3] Biostatistics and Bioinformatics Unit, IMDEA Food Institute, CEI UAM + CSIC, 28049 Madrid, Spain; gonzalo.colmenarejo@imdea.org
[4] Department of Nutrition, Food Science and Physiology, Center for Nutrition Research (CIN), Navarra Institute for Health Research (IdiSNA), 31008 Pamplona, Spain
[5] Center of Biomedical Research in Physiopathology of Obesity and Nutrition (CIBEROBN), Institute of Health Carlos III, 28029 Madrid, Spain
[6] Molecular Oncology and Nutritional Genomics of Cancer, IMDEA-Food Institute, CEI UAM + CSIC, 28049 Madrid, Spain; ana.ramirez@imdea.org
[7] Department of Production and Characterization of Novel Foods, Institute of Food Science Research (CIAL) CEI UAM+CSIC, 28049 Madrid, Spain; guillermo.reglero@imdea.org
[8] Production and Development of Foods for Health, IMDEA-Food Institute, CEI UAM + CSIC, 28049 Madrid, Spain
[*] Correspondence: viviana.loria@imdea.org; Tel.: +34-912796957; Fax: +34-911880756

Received: 31 March 2020; Accepted: 16 April 2020; Published: 19 April 2020

Abstract: The aim of this study was to evaluate the distribution of energy intake and macronutrients consumption throughout the day, and how its effect on nutritional status can be modulated by the presence of the rs3749474 polymorphism of the CLOCK gene in the Cantoblanco Platform for Nutritional Genomics ("GENYAL Platform"). This cross-sectional study was carried out on 898 volunteers between 18 and 69 years old (65.5% women). Anthropometric measurements, social issues and health, dietary, biochemical, genetic, and physical activity data were collected. Subsequently, 21 statistical interaction models were designed to predict the body mass index (BMI) considering seven dietary variables analyzed by three genetic models (adjusted by age, sex, and physical activity). The average BMI was 26.9 ± 4.65 kg/m^2, 62.14% presented an excess weight (BMI > 25 kg/m^2). A significant interaction was observed between the presence of the rs3749474 polymorphism and the evening carbohydrate intake (% of the total daily energy intake [%TEI]) (adjusted $p = 0.046$), when predicting the BMI. Participants carrying TT/CT genotype showed a positive association between the evening carbohydrate intake (%TEI) and BMI ($\beta = 0.3379$, 95% CI = (0.1689,0.5080)) and ($\beta = 0.1529$, 95% CI = (−0.0164,0.3227)), respectively, whereas the wild type allele (CC) showed a negative association ($\beta = −0.0321$, 95% CI = (−0.1505,0.0862)). No significant interaction with the remaining model variables was identified. New dietary strategies may be implemented to schedule the circadian distribution of macronutrients according to the genotype. Clinical Trial number: NCT04067921.

Keywords: dietary parameters; carbohydrate intake; obesity; single nucleotide polymorphism; CLOCK gene; rs3749474

1. Introduction

Overweight and obesity rates are increasing alarmingly, with excess weight being one of today's major health issues. Both conditions are characterized by an excessive accumulation of fat in the adipose tissue, hypertrophying and unbalancing the homeostatic processes in which it is involved [1]. This situation increases the risk of prematurely suffering diseases and pathological conditions, such as hypertension, diabetes, cardiovascular events, osteopathy, and cancer; being one of the main causes of population's morbidity and mortality [2–4].

As for the prevalence worldwide, it is estimated that approximately 30% suffer from obesity and 35% show a body mass index (BMI) in overweight ranges [3]. In Spain, according to the National Statistics Institute, up to 62.5% of men and 46.7% of women have an excess body weight (BMI > 25 kg/m^2) [5].

Obesity treatment strategies traditionally focus on weight loss by combining dietary and physical activity recommendations as well as pharmacological and surgical therapeutic approaches [6]. Although many dietary alternatives are available for weight control, such as low-carbohydrate diet and low-fat diet, it is not known which is the most effective [7]. In recent years, a more detailed study of dietary strategy has been promoted, not only based on the amount of kilocalories or macronutrients provided, but also on their circadian distribution throughout the day. Among the various studies, some show a benefit of carbohydrate consumption mostly displaced towards late-night hours during dinner [8–10]. Others, however, support that the intake of this macronutrient in breakfast or lunch has a metabolic benefit as well as on satiety [11,12]. Indeed, the effectiveness of the distribution of calories and macronutrients over the day is controversial. While some of these discrepancies may be explained by methodological differences in the studies, they may also be due to genetic modulation of the response. In this context, single nucleotide polymorphisms (SNPs) constitute the majority of human genomic variation among individuals. These variations in DNA sequence can affect the response of individuals to drugs, viruses, bacteria, and also to diet [13,14].

Recent research has linked the presence of certain polymorphisms in genes such as *CLOCK* (Circadian Locomotor Output Cycles Kaput) to different responses to caloric and macronutrient intake. Regarding *CLOCK* rs3749474, it has been shown that both TT+CT genotypes tended to have a better weight loss after a treatment involving dietary fat restriction than those homozygous for the wild type allele (CC) [15]. However, the evidence about the mode this SNP could influence the response according to the distribution of different macronutrients throughout the day is unclear [16].

On this basis, the aim of this study was to evaluate the distribution of energy intake and macronutrients throughout the day, and how its effect on nutritional status can be modulated by the presence of the *CLOCK* SNP rs3749474 in the sample of the Cantoblanco Platform for Nutritional Genomics (GENYAL Platform). The final objective was to figure out some of the controversies found in the dietary recommendations and describe new alternatives for a personalized nutritional treatment for obesity.

2. Materials and Methods

2.1. Study Sample and Design

This observational and cross-sectional study was carried out on volunteers belonging to the Cantoblanco Platform for Nutritional Genomics (GENYAL Platform) recruited during the period 2012–2017 in Spain. This platform is a tool at the IMDEA-Food Institute for the study of genome-nutrient interactions on a large scale.

Recruitment was developed through media and included volunteers (men and women), aged between 18 and 69 years of age, who did not suffer from any serious diseases, were pregnant or lactating, who willingly signed the consent participation form.

The study was approved by the Research Ethics Committee of the Autonomous University of Madrid (CEI 27-666) and was adapted to the ethical bases proposed by the Declaration of Helsinki with regard to scientific research studies and valid laws.

2.2. Personal, Social, and Health Data

The information that was collected included date of birth, sex, ethnicity, marital status, level of education, and employment status, as well as health information, such as smoking, alcohol consumption, minor illnesses, medication, and family history of disease.

2.3. Anthropometric Parameters

Anthropometric measurements were performed under standardized procedures. Height (cm) was assessed to the nearest 0.1 cm using a stadiometer (Leicester Biológica Tecnología Médica SL, Barcelona, Spain). Weight (kg), fat mass (%), and muscle mass (%) were evaluated using bioelectrical impedance analysis (Body Composition Monitor BF511-OMRON HEALTHCARE, LT, Kyoto, Japan). Based on these data, the BMI was calculated according to the Quetelet Index ($\frac{Weight\ (kg)}{Height\ (m)^2}$). The World Health Organization's criteria (WHO) [17] was used to classify the subjects and were regrouped as normal weight (NW) when BMI < 25 kg/m^2 and as excess weight (EW) when BMI ≥ 25 kg/m^2. Waist and hip circumference were measured with a flexible Dry 201 metal tape, with measuring range 0–150 cm and 1 mm of precision (Quirumed, Valencia, Spain).

2.4. Dietary Parameters

Regarding the study of the dietary intake, volunteers completed a validated three-day food record prior to the visit (two weekdays and another weekend day) in which they wrote all the food and beverages consumed, as well as the exact or estimated weight in measurements at home (for which they were instructed by trained nutritionists). Subsequently, the extracted data were tabulated and analyzed using the DIAL nutritional software (version 3.7.1.0 (February 2019)-Alce Ingeniería, Madrid, Spain) as described by the supplier [18].

Based on the Spanish food pattern, three main meals (breakfast, lunch, and dinner) and three in-between meals eating occasions (morning, evening, and after dinner snack) were considered. The intake at mid-morning, mid-afternoon, and late-evening occasions in this population is usually a light and easily prepared meal, such as a sugar-sweetened infusion (tea, coffee …), a piece of fruit, a dairy product, cookies and similar products, or reasonable combinations of them.

Multiple dietary variables were considered in the exploratory analyses. The Total Daily Energy Intake (TEI), the Meal Energy Intake (MEI), and the percentages of macronutrients (carbohydrates (CH), proteins (Pr), and lipids (Lip)) of TEI (CH (%TEI), Pr (%TEI), Lip (%TEI), as well as of each meal (CH (%MEI), Pr (%MEI), Lip (%MEI)) were automatically calculated by the DIAL software.

In addition, to account the timing distribution of energy and the macronutrients of the TEI, the following variables were created:

Meal Energy Intake (MEI) as a percentage of TEI:

$$\text{MEI (\%TEI)} = \frac{\text{MEI }(kJ)}{\text{TEI}\left(\frac{kJ}{day}\right)} \times 100.$$

Macronutrients (CH, Pr, Lip) of each MEI over TEI:

$$Meal\ \text{CH (\%TEI)} = \frac{\text{CH } of\ \text{MEI }(kJ)}{\text{TEI}\left(\frac{kJ}{day}\right)} \times 100\ ;\ Meal\ \text{Pr (\%TEI)} = \frac{\text{Pr } of\ \text{MEI }(kJ)}{\text{TEI}\left(\frac{kJ}{day}\right)} \times 100;$$

$$Meal\ \text{Lip (\%TEI)} = \frac{\text{Lip } of\ \text{MEI }(kJ)}{\text{TEI}\left(\frac{kJ}{day}\right)} \times 100.$$

In order to investigate if a higher intake of energy and macronutrient at the beginning or the end of the day may influence the nutritional status, the meals were regrouped into two sub-groups according to the time-of-day: morning intakes (MI) (including breakfast and morning snack) and evening intakes (EI) (including evening snack, dinner, and after dinner snack).

An initial inspection of the data showed a discordance between the information provided by some subjects in the dietary questionnaire and their anthropometric parameters. Thus, some with high BMI, body fat, and high levels of blood pressure, reported lower intake in terms of total calories, as well as healthier and more balanced nutritional profiles. This finding suggested that some of these individuals could be underreporting their dietary data. To overcome this limitation, the method proposed by the European Food Safety Authority (EFSA) [19] and Goldberg and Black [20,21] was followed. This method assumes that, if weight remains stable, energy intake should be equal to energy expenditure. According to this procedure, the volunteers were classified as underreporting, overreporting and correct reporting, creating a final classification of two groups, namely misreporters and plausible reporters. The exploratory analysis of the dietary and intake data was performed with both the total sample and the plausible reporters. However, the statistical models (see below) were only derived for the plausible reporters.

2.5. Physical Activity Parameters

The short version of the International Physical Activity Questionnaire (IPAQ) validated in the Spanish population [22] was used. According to the results, the pattern of activity was classified into three categories: low, moderate, and high, according to the authors proposal [22].

2.6. Biochemical Parameters and Genotyping

A blood sample was collected by venipuncture in the middle cubital vein of the forearm early in the morning, with the subject fasting for at least the previous 12 h.

The CQS Laboratory, following standardized procedures, carried out the analysis of the biochemical determinations. The lipid profile [(triglycerides, total cholesterol, low-density lipoprotein cholesterol (LDL-cholesterol), and high-density lipoprotein cholesterol (HDL-cholesterol)] were determined by enzymatic spectrophotometry. The glycemic profile [glycemia (c. hexokinase mass) and insulin (immunoassay)] and the Homeostasis Model Assessment (HOMA) was calculated regarding the following formula:

$$\text{HOMA} = \frac{Glucose\left(\frac{mg}{dL}\right) \times Insulin\left(\frac{\mu U}{mL}\right)}{405}.$$

The presence of the rs3749474 *CLOCK* polymorphism was determined at the IMDEA-Food Nutritional Genomics Laboratory from these blood samples. This SNP was selected based on its recognized involvement in different parts of the pathogenic processes of obesity and its related phenotypes, and because it has been associated with different genotype response to weight loss in a previous study [15]. Preserved at $-80\ °C$. DNA extraction was performed using the QIAamp DNA Mini Kit, QIAGEN, following the protocol of the commercial company, in which 300 µL of sample was used to obtain 100 µL of DNA, establishing the optimal cut points for DNA concentration at 50 ng/µL and 1.7 of quality (absorbance ratio A260/A280 and A260/A230). Once the DNA had been extracted, genotyping was performed using TaqMan®® SNPs probes with the QuantStudio™ 12K Flex Real-Time PCR System and the AccuFill™ system. Data analysis was performed using TaqMan Genotyper Software v1.3. A quality value of each genotyping of more than 90% was used.

2.7. Statistical Analysis

Categorical data were presented as percentages and absolute frequencies, while quantitative data were expressed as mean ± standard deviation (SD). Linear regression models adjusted by sex and age were used to assess associations between anthropometric, dietary, and physical activity variables. In order to test the interaction between the rs3749474 SNP and nutritional variables in the prediction of BMI, the following seven evening variables were considered: Evening Meal Energy, Carbohydrates, Proteins, and Lipids Intake of the Total Daily Energy Intake [Eve MEI (%TEI), Eve CH (%TEI), Eve Pr (%TEI) and Eve Lip (%TEI)], and Evening Carbohydrates, Proteins, and Lipids Intake of the Meal Energy Intake [Eve CH (%MEI), Eve Pr (%MEI), and Eve Lip (%MEI)]. In addition, three genetic models (additive (ADD), co-dominant (COD), and dominant (DOM)) were considered, giving a total of 21 models, all of them adjusted by sex, age, and physical activity (expressed in metabolic equivalents (METs)). It was finally determined that the ADD model (in which the risk conferred by an allele is increased 2-fold for homozygotes) was the best fit for the data. *p* Values were corrected with Bonferroni's method. All statistical tests were considered as bilateral with a significance level of 0.05. Estimated parameters (Betas) were obtained with 95% confidence intervals. Statistical analyses were performed using R version 3.4 (projects) (www.r-project.org).

Sample size calculations were performed with G*Power 3.1.9.2. for a multiple regression model of 6 predictor variables. For an effect size of 0.01 for adding a new single variable, with a power of 0.8 and a significance level of 0.05, a sample size of 787 was obtained. Assuming a drop-out of 15%, this gave a sample size of 905 individuals.

3. Results

The sample was composed of 898 subjects, of which 65.5% were women and 34.5% were men. The mean age of the sample was 41 ± 12 years and the BMI 26.9 ± 4.65 kg/m^2. About 62.14% of the volunteers presented excess weight (EW, BMI > 25 kg/m^2). Anthropometric, biochemical, physical activity, as well as dietary data of the participants according to their nutritional status is provided in the Appendix A section. A total of 799 subjects had dietary data available and 697 were considered according to Goldberg's method. Finally, 84% were classified as plausible reporters ($n = 585$) and 16% as misreporters ($n = 112$).

When the BMI was regressed on the Total Daily Energy Intake (TEI), a positive and significant association was observed ($\beta = 0.00132$, 95% CI = (0.000505,0.00213); $p = 0.038$). On the other hand, the percentages of macronutrient (CH, Pr, and Lip) as a percentage of TEI showed no significant associations with the BMI of the total participants ($p > 0.05$).

In terms of energy and macronutrient intake per meal (breakfast, morning snack, lunch, evening snack, dinner, and after dinner snack, morning and evening) no significant association with BMI was identified. Nevertheless, morning carbohydrate intake (%TEI) was associated with different markers of the glycemic profile: glycemia ($\beta = -0.419$, 95% CI = (-0.628, -0.21); $p = 0.001$), HOMA ($\beta = -0.0443$, 95% CI = (-0.0707, -0.0178); $p = 0.014$) and insulin ($\beta = -0.181$, 95% CI = (-0.288,-0.0743); $p = 0.012$) and lipid: LDL ($\beta = -1.09$, 95% CI = (-1.69,-0.488); $p = 0.005$) and TG ($\beta = -0.813$, 95% CI = (-1.32,-0.307); $p = 0.022$).

Table 1 describes the main anthropometric and biochemical data of the sample split by genotypes of the rs3749474 polymorphism, as well as the *p* value of a linear model of the SNP as predictor adjusted by sex and age. Table 2 includes dietary data for the total and the time-of-day intakes (breakfast, morning snack, lunch, evening snack, dinner, and after dinner snack, morning and evening intake) in plausible reporters by genotypes. The *p* value of a linear model of the SNP as predictor adjusted by sex and age is also included.

Table 1. Anthropometric and biochemical data by genotypes of the rs3749474.

	Genotype			
Variable	**CC** (*n* = 271)	**CT** (*n* = 283)	**TT** (*n* = 70)	***p* Value**
BMI (kg/m^2)	28.00 (27.44,28.56) [1]	27.41 (26.86,27.96)	28.84 (27.76,29.92)	0.048 *
Total Fat Mass (%)	33.72 (32.81,34.62)	32.93 (32.04,33.82)	34.87 (33.13,36.62)	0.115
WC (cm)	92.59 (91.02,94.15)	91.29 (89.75,92.84)	96.21 (93.18,99.25)	0.016 *
SBP (mm Hg)	124.70 (123.01,126.39)	124.78 (123.11,126.45)	125.62 (122.34,128.91)	0.880
DBP (mm Hg)	77.40 (76.25,78.55)	77.55 (76.41,78.69)	78.84 (76.6,81.07)	0.516
HDL (mg/dL)	51.73 (50.21,53.26)	52.09 (50.59,53.59)	53.37 (50.41,56.33)	0.619
LDL (mg/dL)	127.76 (124.02,131.49)	130.35 (126.66,134.04)	131.09 (123.86,138.32)	0.531
TG (mg/dL)	103.46 (95.97,110.95)	104.23 (96.88,111.58)	115.31 (100.84,129.79)	0.332
LDL/HDL	2.60 (2.5,2.71)	2.64 (2.54,2.74)	2.62 (2.42,2.82)	0.877
TC/HDL	4.08 (3.95,4.21)	4.13 (4.01,4.26)	4.11 (3.87,4.36)	0.818
Log TG/HDL	0.25 (0.22,0.29)	0.25 (0.22,0.28)	0.30 (0.24,0.36)	0.370
Glucose (mg/dL)	87.19 (85.72,88.67)	85.63 (84.17,87.1)	88.22 (85.53,90.92)	0.144
HOMA	2.03 (1.85,2.21)	1.79 (1.61,1.98)	2.08 (1.74,2.43)	0.122
Insulin (μUI/mL)	9.07 (8.38,9.76)	8.34 (7.65,9.03)	9.53 (8.23,10.82)	0.162

Hardy–Weingber equilibrium p = 0.8157. [1] Mean (95% CI). * p < 0.05 is considered as statistically significant. CC: Common Homozygous; CT: Heterozygous; DBP: Diastolic Blood Pressure; HOMA: Homeostatic Model Assessment; SBP: Systolic Blood Pressure; TC: Total Cholesterol; TT: Variant Homozygous; WC: Waist Circumference.

In order to better understand the genetic dependence of the effect of the distribution of meals in the nutritional status, 21 different lineal models were developed to predict BMI. Each of them had an interaction term between the SNP and one of the following 7 evening variables: Evening Meal Energy, Carbohydrates, Proteins, and Lipids Intake, of the Total Daily Energy Intake [Eve MEI (%TEI), Eve CH (%TEI), Eve Pr (%TEI), and Eve Lip (%TEI)], and Evening Carbohydrates, Proteins, and Lipids Intake of the Meal Energy Intake [Eve CH (%MEI), Eve Pr (%MEI) and Eve Lip (%MEI)]. In addition, for each evening variable, three different genetic models for the SNP were tested: additive (ADD), co-dominant (COD), and dominant (DOM)). All the models were adjusted by sex, age, and physical activity (expressed in metabolic equivalents (METs).

After multiple-test correction by Bonferroni method, a significant interaction was observed between the presence of the rs3749474 polymorphism in the additive model and the evening carbohydrate intake (p = 0.046) when predicting the BMI (Figure 1). The heterozygous subjects (CT) showed that for every 1% increase in carbohydrate intake of the total energy intake during evening hours, the BMI increased by 0.1529 kg/m^2 (β = 0.1529, 95% CI = −0.0164,0.3227)) and the homozygotes for the risk allele (TT) showed an even greater variation, increasing the BMI by 0.3379 kg/m^2 for each 1% increase in carbohydrate consumption during this time (β = 0.3379, 95% CI = (0.1689,0.5080)). However, the common homozygous (CC) did not experience statistic increases in the BMI, even showing a slight decrease in it (−0.0321 kg/m^2), (β = −0.0321, 95% CI = (−0.1505,0.0862)). Moreover, no significant interaction with the remaining model variables was identified.

Table 2. Dietary data for the total and the time-of-day intakes by genotype of the rs3749474 (plausible reporters).

	Genotype			
	CC (*n* = 182)	**CT** (*n* = 196)	**TT** (*n* = 49)	
Dietary Data	**X ± SD**	**X ± SD**	**X ± SD**	***p* Value**
Total dietary data				
TEI (kJ/day)	9450 ± 1948	9215 ± 2024	9729 ± 2126	0.159
CH (%TEI)	37.97 ± 6.51	38.44 ± 6.03	36.91 ± 6.67	0.305
Pr (%TEI)	17.59 ± 2.91	17.57 ± 2.81	17.09 ± 2.86	0.717
Lip (%TEI)	40.38 ± 6.00	39.64 ± 5.92	41.05 ± 6.26	0.250
Dietary data: Breakfast				
B MEI (%TEI)	16.71 ± 6.54	16.66 ± 6.12	16.60 ± 8.09	0.994
B CH (%TEI)	8.89 ± 3.77	8.86 ± 3.44	8.23 ± 4.55	0.392
B Pr (%TEI)	2.27 ± 1.14	2.27 ± 1.01	2.33 ± 1.25	0.882
B Lip (%TEI)	5.12 ± 2.95	5.1 ± 3.02	5.61 ± 3.08	0.395
Dietary data: Morning snack				
MS MEI (%TEI)	6.38 ± 5.69	5.94 ± 4.67	7.44 ± 6.07	0.437
MS CH (%TEI)	3.09 ± 2.90	3.01 ± 2.42	3.36 ± 2.70	0.634
MS Pr (%TEI)	0.85 ± 0.97	0.79 ± 0.82	1.09 ± 1.11	0.478
MS Lip (%TEI)	2.08 ± 2.41	1.76 ± 1.98	2.46 ± 2.40	0.320
Dietary data: Lunch				
L MEI (%TEI)	39.77 ± 8.61	40.08 ± 8.88	39.34 ± 9.32	0.745
L CH (%TEI)	13.31 ± 4.49	13.50 ± 4.55	12.62 ± 4.36	0.427
L Pr (%TEI)	7.91 ± 2.24	7.92 ± 2.17	7.37 ± 2.26	0.186
L Lip (%TEI)	16.96 ± 5.62	16.93 ± 5.79	17.32 ± 5.28	0.818
Dietary data: Evening snack				
ES MEI (%TEI)	7.06 ± 6.10	6.60 ± 5.59	6.70 ± 5.07	0.798
ES CH (%TEI)	3.25 ± 2.90	3.17 ± 2.78	3.20 ± 2.37	0.821
ES Pr (%TEI)	0.95 ± 0.98	0.90 ± 0.96	0.87 ± 0.87	0.747
ES Lip (%TEI)	2.58 ± 2.77	2.23 ± 2.51	2.28 ± 2.35	0.512
Dietary Data: Dinner				
D MEI (%TEI)	29.25 ± 8.55	29.59 ± 9.86	29.14 ± 8.39	0.918
D CH (%TEI)	9.14 ± 3.85	9.43 ± 4.11	9.15 ± 3.86	0.888
D Pr (%TEI)	5.45 ± 1.89	5.55 ± 2.00	5.34 ± 2.25	0.660
D Lip (%TEI)	13.31 ± 4.99	13.21 ± 5.48	13.12 ± 4.59	0.967
Dietary data: After dinner snack				
ADS MEI (%TEI)	0.85 ± 2.66	1.14 ± 3.26	0.78 ± 2.74	0.407
ADS CH (%TEI)	0.29 ± 0.87	0.48 ± 1.57	0.35 ± 1.29	0.412
ADS Pr (%TEI)	0.15 ± 0.73	0.13 ± 0.42	0.09 ± 0.33	0.462
ADS Lip (%TEI)	0.33 ± 1.14	0.42 ± 1.51	0.26 ± 0.94	0.565
Grouped dietary data				
Dietary data: Morning				
Morn MEI (%TEI)	23.08 ± 7.45	22.59 ± 7.06	24.04 ± 9.04	0.465
Morn CH (%TEI)	11.98 ± 4.19	11.87 ± 4.02	11.59 ± 4.80	0.884
Morn Pr (%TEI)	3.12 ± 1.44	3.06 ± 1.22	3.43 ± 1.58	0.207
Morn Lip (%TEI)	7.20 ± 3.36	6.86 ± 3.42	8.07 ± 3.69	0.082
Dietary data: Evening				
Eve MEI (%TEI)	37.15 ± 9.18	37.33 ± 10.26	36.62 ± 9.86	0.901
Eve CH (%TEI)	12.68 ± 4.78	13.07 ± 4.58	12.70 ± 4.97	0.678
Eve Pr (%TEI)	6.56 ± 2.07	6.59 ± 2.25	6.30 ± 2.51	0.552
Eve Lip (%TEI)	16.21 ± 5.34	15.85 ± 5.80	15.66 ± 5.09	0.742

ADS: After Dinner Snack; B: Breakfast; CC: Common Homozygous; CH: Carbohydrates; CT: Heterozygous; D: Dinner; ES: Evening Snack; Eve: Evening; kJ: Kilojoules; L: Lunch; Lip: Lipids; MEI: Meal Energy Intake; Morn: Morning; MS: Morning Snack; Pr: Proteins; SD: Standard Deviation; TEI: Total Energy Intake; TT: Variant Homozygous; X: Mean.

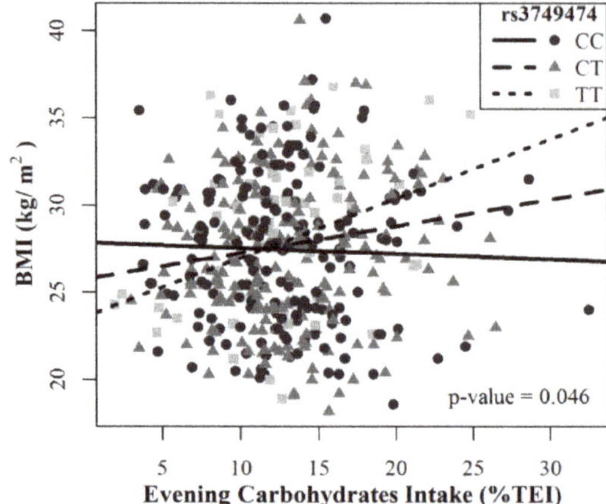

Figure 1. Association between the evening carbohydrates intake as a percentage of Total Daily Energy Intake (TEI) and the body mass index (BMI) according to the presence of the rs3749474. Participants carrying TT and CT genotype showed a positive association between the evening carbohydrate intake (%TEI) and the BMI (β = 0.3379, 95% CI = (0.1689,0.5080)) and (β = 0.1529, 95% CI = (−0.0164,0.3227)) respectively, whereas the common homozygous (CC) showed a negative association (β = −0.0321, 95% CI = (−0.1505,0.0862)) (adjusted p = 0.046).

4. Discussion

The relationship between the food intake at different times of the day and the nutritional status in a group of volunteers belonging to the GENYAL Platform sample was studied in the present research. A significant interaction between evening intake of carbohydrates of the volunteers and the rs3747494 *CLOCK* polymorphism when predicting BMI was identified, suggesting new alternatives of personalized nutritional treatment for obesity.

In this research, excess weight people had defective lipid and glycemic profiles in relation to normal weight people. Furthermore, this group showed lower levels of physical activity (these results are provided in the Appendix A section). These features of excess weight people have been well described in the scientific literature [4,23,24] and have been associated with an increase of concomitant diseases, such as type II diabetes mellitus, cancer, etc. [2,11,25], and thus, emphasize the importance of managing weight control in order to avoid secondary clinical complications.

The total daily energy intake has been postulated as a regulator of the nutritional status, especially if it is not combined with physical exercise according to the energy intake [26,27]. This positive association between the energy intake and BMI was observed, but only after considering the plausible reporters. This highlights the importance of verifying the reliability of dietary intake data using the proposed methods by Goldberg and Black in order to minimize the bias generated by participants reporting [28].

Furthermore, we observed that morning carbohydrate intake was associated with better lipid and glycemic profiles. This finding corroborates previous studies [11,12] that support better nutritional and health status when a higher proportion of carbohydrates is made in the first half of the day. These studies suggest that a higher intake in the morning hours translates into a lower overall intake compared to main meals eaten later. It has also been suggested that this behavior could be influenced by the levels of ghrelin, a hormone that is involved in controlling appetite, since elevated levels of ghrelin were found during the early part of the night, decreasing in the morning before awakening [29]. In addition, according to these studies, insulin sensitivity and glucose tolerance decrease progressively throughout the day [11] with a negative effect on weight management.

Conversely, other authors reported better results in weight control when the highest intake of carbohydrates is made in late hours [8–10]; however, this was not observed in this study. The authors who support late high carbohydrates intake postulate a benefit in maintaining the levels of leptin with a consequent effect on satiety. Moreover, they argue that the consumption of nocturnal carbohydrates is related to higher levels of adiponectin, which results in better insulin sensitivity and a better inflammatory profile [9].

Given the controversy about the association between carbohydrate consumption during the evening and the nutritional status, it was planned to investigate this association by considering the SNP rs3749474 of the *CLOCK* gene as a modulator. A significant interaction was observed between the intake of carbohydrates displaced at evening hours (evening snack, dinner, and after dinner snack) and the SNP genotype when predicting the BMI. The higher intake of carbohydrates in the evening was associated with a higher BMI in volunteers with one or more risk alleles, a situation not observed in wild type. The *CLOCK* gene is a regulator of circadian rhythms that modulates the expression of PPAR (Peroxisome Proliferator-Activated Receptor), which corresponds to a family of transcription factors involved in cellular lipid metabolism (lipolysis and lipogenesis), whose activity is regulated following circadian cycles [30,31]. The presence of the polymorphism of the *CLOCK* rs3749474 gene implies the exchange of a cytosine (C) for a thymine (T) on the 3'-non-coding region of the gene. This situation affects the folding and stability of its mRNA, compromising its functionality and the factors it regulates, such as PPAR [15,32] so that the processes of lipogenesis and lipolysis are altered and lose effectiveness, influencing fat deposition.

Several studies have investigated the relationship between fat-rich diets and the deregulation of circadian rhythms [31,33]. It has been observed that the presence of alterations in the *CLOCK* gene implies a modification in the normal patterns of secretion of neuropeptides involved in the appetite/satiety pathways. Consequently, the presence of the polymorphism rs3749474 has been associated with higher energy intake [34]. However, the influence that this polymorphism can have in the context of a higher consumption of carbohydrates in late hours has been less studied, as is shown by the interaction found in this work.

The metabolism of carbohydrates and lipids is strongly linked in a context of the overabundance of glucose (e.g., high-carbohydrate diet), excess of glucose is metabolized to lipids for storage [35]. In this way, a high-calorie diet based on carbohydrates can lead to an increase in total body fat. In addition, the presence of polymorphism in the *CLOCK* rs3749474 gene has been associated with higher levels of orexigenic hormones, such as ghrelin, and lower levels of anorexigenic hormones, such as leptin, leading to physiological disorders in circadian patterns [33]. Under normal conditions, which one release is increased during the late hours. Nevertheless, this polymorphism influences maintaining low levels of leptin during this period inducing greater appetite. Deregulation of insulin secretion has also been observed in previous studies when *CLOCK* is altered [31].

We hypothesized then that all these altered mechanisms described in previous studies could contribute to a poor energy management in the presence of the SNP and explain in part the negative effect on the nutritional status. The evening energy and carbohydrate intake showed no difference due to genotype. These data suggest that the effect on BMI may be more related to the metabolic disorders associated with the presence of the SNP than to a greater appetite.

Given that the volunteers carrying TT alleles presented a worse nutritional status, we suggest that controlling the amount of carbohydrates consumed in the evening could be an approach for personalizing nutrition strategies in this group. Individuals with CT and TT genotypes would be recommended carbohydrate intakes mostly at morning hours in order to avoid the negative effect on the nutritional state. This finding highlights the interest of studies on gene-diet interaction, to contribute to dietary interventions aimed at the normalization of nutritional status.

This work is an observational study based on self-reported dietary surveys; therefore, this data is subject to under/overreporting. In order to overcome this limitation, people whose weight stability matched their reported energy intake were included and described as 'plausible reporters'.

More research is needed to confirm these findings, through prospective studies that include different interventions after genotype and compare different dietary approaches to the carbohydrate distribution throughout the day, taking into account the presence of this SNP.

5. Conclusions

Evening carbohydrate intake in the presence of the rs3747494 polymorphism in homozygosis or heterozygosis is associated with a higher BMI. In accordance with this evidence, the response to late-time carbohydrate consumption should consider a sufficient knowledge of genetic factors. It seems reasonable to hypothesize that the controversy regarding the correct pattern of intake could be directly related to the genotype studied, since previous studies have not considered the influence of SNPs. As a new nutritional intervention tool, dietary recommendations for SNP CLOCK carriers should be aimed at distributing carbohydrates mainly at morning hours for a personalized circadian nutrition.

Author Contributions: The authors' contributions were as follows: Conceptualization, V.L.-K.; methodology, V.L.-K. and G.C.; formal analysis, G.C.; investigation, E.B.-R. and E.A.-A.; writing—original draft preparation, M.C.M.; writing—review and editing, V.L.-K.; E.A.-A.; J.A.M.; A.R.D.M. and G.R.; visualization, M.C.M. and E.A.-A.; supervision, V.L.-K.; funding acquisition, A.R.D.M. and G.R. All authors have read and agreed to the published version of the manuscript.

Funding: This research received no external funding.

Acknowledgments: The authors would like to thank the participants of the Cantoblanco Platform for Nutritional Genomics ("GENYAL Platform") and the contribution of the rest of the members of the Nutrition and Clinical Trials Unit and Genomic Laboratory of IMDEA-Food who contributed to its successful completion.

Conflicts of Interest: The authors declare no conflict of interest.

Appendix A

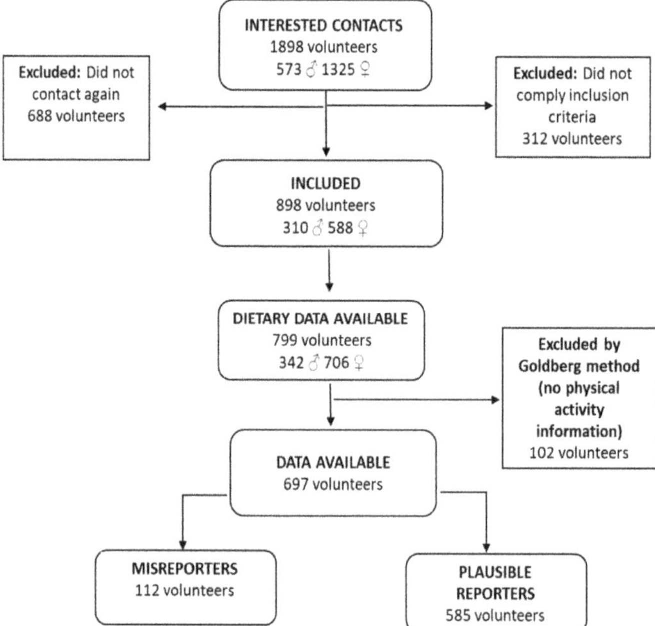

Figure A1. Study's flow chart.

Table A1. Anthropometric and biochemical data distributed according to nutritional state (total sample).

		Total Sample		
		NW [1]		EW
Anthropometry	*N*	X ± SD [2]	*N*	X ± SD
BMI (kg/m^2)	340	22.30 ± 1.78	558	29.91 ± 3.31 †
Total Fat Mass (%)	339	27.19 ± 8.05	555	38.83 ± 8.49 †
Total Muscular Mass (%)	336	31.83 ± 6.10	555	27.16 ± 4.91 †
Visceral Fat (%)	339	4.73 ± 1.68	555	10.34 ± 3.62 †
BME (kJ)	329	5853 ± 744	550	6799 ± 1020 †
Waist Circumference (cm)	324	76.77 ± 7.02	551	97.10 ± 11.35 †
Hip Circumference (cm)	232	97.54 ± 5.19	389	112.67 ± 8.16 †
Waist-Hip Ratio	232	0.79 ± 0.07	389	0.87 ± 0.09 †
Biochemistry				
TC (mg/dL)	177	193.36 ± 34.33	484	207.66 ± 36.59 †
HDL (mg/dL)	177	57.61 ± 12.32	482	52.47 ± 12.35 †
LDL (mg/dL)	177	117.90 ± 31.05	481	132.17 ± 32.20 †
TG (mg/dL)	177	75.43 ± 32.27	482	109.13 ± 60.79 †
LDL/HDL	177	2.13 ± 0.74	481	2.65 ± 0.89 †
TC/ HDL	177	3.48 ± 0.85	482	4.14 ± 1.11 †
TG/HDL	177	1.40 ± 0.79	482	2.29 ± 1.71 †
Glucose (mg/dL)	115	81.75 ± 9.10	472	87.23 ± 12.76 †
Insulin (µUI/mL)	96	5.74 ± 2.30	410	9.29 ± 4.83 †
HOMA	96	1.17 ± 0.54	410	2.03 ± 1.30 †

[1] Differences between NW (Normal Weight) and EW (Excess Weight): † $p < 0.01$. [2] X: Mean; SD: Standard Deviation. BME: Basal Metabolic Expenditure; HOMA: Homeostasis Model Assessment; kJ: Kilojoules; TG: Triglycerides; TC: Total Cholesterol.

Table A2. Physical activity data distributed according to nutritional state (total sample).

		Total Sample		
		NW [1]		EW
Physical Activity (METs)	*N*	X ± SD [2]	*N*	X ± SD
Low	186	774.37 ± 894.93	452	707.50 ± 919.64
Moderate	187	365.29 ± 666.62	452	299.00 ± 528.05 *
High	187	433.66 ± 877.82	452	269.50 ± 711.81 †

[1] Differences between NW (Normal Weight) and EW (Excess Weight): * $p < 0.05$, † $p < 0.01$. [2] X: Mean; SD: Standard Deviation. MET: Metabolic Equivalent.

Table A3. Dietary data according to nutritional state (total sample and plausible reporters).

		Total Sample				Plausible Reporters		
		NW [1]		EW		NW		EW
Dietary Data	*N*	X ± SD [2]	*N*	X ± SD	*N*	X ± SD	*N*	X ± SD
Total dietary data								
TEI (kJ/day)	292	9297 ± 2319	507	8955 ± 2336	226	9222 ± 1810	359	9470 ± 2053
CH (%TEI)	292	39.07 ± 6.48	507	37.79 ± 6.47 †	226	38.98 ± 6.13	359	37.75 ± 6.23 *
Pr (%TEI)	292	17.19 ± 3.09	507	17.84 ± 3.05 †	226	17.04 ± 2.81	359	17.60 ± 2.90 *
Lip (%TEI)	292	39.68 ± 6.53	507	39.78 ± 6.04	226	39.99 ± 6.28	359	40.02 ± 6.05
Dietary data: Breakfast								
B MEI (%TEI)	292	17.68 ± 6.97	507	16.50 ± 6.63 *	226	17.42 ± 6.57	359	16.06 ± 6.55 *
B CH (%TEI)	292	9.45 ± 4.04	507	8.57 ± 3.69 †	226	9.24 ± 3.75	359	8.39 ± 3.69 *
B Pr (%TEI)	292	2.39 ± 1.10	507	2.32 ± 1.15	226	2.39 ± 1.08	359	2.20 ± 1.07 *
B Lip (%TEI)	292	5.40 ± 3.63	507	5.17 ± 2.99	226	5.39 ± 3.5	359	5.05 ± 2.93

Table A3. *Cont.*

| | Total Sample | | | | Plausible Reporters | | | |
| | NW [1] | | EW | | NW | | EW | |
Dietary Data	*N*	X ± SD [2]	*N*	X ± SD	*N*	X ± SD	*N*	X ± SD
Dietary data: Morning snack								
MS MEI (%TEI)	292	6.07 ± 5.33	507	6.64 ± 5.47	226	5.78 ± 5.21	359	6.41 ± 5.41
MS CH (%TEI)	292	3.15 ± 2.75	507	3.23 ± 2.75	226	3.02 ± 2.69	359	3.07 ± 2.71
MS Pr (%TEI)	292	0.78 ± 0.91	507	0.92 ± 0.96 *	226	0.73 ± 0.90	359	0.89 ± 0.93 *
MS Lip (%TEI)	292	1.80 ± 2.09	507	2.07 ± 2.35	226	1.72 ± 1.98	359	2.04 ± 2.32
Dietary data: Lunch								
L MEI (%TEI)	292	38.93 ± 8.64	507	39.51 ± 8.88	226	39.23 ± 8.62	359	39.74 ± 8.93
L CH (%TEI)	292	13.33 ± 4.59	507	13.01 ± 4.56	226	13.41 ± 4.50	359	13.19 ± 4.32
L Pr (%TEI)	292	7.55 ± 2.30	507	7.95 ± 2.36 *	226	7.56 ± 2.19	359	7.86 ± 2.26
L Lip (%TEI)	292	16.55 ± 5.32	507	16.75 ± 5.68	226	16.8 ± 5.30	359	16.85 ± 5.80
Dietary data: Evening snack								
ES MEI (%TEI)	292	6.43 ± 5.51	507	7.02 ± 5.82	226	6.60 ± 5.55	359	7.02 ± 5.74
ES CH (%TEI)	292	3.24 ± 2.81	507	3.34 ± 2.88	226	3.31 ± 2.81	359	3.28 ± 2.80
ES Pr (%TEI)	292	0.86 ± 0.91	507	0.94 ± 0.97	226	0.87 ± 0.91	359	0.93 ± 0.97
ES Lip (%TEI)	292	2.02 ± 2.27	507	2.41 ± 2.63 *	226	2.09 ± 2.31	359	2.52 ± 2.63 *
Dietary data: Dinner								
D MEI (%TEI)	292	30.30 ± 8.93	507	29.35 ± 9.07	226	30.41 ± 8.84	359	29.63 ± 9.03
D CH (%TEI)	292	9.66 ± 3.97	507	9.27 ± 3.84	226	9.77 ± 4.01	359	9.40 ± 3.78
D Pr (%TEI)	292	5.54 ± 2.02	507	5.59 ± 2.15	226	5.41 ± 1.91	359	5.57 ± 2.02
D Lip (%TEI)	292	13.74 ± 5.24	507	13.03 ± 5.10	226	13.84 ± 5.13	359	13.15 ± 5.10
Dietary data: After dinner snack								
ADS MEI (%TEI)	292	0.55 ± 1.90	507	0.98 ± 2.93 *	226	0.56 ± 1.56	359	1.14 ± 3.26
ADS CH (%TEI)	292	0.23 ± 0.81	507	0.37 ± 1.24 *	226	0.24 ± 0.72	359	0.42 ± 1.38
ADS Pr (%TEI)	292	0.06 ± 0.23	507	0.12 ± 0.52	226	0.07 ± 0.26	359	0.15 ± 0.6
ADS Lip (%TEI)	292	0.14 ± 0.51	507	0.34 ± 1.28 *	226	0.15 ± 0.52	359	0.41 ± 1.4
Grouped dietary data								
Dietary data: Morning								
Morn MEI (%TEI)	292	23.79 ± 8.17	507	23.14 ± 7.39	226	23.20 ± 7.97	359	22.47 ± 7.27
Morn CH (%TEI)	292	12.60 ± 4.68	507	11.80 ± 4.12 *	226	12.25 ± 4.42	359	11.46 ± 4.10 *
Morn Pr (%TEI)	292	3.18 ± 1.38	507	3.25 ± 1.38	226	3.12 ± 1.40	359	3.09 ± 1.34
Morn Lip (%TEI)	292	7.23 ± 3.99	507	7.25 ± 3.50	226	7.12 ± 3.81	359	7.09 ± 3.35
Dietary data: Evening								
Eve MEI (%TEI)	292	37.28 ± 9.87	507	37.35 ± 9.62	226	37.57 ± 9.56	359	37.79 ± 9.73
Eve CH (%TEI)	292	13.13 ± 4.76	507	12.98 ± 4.76	226	13.32 ± 4.66	359	13.10 ± 4.67
Eve Pr (%TEI)	292	6.46 ± 2.22	507	6.65 ± 2.25	226	6.36 ± 2.12	359	6.65 ± 2.20
Eve Lip (%TEI)	292	15.9 ± 5.59	507	15.78 ± 5.40	226	16.08 ± 5.50	359	16.07 ± 5.52

[1] Differences between NW (Normal Weight) and EW (Excess Weight): * $p < 0.05$; † $p < 0.01$. [2] X: Mean; SD: Standard Deviation; ADS: After Dinner Snack; B: Breakfast; CH: Carbohydrates; D: Dinner; Eve: Evening; ES: Evening Snack; kJ: Kilojoules; Lip: Lipids; L: Lunch; MEI: Meal Energy Intake; Morn: Morning; MS: Morning Snack; Pr: Proteins; TEI: Total Energy Intake.

References

1. World Health Organization (WHO). Available online: www.who.int (accessed on 9 February 2019).
2. Barberio, A.M.; Alareeki, A.; Viner, B.; Pader, J.; Vena, J.E.; Arora, P.; Friedenreich, C.M.; Brenner, D.R. Central body fatness is a stronger predictor of cancer risk than overall body size. *Nat. Commun.* **2019**, *10*, 383. [CrossRef] [PubMed]
3. Ladhani, M.; Craig, J.C.; Irving, M.; Clayton, P.A.; Wong, G. Obesity and the risk of cardiovascular and all-cause mortality in chronic kidney disease: A systematic review and meta-analysis. *Nephrol. Dial. Transpl.* **2017**, *32*, 439–449. [CrossRef] [PubMed]

4. Wadolowska, L.; Hamulka, J.; Kowalkowska, J.; Kostecka, M.; Wadolowska, K.; Biezanowska-Kopec, R.; Czarniecka-Skubina, E.; Kozirok, W.; Piotrowska, A. Prudent-Active and Fast-Food-Sedentary Dietary-Lifestyle Patterns: The Association with Adiposity, Nutrition Knowledge and Sociodemographic Factors in Polish Teenagers-The ABC of Healthy Eating Project. *Nutrients* **2018**, *10*, 1988. [CrossRef]

5. Instituto Nacional de Estadística (INE) Encuesta Nacional de Salud 2017. 4.6 Determinantes de Salud (Sobrepeso, Consumo de Fruta y Verdura, Tipo de Lactancia, Actividad Física). Available online: http://www.ine.es/ss/Satellite?L=es_ES&c=INESeccion_C&cid=1259926457058&p=1254735110672& pagename=ProductosYServicios%2FPYSLayout¶m1=PYSDetalle¶m3=1259924822888 (accessed on 9 February 2019).

6. Susmallian, S.; Raziel, A.; Barnea, R.; Paran, H. Bariatric surgery in older adults: Should there be an age limit? *Medicine* **2019**, *98*, e13824. [CrossRef]

7. Witjaksono, F.; Jutamulia, J.; Annisa, N.G.; Prasetya, S.I.; Nurwidya, F. Comparison of low calorie high protein and low calorie standard protein diet on waist circumference of adults with visceral obesity and weight cycling. *BMC Res. Notes* **2018**, *11*, 674. [CrossRef]

8. Sofer, S.; Eliraz, A.; Kaplan, S.; Voet, H.; Fink, G.; Kima, T.; Madar, Z. Greater weight loss and hormonal changes after 6 months diet with carbohydrates eaten mostly at dinner. *Obesity (Silver Spring)* **2011**, *19*, 2006–2014. [CrossRef]

9. Sofer, S.; Eliraz, A.; Kaplan, S.; Voet, H.; Fink, G.; Kima, T.; Madar, Z. Changes in daily leptin, ghrelin and adiponectin profiles following a diet with carbohydrates eaten at dinner in obese subjects. *Nutr. Metab. Cardiovasc. Dis.* **2013**, *23*, 744–750. [CrossRef]

10. Alves, R.D.M.; de Oliveira, F.C.E.; Hermsdorff, H.H.M.; Abete, I.; Zulet, M.A.; Martínez, J.A.; Bressan, J. Eating carbohydrate mostly at lunch and protein mostly at dinner within a covert hypocaloric diet influences morning glucose homeostasis in overweight/obese men. *Eur. J. Nutr.* **2014**, *53*, 49–60. [CrossRef]

11. Jakubowicz, D.; Barnea, M.; Wainstein, J.; Froy, O. High caloric intake at breakfast vs. dinner differentially influences weight loss of overweight and obese women. *Obesity (Silver Spring)* **2013**, *21*, 2504–2512. [CrossRef]

12. Madjd, A.; Taylor, M.A.; Delavari, A.; Malekzadeh, R.; Macdonald, I.A.; Farshchi, H.R. Beneficial effect of high energy intake at lunch rather than dinner on weight loss in healthy obese women in a weight-loss program: A randomized clinical trial. *Am. J. Clin. Nutr.* **2016**, *104*, 982–989. [CrossRef]

13. de Toro-Martín, J.; Arsenault, B.J.; Després, J.-P.; Vohl, M.-C. Precision Nutrition: A Review of Personalized Nutritional Approaches for the Prevention and Management of Metabolic Syndrome. *Nutrients* **2017**, *9*, 913. [CrossRef]

14. Ordovas, J.M.; Ferguson, L.R.; Tai, E.S.; Mathers, J.C. Personalised nutrition and health. *BMJ* **2018**, *361*, bmj.k2173. [CrossRef]

15. Loria-Kohen, V.; Espinosa-Salinas, I.; Marcos-Pasero, H.; Lourenço-Nogueira, T.; Herranz, J.; Molina, S.; Reglero, G.; Ramirez de Molina, A. Polymorphism in the CLOCK gene may influence the effect of fat intake reduction on weight loss. *Nutrition* **2016**, *32*, 453–460. [CrossRef]

16. Garaulet, M.; Gómez-Abellán, P.; Alburquerque-Béjar, J.J.; Lee, Y.-C.; Ordovás, J.M.; Scheer, F.A.J.L. Timing of food intake predicts weight loss effectiveness. *Int. J. Obes. (Lond.)* **2013**, *37*, 604–611. [CrossRef]

17. World Health Organization. Obesity: Preventing and managing the global epidemic. Report of a WHO consultation. *World Health Organ. Tech. Rep. Ser.* **2000**, *894*, 1–253.

18. Ortega, R.M.; López-Sobaler, A.M.; Andrés, P.; Requejo, A.M.; Aparicio, A.; Molinero, L.M. *Programa DIAL Para Valoración de Dietas y Cálculos de Alimentación*; Departamento de Nutrición (UCM) y ALCE Ingeniería: Madrid, Spain, 2016.

19. European Food Security Authority (EFSA) EU Menu Guidance. Appendix 8.2.1. Example of a Protocol for Identification of Misreporting (under- and Overreporting of Energy Intake) Based on the PILOT-PANEU Project. 2013. Available online: http://www.efsa.europa.eu/sites/default/files/efsa_rep/blobserver_assets/ 3944A-8-2-1.pdf (accessed on 30 October 2018).

20. Goldberg, G.R.; Black, A.E.; Jebb, S.A.; Cole, T.J.; Murgatroyd, P.R.; Coward, W.A.; Prentice, A.M. Critical evaluation of energy intake data using fundamental principles of energy physiology: 1. Derivation of cut-off limits to identify under-recording. *Eur. J. Clin. Nutr.* **1991**, *45*, 569–581.

21. Black, A.E. Critical evaluation of energy intake using the Goldberg cut-off for energy intake: Basal metabolic rate. A practical guide to its calculation, use and limitations. *Int. J. Obes. Relat. Metab. Disord.* **2000**, *24*, 1119–1130. [CrossRef]

22. International Physical Activity Questionnaire (IPAQ). Group USA Spanish Version Translated 3/2003-Short Last 7 Days Slef-Administered Version of the IPAQ–Revised August 2002. Available online: www.ipaq.ki.se (accessed on 5 August 2019).

23. Dankel, S.J.; Loenneke, J.P.; Loprinzi, P.D. Health Outcomes in Relation to Physical Activity Status, Overweight/Obesity, and History of Overweight/Obesity: A Review of the WATCH Paradigm. *Sports Med.* **2017**, *47*, 1029–1034. [CrossRef]

24. Müller, M.J.; Lagerpusch, M.; Enderle, J.; Schautz, B.; Heller, M.; Bosy-Westphal, A. Beyond the body mass index: Tracking body composition in the pathogenesis of obesity and the metabolic syndrome. *Obes. Rev.* **2012**, *13*, 6–13. [CrossRef]

25. Thom, G.; Lean, M. Is There an Optimal Diet for Weight Management and Metabolic Health? *Gastroenterology* **2017**, *152*, 1739–1751. [CrossRef]

26. Brown, R.E.; Sharma, A.M.; Ardern, C.I.; Mirdamadi, P.; Mirdamadi, P.; Kuk, J.L. Secular differences in the association between caloric intake, macronutrient intake, and physical activity with obesity. *Obes. Res. Clin. Pract.* **2016**, *10*, 243–255. [CrossRef]

27. Buscemi, J.; Rybak, T.M.; Berlin, K.S.; Murphy, J.G.; Raynor, H.A. Impact of food craving and calorie intake on body mass index (BMI) changes during an 18-month behavioral weight loss trial. *J. Behav. Med.* **2017**, *40*, 565–573. [CrossRef]

28. Leech, R.M.; Worsley, A.; Timperio, A.; McNaughton, S.A. The role of energy intake and energy misreporting in the associations between eating patterns and adiposity. *Eur. J. Clin. Nutr.* **2018**, *72*, 142–147. [CrossRef]

29. Westerterp-Plantenga, M.S. Sleep, circadian rhythm and body weight: Parallel developments. *Proc. Nutr. Soc.* **2016**, *75*, 431–439. [CrossRef]

30. Yang, X.; Downes, M.; Yu, R.T.; Bookout, A.L.; He, W.; Straume, M.; Mangelsdorf, D.J.; Evans, R.M. Nuclear receptor expression links the circadian clock to metabolism. *Cell* **2006**, *126*, 801–810. [CrossRef]

31. Kohsaka, A.; Laposky, A.D.; Ramsey, K.M.; Estrada, C.; Joshu, C.; Kobayashi, Y.; Turek, F.W.; Bass, J. High-fat diet disrupts behavioral and molecular circadian rhythms in mice. *Cell Metab.* **2007**, *6*, 414–421. [CrossRef]

32. Garaulet, M.; Lee, Y.-C.; Shen, J.; Parnell, L.D.; Arnett, D.K.; Tsai, M.Y.; Lai, C.-Q.; Ordovas, J.M. CLOCK genetic variation and metabolic syndrome risk: Modulation by monounsaturated fatty acids. *Am. J. Clin. Nutr.* **2009**, *90*, 1466–1475. [CrossRef]

33. Garaulet, M.; Lee, Y.-C.; Shen, J.; Parnell, L.D.; Arnett, D.K.; Tsai, M.Y.; Lai, C.-Q.; Ordovas, J.M. Genetic variants in human CLOCK associate with total energy intake and cytokine sleep factors in overweight subjects (GOLDN population). *Eur. J. Hum. Genet.* **2010**, *18*, 364–369. [CrossRef]

34. Turek, F.W.; Joshu, C.; Kohsaka, A.; Lin, E.; Ivanova, G.; McDearmon, E.; Laposky, A.; Losee-Olson, S.; Easton, A.; Jensen, D.R.; et al. Obesity and metabolic syndrome in circadian Clock mutant mice. *Science* **2005**, *308*, 1043–1045. [CrossRef]

35. Gil Hernández, Á. *Nutrition Treatise: Human Nutrition in Health Status*, 2nd ed.; Editorial Médica Panamericana: Madrid, Spain, 2010; ISBN 978-84-9835-239-9.

Article

Identification of Genetic Factors Underlying the Association between Sodium Intake Habits and Hypertension Risk

Yu-Jin Kwon [1,†], Jung Oh Kim [2,†], Jae-Min Park [3], Ja-Eun Choi [2], Da-Hyun Park [2], Youhyun Song [3], Seong-Jin Kim [2], Ji-Won Lee [3,*] and Kyung-Won Hong [2,*]

[1] Department of Family Medicine, Yongin Severance Hospital, Yonsei University College of Medicine, 363, Dongbaekjukjeon-daero, Giheung-gu, Yongin-si 16995, Korea; digda3@yuhs.ac
[2] Theragen Bio Co., Ltd., Suwon 16229, Korea; jungoh.kim@therabio.kr (J.O.K.); jaeun.choi@therabio.kr (J.-E.C.); dahyun.park@therabio.kr (D.-H.P.); jasonsjkim@therabio.kr (S.-J.K.)
[3] Department of Family Medicine, Gangnam Severance Hospital, Yonsei University College of Medicine, 211 Eonju-ro, Gangnam-gu, Seoul 06273, Korea; milkcandy@yuhs.ac (J.-M.P.); WLGMEO_O@yuhs.ac (Y.S.)
*   Correspondence: indi5645@yuhs.ac (J.-W.L.); kyungwon.hong@theragenetex.com (K.-W.H.)
†   These authors are co-first authors who equally contribute to this work.

Received: 5 August 2020; Accepted: 22 August 2020; Published: 25 August 2020

Abstract: The role of sodium in hypertension remains unresolved. Although genetic factors have a significant impact on high blood pressure, studies comparing genetic susceptibility between people with low and high sodium diets are lacking. We aimed to investigate the genetic variations related to hypertension according to sodium intake habits in a large Korean population-based study. Data for a total of 57,363 participants in the Korean Genome and Epidemiology Study Health Examination were analyzed. Sodium intake was measured by a semi-quantitative food frequency questionnaire. We classified participants according to sodium intake being less than or greater than 2 g/day. We used logistic regression to test single-marker variants for genetic association with a diagnosis of hypertension, adjusting for age, sex, body mass index, exercise, alcohol, smoking, potassium intake, principal components 1, and principal components 2. Significant associations were defined as $p < 5 \times 10^{-8}$. In participants whose sodium intake was greater than 2 g/day, chromosome 6 open reading frame 10 (C6orf10)-human leukocyte antigen (HLA)-DQB1 rs6913309, ring finger protein (RNF)213 rs112735431, glycosylphosphatidylinositol anchored molecule-like (GML)- cytochrome P450 family 11 subfamily B member 1(CYP11B1) rs3819496, myosin light chain 2 (MYL2)-cut like homeobox 2 (CUX2) rs12229654, and jagged1 (JAG1) rs1887320 were significantly associated with hypertension. In participants whose intake was less than 2 g/day, echinoderm microtubule-associated protein-like 6(EML6) rs67617923 was significantly associated with hypertension. Genetic susceptibility associated with hypertension differed according to sodium intake. Identifying gene variants that contribute to the dependence of hypertension on sodium intake status could make possible more individualized nutritional recommendations for preventing cardiovascular diseases.

Keywords: sodium intake; hypertension; single-nucleotide polymorphism

---

1. Introduction

Sodium is the most important electrolyte for maintaining extracellular fluid volume and regulating cellular membrane potential [1]. The importance of dietary sodium in regulating blood pressure (BP) has received much attention in the past. Hypertension has been the most important global risk factor for all-cause mortality and for cardiovascular mortality [2]. Many studies have demonstrated the association of sodium consumption with hypertension and risk of cardiovascular diseases (CVD) [3–5].

Therefore, the World Health Organization (WHO) recommends sodium intake of less than 2 g/day to reduce BP and the risk of CVD [6].

Under normal physiological adaptation to sodium intake, the pressure natriuresis curve is regulated by the renin–angiotensin system and renal sympathetic nerve activity [7]. Increased sodium intake suppressed angiotensin II and led to pressure natriuresis cure shifting, which increased renal sodium excretion [7]. Both epithelial sodium transporter and aldosterone level are also involved in adapting the dietary sodium intake. In patients with salt-sensitive increased BP, enhanced sodium reabsorption, changes in pressure natriuresis curve, a suppressed renin–angiotensin system, and gene polymorphisms in voltage-dependent Ca^{2+} channels and sodium-bicarbonate cotransporter were noted [8,9]. Furthermore, the levels of natriuretic peptides could be affected by excessive sodium intake, decreased potassium and magnesium intake, and metabolic diseases such as obesity [10–12].

However, the relationship between sodium intake and BP remains unresolved. A meta-analysis of 13 prospective studies with 177,035 participants reported that high salt intake is associated with significantly increased risk of stroke and total CVD [3]. Another meta-analysis that included 22 trials in hypertensive patients and 12 trials in normotensive participants reported that a salt reduction of 4.4 g/day led to a mean systolic blood pressure (SBP) change of −4.18 mm Hg (95% confidence interval [CI] −5.18 to −3.18, I2 = 75%), and a diastolic blood pressure (DBP) change of −2.06 mm Hg (CI, −2.67 to −1.45, I2 = 68%) [4]. Conversely, several studies reported an inverse association between sodium intake and CVD. Stolarz-Skrzypek et al. [13] found that SBP, but not DBP, was significantly correlated with 24 h urinary sodium excretion; however, the incidence of hypertension did not increase, and CVD risk decreased with increasing sodium excretion tertiles. Interestingly, some studies have shown a J-shaped association between sodium intake and CVD. Martin et al. [14] showed that sodium excretion rates greater than 7 g/day or less than 3 g/day were associated with increased risk of all CV events as well as CVD mortality, compared to sodium excretion of 4 to 5.99 g/day, using the two-cohort data. The same author [15] reaffirmed that estimated sodium intake of 3 to 6 g/day was associated with a low risk of CVD among 101,945 persons in 17 countries. These conflicting findings are due not only to differences among studies but also to the complexity of traits of hypertension. Essential hypertension, with varying or unknown pathology, accounts for 95% of all hypertension cases [16]. High BP is known to result from interaction among multiple factors, including genetic susceptibility, obesity, aging, sedentary life style, alcohol consumption, high salt intake (especially in salt-sensitive persons), and low potassium intake [16]. Genetic elements were reported to make a 30–70% contribution to BP variation [17,18]. Under similar environmental conditions, some individuals develop hypertension and others do not.

Single-nucleotide polymorphisms (SNPs), single base substitutions within the deoxyribonucleic acid (DNA) sequence, are the most common type of human genetic variation [19]. Inter-individual genetic variation is an important determinant of human nutritional requirements [20]. However, studies comparing genetic susceptibility associated with hypertension between people with low versus high sodium diets have been limited. Identifying gene variants that contribute to the association of hypertension with sodium intake could contribute to better understanding of the pathophysiology of hypertension, and offer opportunities to determine optimal nutrition status for individuals.

Therefore, we aimed to investigate the genetic variations involved in the relationship between hypertension and sodium intake, in a large Korean population-based study.

2. Materials and Methods

2.1. Study Population

The Korean Genome and Epidemiology Study (KoGES) is a large cohort study to find genetic and environmental factors, and their interactions, in non-communicable diseases, with government funding [21]. KoGES Health Examination (KoGES_HEXA), one of the subset cohorts of KoGES, consists of community dwellers and participants, aged ≥ 40 years at baseline recruited from the national health examinee registry.

In the current study, we included the total 58,701 participants who participated in KoGES_HEXA. We excluded participants in KoGES_HEXA from the present study if values were missing for BP, body mass index (BMI), waist circumference (WC), heart rate (HR), alcohol, smoking, or exercise (n = 1338). A total of 57,363 participants were included in the current study. Hypertensive patients (n = 15,245) were defined as those with SBP ≥ 140 mm Hg, DBP ≥ 90 mm Hg, or a history of hypertension or taking antihypertensive medication. Controls (n = 42,114) were defined as those without hypertension or taking anti-hypertensive drug or cardiovascular diseases.

Figure 1 shows a flow chart describing this study. We treated the three analyses set. In analysis 1, we compared with control (n = 42,114) and hypertension patients (n = 15,245). In analysis 2, we compared with controls (n = 17,869) and hypertension patients (n = 6,546) in the participants with <2 g/day (n = 24,415). In analysis 3, we compared with controls (n = 24,245) and hypertension patients (n = 8699) in the participants with ≥2 g/day (n = 32,994). The study was approved by the institutional review board of Theragen Bio Co., Ltd. (approval number: 700062-20190819-GP-006-02).

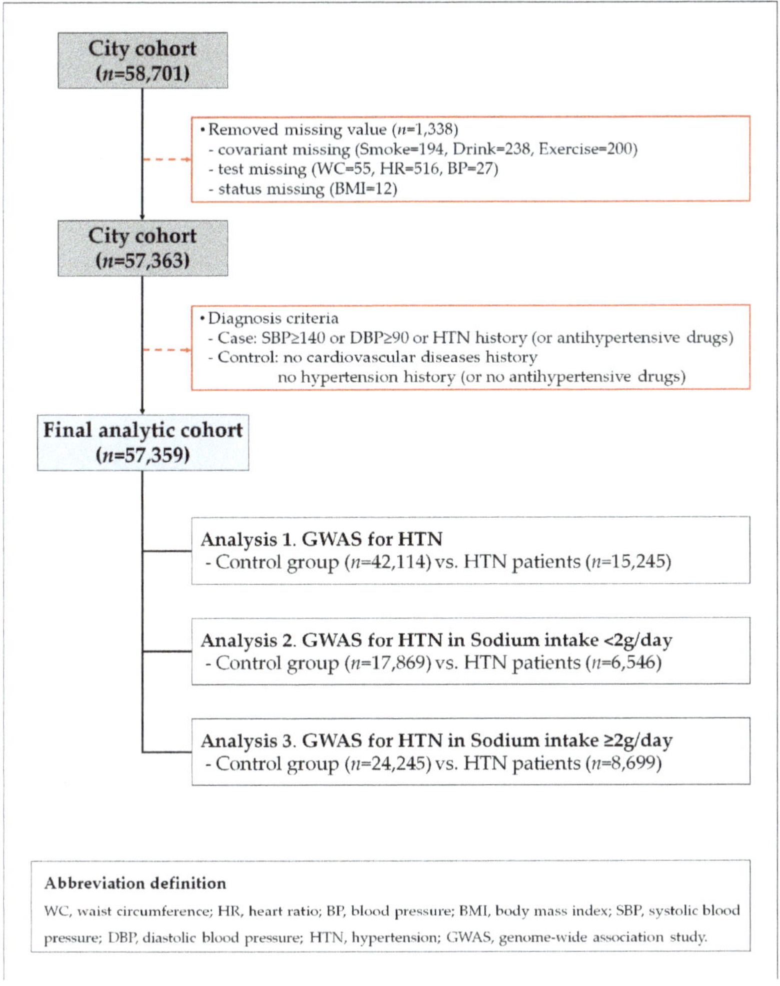

Figure 1. Flow chart of study population.

2.2. Assessment of Dietary Sodium and Potassium Intake and Covariates

For dietary assessment, a semi-quantitative food frequency questionnaire (FFQ) involving 103 items was developed for the KoGES. Participants reported the frequency and amount of foods eaten over the past year. The results of the questionnaire were analyzed, with reference to a food composition database, to estimate intakes. FFQs are widely used as the primary dietary assessment tool in epidemiological studies [22]. We classified participants based on sodium intake, according to the WHO recommendation of 2 g/day [6]. The WC was measured midway between the bottom rib and the iliac crest. Body mass index (BMI) was calculated as weight (kg) divided by height (m) squared. Blood pressures in the seated position were measured twice, using a mercury sphygmomanometer. Smoking status was classified into three categories: non-smokers, ex-smokers and current smokers. Drinking status was classified into three groups: non-drinkers (those who drink alcohol fewer than 12 times a year, with one drink not exceeding one cup), ex-drinkers, and current drinkers. Exercise was defined as regular exercise sufficient to cause perspiration.

2.3. Genotyping

Fasting blood samples were collected into one serum separator tube and two ethylenediaminetetraacetic acid tubes. Blood DNA samples were prepared, and all samples were then transported to the National Biobank of Korea. The SNP genotypes of participants were extracted from the Korea Biobank array (referred to as KoreanChip), which was optimized for the Korean population and to demonstrate findings of genome-wide association study (GWAS) of blood biochemical traits. The KoreanChip comprised >833,000 markers, including >247,000 rare or functional variants, derived from sequencing data for over 2500 Koreans [23]. Detailed information about the KoreanChip was described in a previous study [23]. We applied the following criteria in the analysis of KoreanChip data, to control the quality of genotyping results: call rate >97%, minor allele frequency >0.01, missing genotype >0.01, Hardy-Weinberg equilibrium $p > 0.000001$. In addition, the genotype used in the analysis is genome data which imputed data from a dataset of 1000 genome phase 1 and 2 Asian panels.

2.4. Statistical Analysis

The data were presented either as mean ± standard deviation or as numbers (percentage). To compare participants with and without hypertension, we used two-tailed Student's t-tests for continuous variables, or chi-squared tests for categorical variables. In addition, we performed principal component analysis (PCA) to reduce bias of genomic data according to the region where samples were collected, and used principal component (PC)1 and PC2 as covariates in statistical analyses. We used logistic regression to test single-marker variants for genetic association with a diagnosis of hypertension, while adjusting for age, sex, BMI, exercise, alcohol, smoking, potassium intake, PC1 and PC2. All statistical tests were based on an adjusted model using PLINK (ver. 1.07). p values $< 5 \times 10^{-8}$ were considered as statistically significant.

3. Results

3.1. General Characteristics of the Study Population

Table 1 shows the general characteristics of participant categorized according to sodium intake. There were 24,415 (42.6%) and 32,944 (57.4%) participants with sodium intakes <2 g/day or ≥2 g/day, respectively, and with respective mean ages of 54.1 and 53.6 years. The proportion of men was significantly higher among participants with sodium intake ≥2 g/day. SBP and DBP were also significantly higher in this group. The mean total cholesterol (TC) and low-density lipoprotein (LDL) cholesterol level were not different between two groups. The mean level of triglyceride (TG) was significantly higher in participants who intake sodium intake was ≥2 g/day, while the mean level of high-density lipoprotein (HDL) cholesterol was significantly lower in this group. The mean level of C-reactive protein (CRP) was not different between two groups. The proportions of exercise, drinking, and smoking were higher among participants with sodium intake ≥ 2 g/day. Sodium and potassium consumptions were significantly higher in participants with sodium intake ≥ 2 g/day (all $p < 0.001$). The sodium to potassium (Na/K) ratio was also significantly higher in participants with sodium intake ≥ 2 g/day ($p < 0.001$). Table 1 also presents characteristics of participants subcategorized according to the presence of hypertension (HTN). The mean age, BMI, WC, SBP, and DBP were significantly higher in hypertensive patients than controls, whether sodium intake was <2 or ≥ 2 g/day. The mean TC, HDL, LDL were significantly lower in hypertensive patients than controls, whether sodium intake was < or ≥ 2 g/day. The TG and CRP were significantly higher in hypertensive patients than controls, whether sodium intake was <2 or ≥ 2 g/day. Among participants with sodium intake <2 g/day, sodium and potassium intakes were lower for hypertensive patients ($p = 0.038$, and $p < 0.001$), whereas mean Na/K was higher in hypertensive patients than in controls ($p < 0.001$). Among participants with sodium intake ≥ 2 g/day, sodium intake was similar between hypertensives and controls ($p = 0.433$), while potassium intake was significantly lower ($p < 0.001$) and Na/K was higher in hypertensive patients ($p < 0.001$) than in controls.

3.2. SNPs Associated with Hypertension Based on Sodium Intake

Table 2 shows the SNPs most strongly associated or clustered with hypertension in the Korean subjects, according to their sodium intake. Odds ratios (OR) and 95% CIs were calculated using logistic regression analysis after adjusting for age, sex, BMI, alcohol consumption, smoking, physical activity, and potassium intake. SNPs rs16998073 and rs12509595 demonstrated significant association with hypertension risk both in participants with sodium intake <2 g/day and those with intake ≥ 2 g/day. SNPs rs1191582, rs11105378, and rs140473396 were significantly associated with a decreased risk of hypertension, both in participants with sodium intake <2 g/day and those with intake ≥ 2g/day. SNP rs67617923 was significantly associated with increased risk of hypertension only in participants with sodium intake <2 g/day (OR = 1.294 [1.187–1.410], $p = 4.29 \times 10^{-9}$). SNPs rs6913309 and rs112735431 were significantly associated with hypertension only in participants with sodium intake ≥ 2 g/day (OR = 1.145 [1.094–1.197], $p = 4.23 \times 10^{-9}$; and OR =1.706 [1.446–2.012], $p = 2.38 \times 10^{-10}$, respectively). SNPs rs3819496, rs12229654, and rs1887320 were significantly associated with decreased risk of hypertension in participants with sodium intake ≥ 2 g/day (OR = 0.892 [0.857–0.929], $p = 3.73 \times 10^{-8}$; OR = 0.834 [0.787–0.883], $p = 5.25 \times 10^{-10}$; and OR = 0.892 [0.859–0.925], $p = 1.45 \times 10^{-9}$, respectively). All SNPs that were found to be significantly related to hypertension are described in the Supplementary Tables (Table S1, SNPs significantly related to hypertension; Table S2, SNPs significantly related to hypertension in participants with sodium intake <2 g/day; Table S3, SNPs significantly related to hypertension in participants with sodium intake ≥ 2 g/day).

Table 1. General characteristics of study population based on sodium intake and according to prevalence of hypertension.

Characteristics	Sodium Intake		$p^{1,\dagger}$	$p^{2,\dagger}$	$p^{3,\dagger}$	Sodium Intake <2 g/Day			Sodium Intake ≥2 g/Day		
	<2 g/Day (n = 24,415)	≥2 g/Day (n = 32,944)				Without HTN (n = 17,869)	With HTN (n = 6546)	$p^{\dagger}$	Without HTN (n = 24,245)	With HTN (n = 8699)	$p^{\dagger}$
Age (years,)	54.05 ± 7.90	53.58 ± 8.07	<0.0001	<0.0001	0.002	52.70 ± 7.70	57.73 ± 7.25	<0.0001	52.23 ± 7.85	57.36 ± 7.44	<0.0001
Sex (male, %)	7453 (30.5)	12,348 (37.5)	<0.0001 ‡	<0.0001 ‡	<0.0001 ‡	4929 (27.6)	2524 (38.6)	<0.0001 ‡	8264 (34.1)	4084 (46.9)	<0.0001 ‡
BMI (kg/m2)	24.02 ± 46.63	24.20 ± 34.68	0.606	0.743	<0.0001	23.29 ± 2.70	24.91 ± 2.92	<0.0001	23.59 ± 2.74	25.17 ± 2.95	<0.0001
WC (cm,)	80.08 ± 8.54	81.26 ± 8.65	<0.0001	<0.0001	<0.0001	78.72 ± 8.23	83.78 ± 8.29	<0.0001	79.93 ± 8.35	84.97 ± 8.40	<0.0001
SBP (mmHg)	122.14 ± 14.80	122.57 ± 14.73	0.001	<0.0001	0.042	117.57 ± 11.73	134.62 ± 15.12	<0.0001	118.07 ± 11.65	135.12 ± 15.14	<0.0001
DBP (mmHg)	75.44 ± 9.71	75.94 ± 9.72	<0.0001	<0.0001	0.001	72.89 ± 8.18	82.43 ± 10.11	<0.0001	73.42 ± 8.21	82.96 ± 10.15	<0.0001
HR (bpm)	69.30 ± 9.28	69.06 ± 9.00	0.002	0.012	0.09	68.94 ± 8.93	70.27 ± 10.11	<0.0001	68.73 ± 8.70	70.00 ± 9.74	<0.0001
TC (mg/dL)	197.19 ± 36.00	197.50 ± 35.45	0.307	0.424	0.558	197.88 ± 35.46	195.33 ± 37.35	<0.0001	198.16 ± 35.11	195.68 ± 36.30	<0.0001
TG (mg/dL)	122.06 ± 83.51	127.35 ± 87.03	<0.0001	<0.0001	0.0001	115.17 ± 77.32	140.86 ± 95.93	<0.0001	120.52 ± 82.17	146.35 ± 96.79	<0.0001
HDL-C (mg/dL)	54.23 ± 13.30	53.43 ± 13.04	<0.0001	<0.0001	<0.0001	55.05 ± 13.34	52.00 ± 12.91	<0.0001	54.23 ± 13.14	51.19 ± 12.49	<0.0001
LDL-C (mg/dL)	119.14 ± 32.36	119.40 ± 31.98	0.344	0.531	0.52	120.28 ± 31.72	116.00 ± 33.89	<0.0001	120.48 ± 31.46	116.36 ± 33.21	<0.0001
CRP (mg/dL)	0.14 ± 0.45	0.14 ± 0.34	0.333	0.432	0.639	0.13 ± 0.46	0.17 ± 0.41	<0.0001	0.13 ± 0.34	0.16 ± 0.34	<0.0001
Exercise status											
Yes (%)	13,103 (53.7)	18,250 (55.4)	<0.0001 ‡	<0.0001 ‡	0.441	9422 (52.7)	3681 (56.2)	<0.0001 ‡	13,304 (54.9)	4946 (56.9)	<0.0001 ‡
No (%)	11,312 (46.3)	14,694 (44.6)				8447 (47.3)	2865 (43.8)		10,941 (45.1)	3753 (43.1)	
Smoking status											
Non-smokers (%)	18,638 (76.3)	23,386 (71.0)	<0.0001 ‡	<0.0001 ‡	<0.0001 ‡	13,957 (78.1)	4681 (71.5)	<0.0001 ‡	17,779 (73.3)	5607 (64.5)	<0.0001 ‡
Ex-smokers (%)	3484 (14.3)	5586 (17.0)				2235 (12.5)	1249 (19.1)		3562 (14.7)	2024 (23.3)	
Current smokers (%)	2293 (9.4)	3972 (12.1)				1677 (9.4)	616 (9.4)		2904 (12.0)	1068 (12.3)	
Drinking status											
Non-drinker (%)	13,411 (54.9)	16,369 (49.7)	<0.0001 ‡	<0.0001 ‡	<0.0001 ‡	9880 (55.3)	3531 (53.9)	<0.0001 ‡	12,333 (50.9)	4036 (46.4)	<0.0001 ‡
Ex-drinker (%)	921 (3.8)	1212 (3.7)				605 (3.4)	316 (4.8)		767 (3.2)	445 (5.1)	
Current drinker (%)	10083 (41.3)	15363 (46.6)				7384 (41.3)	2699 (41.2)		11145 (46.0)	4218 (48.5)	
Total intake energy (kcal/day)	1498.32 ± 394.32	1923.57 ± 577.56	<0.0001	<0.0001	<0.0001	1502.56 ± 397.52	1486.77 ± 385.27	0.0056	1938.06 ± 588.37	1883.05 ± 544.22	<0.0001
Sugar (g/day)	275.09 ± 74.34	335.71 ± 92.28	<0.0001	<0.0001	<0.0001	274.83 ± 74.99	275.80 ± 72.54	0.3667	337.21 ± 93.79	331.51 ± 87.81	<0.0001
Fat (g/day)	21.03 ± 10.72	32.85 ± 19.94	<0.0001	<0.0001	<0.0001	21.54 ± 10.94	19.62 ± 9.96	<0.0001	33.57 ± 20.32	30.81 ± 18.71	<0.0001
Protein (g/day)	46.51 ± 14.39	68.25 ± 27.58	<0.0001	<0.0001	<0.0001	46.83 ± 14.50	45.64 ± 14.03	<0.0001	68.90 ± 28.03	66.45 ± 26.22	<0.0001
Sugar ratio	73.48 ± 6.45	70.41 ± 7.05	<0.0001	<0.0001	<0.0001	12.53 ± 2.23	12.32 ± 2.12	<0.0001	14.12 ± 2.55	14.02 ± 2.69	<0.0001
Fat ratio	5.58 ± 2.28	6.60 ± 2.40	<0.0001	<0.0001	<0.0001	5.71 ± 2.32	5.24 ± 2.12	<0.0001	6.70 ± 2.39	6.32 ± 2.40	<0.0001
Protein ratio	12.47 ± 2.20	14.09 ± 2.59	<0.0001	<0.0001	<0.0001	73.19 ± 6.57	74.28 ± 6.05	<0.0001	70.20 ± 7.01	71.01 ± 7.12	<0.0001
Na (mg/day)	1323.39 ± 435.98	3254.94 ± 1251.19	<0.0001	<0.0001	<0.0001	1328.30 ± 433.08	1310.05 ± 443.58	0.0038	3258.11 ± 1258.90	3245.85 ± 1229.30	0.4329
K (mg/day)	1591.80 ± 561.36	2694.00 ± 1046.87	<0.0001	<0.0001	<0.0001	1607.78 ± 564.81	1548.20 ± 549.57	<0.0001	2719.81 ± 1069.77	2621.77 ± 976.41	<0.0001
Na/K ratio	0.87 ± 0.28	1.26 ± 0.34	<0.0001	<0.0001	<0.0001	0.86 ± 0.28	0.88 ± 0.29	<0.0001	1.25 ± 0.34	1.29 ± 0.35	<0.0001

Data are presented either as mean ± standard deviation or as numbers (percentage). p values are calculated by two-tail Student's t-test † or Chi-squared test ‡. [1] The p-value comparing the baseline characteristic between the sodium intake <2 g/day group and the sodium intake ≥2 g/day group, among all participants. [2] The p-value comparing the baseline characteristic between the sodium intake <2 g/day group and sodium intake ≥2 g/day group, among participants without HTN. [3] The p-value comparing the baseline characteristic between the sodium intake <2 g/day group and sodium intake ≥2 g/day group, among participants with HTN. Sugar intake ratio = total sugar (gram) × 4 kcal/total energy intake (kcal) × 100; protein intake ratio = protein (gram) × 4 kcal/total energy intake (kcal) × 100; fat intake ratio (gram) × 9 kcal/total energy intake (kcal) × 100. BMI, body mass index; WC, waist circumference; SBP, systolic blood pressure; DBP, diastolic blood pressure; HR, heart rate; TC, total cholesterol; TG, triglyceride; HDL, high-density lipoprotein cholesterol; LDL, low density lipoprotein cholesterol; CRP, C-reactive protein. HTN, hypertension.

Table 2. Single-nucleotide polymorphisms (SNPs) most strongly associated with hypertension susceptibility loci in the Korean population, according to sodium intake.

SNP	Chr:BP	A1	MAF				Gene	Feature	Cluster SNP *	OR (95% CI)	p
			Present Study	EAS	EUR	AMR					
Participants with sodium intake <2 g/day											
rs67617923	2:54968517	A	0.063	0.074	0.160	0.098	*EML6*	intron variant	rs72806698; rs67246257; rs67514855	1.294 (1.187–1.410)	4.29×10^{-9}
rs16998073	4:81184341	T	0.347	0.360	0.268	0.267	*FGF5*	upstream gene variant	rs12509595; rs10857147	1.245 (1.190–1.302)	1.14×10^{-21}
rs11191582	10:104913653	A	0.227	0.265	0.089	0.193	*NT5C2*	intron variant	rs11191479; rs11191484; rs72050190; rs145010450; rs10883815	0.849 (0.806–0.895)	1.08×10^{-9}
rs11105378	12:90090741	T	0.372	0.310	0.141	0.112	*ATP2B1–LINC00936*	intergenic region	rs2681485; rs7136259; rs11105377; rs1401982; rs1689040	0.874 (0.836–0.915)	4.67×10^{-9}
Participants with sodium intake ≥2 g/day											
rs12509595	4:81182554	C	0.347	0.361	0.267	0.267	*PRDM8–FGF5*	intergenic region	rs16998073; rs10857147	1.228 (1.181–1.277)	7.46×10^{-25}
rs6913309	6:32339840	A	0.212	0.139	0.310	0.244	*C6orf10–HLA-DQB1*	upstream gene variant	N/A	1.145 (1.094–1.197)	4.23×10^{-9}
rs112735431	17:78358945	A	0.011	0.002	0.000	0.000	*RNF213*	missense variant	rs138309870	1.706 (1.446–2.012)	2.38×10^{-10}
rs3819496	8:143923891	G	0.312	0.321	0.421	BP0.442	*GML–CYP11B1*	intron variant	rs3753123; rs143247792; rs4527848; rs4606038; rs28524031	0.892 (0.857.0.929)	3.73×10^{-8}
rs140473396	10:104795885	GAC	0.247	0.285	0.097	0.197	*CNNM2–NT5C2*	intron variant	rs11191479; rs11191484; rs72050190; rs145010450; rs10883815	0.836 (0.800–0.873)	1.11×10^{-15}
rs12229654	12:111414461	G	0.141	0.159	0.000	0.000	*MYL2–CUX2*	intergenic region	rs149607519; rs148177611; rs2188380; rs12227162	0.834 (0.787–0.883)	5.25×10^{-10}
rs1887320	20:10965998	G	0.478	0.540	0.461	0.264	*JAG1*	intergenic region	rs6108787; rs1327235; rs6108789; rs913220	0.892 (0.859–0.925)	1.45×10^{-9}

SNP, single-nucleotide polymorphism; Chr, chromosome; BP, base pair; EAS, East Asian; EUR, European; AMR, American; N/A, not applicable; MAF, major allele frequency; A1, minor allele; OR, odds ratio; 95% CI, 95% confidence interval. * The cluster SNP is the top five SNPs with an R^2 value of 0.8 or higher, and within a ±200 kb range.

A Miami plot shows *p*-values for the SNP associations with hypertension in participants whose sodium intake was either <2 g/day or ≥2 g/day (Figure 2).

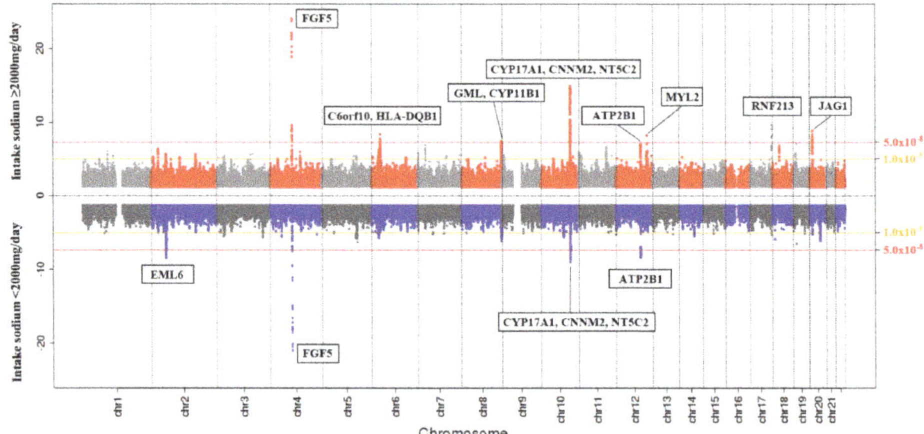

Figure 2. A Miami plot shows *p*-values for the SNP associations with hypertension in participants whose sodium intake was either <2 g/day or ≥2 g/day.

4. Discussion

This study identified both shared loci and sodium intake-specific loci related to hypertension. Fibroblast growth factor 5 (FGF5), PR domain zinc finger protein 8(PRDM8)-FGF5, 5'-nucleotidase, cytosolic II(NT5C2), ATPase plasma membrane Ca2+ transporting 1(ATP2B1), long intergenic non-protein coding RNA 936(LINC00936), and cyclin and CBS domain divalent metal cation transport mediator 2(CNNM2)-NT5C2 were commonly identified loci both in participants whose intakes were less than 2 g/day and in those with intakes greater than 2 g/day. Chromosome 6 open reading frame 10(C6orf10), human leukocyte antigen (HLA)-DQB1, ring finger protein (RNF)213, glycosylphosphatidylinositol anchored molecule-like (GML), cytochrome P450 family 11 subfamily B Member 1(CYP11B1), myosin light chain 2 (MYL2), cut like homeobox 2 (CUX2), and jagged1(JAG1) were significantly associated with hypertension in participants whose sodium intake was greater than 2 g/day, while loci in echinoderm microtubule-associated protein-like 6 (EML6) were significantly associated with hypertension in participants whose sodium intake was less than 2 g/day.

Guyton [24] established that long-term elevation of blood pressure is caused by vasoconstriction including the renal arteries or excess sodium retention through the kidney. The role of the kidney in BP control had been discovered by hypotension or hypertension caused by gene mutations which affect net renal sodium reabsorption [25]. For example, the loss of function mutations of the thiazide-sensitive NaCl symporter (e.g., Gitelman syndrome) impairs sodium reabsorption in the distal convoluted tubes and this results in a loss of sodium, potassium, and magnesium and a decrease in BP [25]. Enhanced tubular reabsorption of salt is important in the pathogenesis of obesity-related hypertension by regulating phosphorylation of Na^+-K^+-$2Cl^-$ cotransporter and regulation of STE20/SPS1-related proline/alanine-rich kinase (SPAK)/oxidative-stress-responsive kinase-1 (OSR1) by AMP-activated protein kinase [26]. Recently, it was discovered that genetic variations at a number of loci increases susceptibility to hypertension in the context of environmental exposures through a variety of physiological mechanisms. Salt sensitivity has been more frequently observed in black people than white people and hypertensive persons than normotensive persons [27,28]. Therefore, races and individuals' circumstance should be considered in salt intake and gene interaction studies [29,30]. In the current study, FGF5 rs16998073 and PRDM8-FGF5 rs12509595 were significantly associated with an increased risk of hypertension both in participants with sodium intake < 2 g/day and in those with intake ≥2 g/day. FGF5 rs16998073 was a well noted polymorphism in the largest GWAS performed by the Global Blood Pressure Genetics Consortium [31]. FGF5, a member of the fibroblast growth factor family, stimulates cell growth and proliferation of cardiac myocytes and promotes angiogenesis [32]. The association between FGF5 rs16998073 and hypertension was also recapitulated in a study of East Asians [33], and this polymorphism was shown to be associated with salt sensitivity in Koreans [34].

NT52C rs1191582, ATP2B1-LINC00936 rs11105378, and CNNM2-NT5C2 rs140473396 were significantly associated with decreased risk of hypertension, whether sodium intake was <2 or ≥2 g/day. NT52C rs11191582 is located in the gene-rich region near CYP17A1-CNNM2-NT5C2, which in GWAS was reported to contain a number of regulatory polymorphisms related to CVD [35,36]. Our study is the first to note the association of this polymorphism with hypertension. The association of ATP2B1 rs11105378 with hypertension was reported in European, Japanese and Korean studies [37]. ATP2B1 encodes the plasma membrane calcium transporting ATPase isoform 1, which plays a critical role in regulating blood pressure through alteration of intracellular calcium homeostasis and vasoconstriction in vascular smooth muscle cells [37,38]. CNNM2-NT5C2 rs140473396 was recently noted in a large, trans-ethnic study that included 776,078 participants from the Million Veteran Program, and in collaborating studies to identify the common variants, rare variants, and genetically predicted expression across multiple tissues of genes associated with blood pressure [39].

In the participants with sodium intake <2 g/day, we found significant association of rs67617923 in EML6 with increased risk of hypertension. While associations of several genetic variants in EML6 (e.g., rs17046380, rs72806698) with hypertension have been noted previously [40], rs67617923 is a novel genetic variant that was newly discovered in our study. Future studies to replicate this polymorphism association, and efforts to uncover the role of EML6 in blood pressure, are needed.

In participants with sodium intake ≥2 g/day, C6orf10-HLA-DQB1 rs6913309, and RNF213 rs112735431 were associated with increased risk of hypertension. C6orf10-HLA-DQB1 rs6913309 is another novel genetic variant that this study has newly discovered. An allele of HLA-DQB1 (which encodes a class II molecule expressed in antigen-presenting cells) increases the production of autoantibodies against angiotensin AT1 receptors, which was associated with essential hypertension in Chinese patients [41]. However, the exact role of HLA-DQB1 remains unclear. Lie et al. [42] found that the rs112735431 polymorphism of RNF213 was strongly associated with moyamoya disease in East Asian populations, including Chinese, Japanese, and Korean. This polymorphism has also been found to be related to intracranial artery steno-occlusive disease and moyamoya disease in Koreans [43]. The prevalence of moyamoya disease is 10 times higher in Japan and Korea than in Europe [44]. A previous study, which investigated the moyamoya disease susceptibility polymorphisms, reported that p.R4810K in RNF213 was found in the East Asian population but not in Southeast Asians [45]. Interestingly, the minor allele frequencies of the rs112735431 polymorphism were specified only in the East Asian population and in the present study. Although the physiologic function of RNF213 is not yet clear, previous studies found it to be involved in a novel signaling pathway in intracranial angiogenesis, and in the proliferation and maintenance of endothelial cells [42,46]. Ohkubo et al. [47] suggested that RNF213 promotes endothelial cell proliferation in response to inflammatory signals from the environment. Excess salt intake promotes vasoconstriction by decreasing nitric oxide production and increasing endothelial cell stiffness [48]. Furthermore, sodium intake is associated with systemic inflammation [49]. We may assume that excess sodium intake could be a provoking factor for the genetic effect of RNF213 on hypertension. Koizumi et al. [50] revealed that RNF213 was significantly associated with high BP in Japanese populations. Park et al. [43] also reported that the proportion of hypertension was higher in moyamoya diseases patients with the rs112735431 polymorphism of RNF213 than in those with wild type. This Korean GWAS was the first to note the association of this polymorphism with hypertension. Further studies to find association between the rs112735431 polymorphism in RNF213 and hypertension in other races/ethnicities are also needed.

We also found GML-CYP11B1 rs3819496, MYL2-CUX2 rs12229654, and JAG1 rs1887320 to be significantly associated with decreased risk of hypertension in participants with sodium intake ≥2 g/day. SNP rs3819496 represents a novel genetic variant, which was newly discovered in this study. Although MYL2-CUX2 rs12229654 and its association with hypertension were first reported in this study, a strong association of genetic variants of MYL2-CUX2 with high-density lipoprotein cholesterol was shown in a Korean GWAS meta-analysis, and it was replicated in a BioBank Japan GWAS, Health 2, and Shanghai Jiao Tong University cohort [51]. Another study conducted in Korea found that rs1229654 was also associated with dyslipidemia and diabetes [52]. Metabolic alteration due to rs1229654 might lead to the development of hypertension. Interestingly, the frequency of this polymorphism was determined only in the present study and in the East Asian population. Furthermore, persons carrying mutations in MYL-2, encoding slow cardiac myosin regulatory light chain 2, developed hypertrophic cardiomyopathy in the presence of hypertension or other risk factors for hypertrophy [53]. We may cautiously assume that excess salt intake might be an additional risk factor for hypertension or CVD in individuals with this genetic susceptibility. Association of JAG1 rs1887320 with hypertension and CVD risk was reported in Chinese cohorts [54,55].

Our study has certain limitations. We investigated the hypertension-related SNPs according to dietary sodium intake, as measured by FFQ. Although FFQ is a practical method to assess intake in large cohort studies, such questionnaires use a limited list of food items and cannot accurately consider additional salt intake via seasoning. Recall bias is another important limitation with FFQ. Second, we could not exclude the possibility of secondary hypertension due to a lack of information about it. Nevertheless, this is the first study to investigate hypertension-related SNPs according to sodium intake in a large population-based study. The current study identified previously well-reported SNPs related to hypertension. Furthermore, we identified several novel genetic variants associated with hypertension according to sodium intake.

5. Conclusions

In this large population-based study, we identified genetic susceptibility differences between participants whose sodium intake was less than 2 g/day and those whose intake was greater than 2 g/day. Discovering genetic predisposition for different sodium intakes would be helpful to establish the individualized medical nutrition therapy for disease management, and better targeted public health nutrition interventions. In further study, the effects and contributions of other confounding and interaction factors such as smoking, alcohol, and environmental factors on hypertension should be considered comprehensively.

Supplementary Materials: The following are available online at http://www.mdpi.com/2072-6643/12/9/2580/s1, Table S1: The hypertension susceptibility loci identified by GWAS signal in the Korean, Table S2: The significant association SNPs with hypertension susceptibility loci according to sodium intake (<2 g) in the Korean, Table S3: The significant association SNPs with hypertension susceptibility loci according to sodium intake greater than 2g per day in the Korean.

Author Contributions: Y.-J.K., J.O.K., J.-W.L., K.-W.H.: conception and design of data, analysis and interpretation of data, drafting the article or revising it critically for important intellectual content and final approval of the version to be published. J.-M.P., J.-E.C., D.-H.P., Y.S., S.-J.K.: analysis and interpretation of data. All authors have read and agreed to the published version of the manuscript.

Funding: This work was supported by the Bio and Medical Technology Development Program through the National Research Foundation of Korea, funded by the Ministry of Science, ICT, and Future Planning (NRF2018R1D1A1B07049223), and by the Technology Innovation Program (20002781, A Platform for Prediction and Management of Health Risk Based on Personal Big Data and Lifelogging) funded by the Ministry of Trade, Industry and Energy (MOTIE), Korea.

Acknowledgments: This study was conducted with bioresources from National Biobank of Korea, the Centers of Disease Control and Prevention, Republic of Korea (2019-059).

Conflicts of Interest: The authors declare no conflict of interest.

References

1. Shrimanker, I.; Bhattarai, S. Electrolytes. In *StatPearls*; StatPearls Publishing: Treasure Island, FL, USA, 2020.
2. GBD 2017 Risk Factor Collaborators. Global, regional, and national comparative risk assessment of 84 behavioural, environmental and occupational, and metabolic risks or clusters of risks for 195 countries and territories, 1990–2017: A systematic analysis for the Global Burden of Disease Study 2017. *Lancet* **2018**, *392*, 1923–1994.
3. Strazzullo, P.; D'Elia, L.; Kandala, N.B.; Cappuccio, F.P. Salt intake, stroke, and cardiovascular disease:meta-analysis of prospective studies. *BMJ* **2009**, *339*, b4567. [CrossRef]
4. He, F.J.; Li, J.; Macgregor, G.A. Effect of longer term modest salt reduction on blood pressure: Cochrane systematic review and meta-analysis of randomised trials. *BMJ* **2013**, *346*, f1325. [CrossRef] [PubMed]
5. Khaw, K.T.; Bingham, S.; Welch, A.; Luben, R.; O'Brien, E.; Wareham, N.; Day, N. Blood pressure and urinary sodium in men and women: The Norfolk Cohort of the European Prospective Investigation into Cancer (EPIC-Norfolk). *Am. J. Clin. Nutr.* **2004**, *80*, 1397–1403. [CrossRef] [PubMed]
6. World Health Organization. *Guideline: Sodium Intake for Adults and Children*; World Health Organization: Geneva, Switzerland, 2012.
7. Ivy, J.R.; Bailey, M.A. Pressure natriuresis and the renal control of arterial blood pressure. *J. Physiol.* **2014**, *592*, 3955–3967. [CrossRef]
8. Suzuki, H.; Kimmel, P.L. *Nutrition and Kidney Disease: A New Era*; Karger Medical and Scientific Publishers: Basel, Switzerland, 2007; Volume 155.
9. Huang, L.; Chu, Y.; Huang, X.; Ma, S.; Lin, K.; Huang, K.; Sun, H.; Yang, Z. Association between gene polymorphisms of voltage-dependent Ca^{2+} channels and hypertension in the dai people of china: A case-control study. *BMC Med. Genet.* **2020**, *21*, 44. [CrossRef]
10. Mervaala, E.M.; Himberg, J.J.; Laakso, J.; Tuomainen, P.; Karppanen, H. Beneficial effects of a potassium- and magnesium-enriched salt alternative. *Hypertension* **1992**, *19*, 535–540. [CrossRef] [PubMed]
11. Kathryn, L.M. *Pathophysiology: The Biologic Basis for Disease in Adults and Children*, 7th ed.; Elsevier, Mosby: St. Louis, MO, USA, 2014.
12. Asferg, C.L.; Nielsen, S.J.; Andersen, U.B.; Linneberg, A.; Møller, D.V.; Hedley, P.L.; Christiansen, M.; Goetze, J.P.; Esler, M.; Jeppesen, J.L. Relative atrial natriuretic peptide deficiency and inadequate renin and angiotensin ii suppression in obese hypertensive men. *Hypertension* **2013**, *62*, 147–153. [CrossRef] [PubMed]
13. Stolarz-Skrzypek, K.; Kuznetsova, T.; Thijs, L.; Tikhonoff, V.; Seidlerová, J.; Richart, T.; Jin, Y.; Olszanecka, A.; Malyutina, S.; Casiglia, E.; et al. Fatal and nonfatal outcomes, incidence of hypertension, and blood pressure changes in relation to urinary sodium excretion. *JAMA* **2011**, *305*, 1777–1785. [CrossRef]
14. O'Donnell, M.J.; Yusuf, S.; Mente, A.; Gao, P.; Mann, J.F.; Teo, K.; McQueen, M.; Sleight, P.; Sharma, A.M.; Dans, A.; et al. Urinary sodium and potassium excretion and risk of cardiovascular events. *JAMA* **2011**, *306*, 2229–2238. [CrossRef]
15. O'Donnell, M.; Mente, A.; Rangarajan, S.; McQueen, M.J.; Wang, X.; Liu, L.; Yan, H.; Lee, S.F.; Mony, P.; Devanath, A.; et al. Urinary sodium and potassium excretion, mortality, and cardiovascular events. *N. Engl. J. Med.* **2014**, *371*, 612–623. [CrossRef] [PubMed]
16. Carretero, O.A.; Oparil, S. Essential hypertension. Part I: Definition and etiology. *Circulation* **2000**, *101*, 329–335. [CrossRef] [PubMed]
17. Doris, P.A. The genetics of blood pressure and hypertension: The role of rare variation. *Cardiovasc. Ther.* **2011**, *29*, 37–45. [CrossRef] [PubMed]
18. Levy, D.; DeStefano, A.L.; Larson, M.G.; O'Donnell, C.J.; Lifton, R.P.; Gavras, H.; Cupples, L.A.; Myers, R.H. Evidence for a gene influencing blood pressure on chromosome 17. Genome scan linkage results for longitudinal blood pressure phenotypes in subjects from the framingham heart study. *Hypertension* **2000**, *36*, 477–483. [CrossRef] [PubMed]
19. Wang, D.G.; Fan, J.B.; Siao, C.J.; Berno, A.; Young, P.; Sapolsky, R.; Ghandour, G.; Perkins, N.; Winchester, E.; Spencer, J.; et al. Large-scale identification, mapping, and genotyping of single-nucleotide polymorphisms in the human genome. *Science* **1998**, *280*, 1077–1082. [CrossRef] [PubMed]
20. Stover, P.J. Influence of human genetic variation on nutritional requirements. *Am. J. Clin. Nutr.* **2006**, *83*, 436S–442S. [CrossRef]

21. Kim, Y.; Han, B.G. Cohort Profile: The Korean Genome and Epidemiology Study (KoGES) Consortium. *Int. J. Epidemiol.* **2017**, *46*, 1350. [CrossRef]

22. Shim, J.S.; Oh, K.; Kim, H.C. Dietary assessment methods in epidemiologic studies. *Epidemiol. Health* **2014**, *36*, e2014009. [CrossRef]

23. Moon, S.; Kim, Y.J.; Han, S.; Hwang, M.Y.; Shin, D.M.; Park, M.Y.; Lu, Y.; Yoon, K.; Jang, H.M.; Kim, Y.K.; et al. The Korea Biobank Array: Design and Identification of Coding Variants Associated with Blood Biochemical Traits. *Sci. Rep.* **2019**, *9*, 1382. [CrossRef]

24. Guyton, A.C. Blood pressure control—Special role of the kidneys and body fluids. *Science* **1991**, *252*, 1813–1816. [CrossRef]

25. Gug, C.; Mihaescu, A.; Mozos, I. Two mutations in the thiazide-sensitive nacl co-transporter gene in a romanian gitelman syndrome patient: Case report. *Ther. Clin. Risk Manag.* **2018**, *14*, 149–155. [CrossRef] [PubMed]

26. Davies, M.; Fraser, S.A.; Galic, S.; Choy, S.W.; Katerelos, M.; Gleich, K.; Kemp, B.E.; Mount, P.F.; Power, D.A. Novel mechanisms of na+ retention in obesity: Phosphorylation of nkcc2 and regulation of spak/osr1 by ampk. *Am. J. Physiol. Renal Physiol.* **2014**, *307*, F96–F106. [CrossRef] [PubMed]

27. Weinberger, M.H.; Miller, J.Z.; Luft, F.C.; Grim, C.E.; Fineberg, N.S. Definitions and characteristics of sodium sensitivity and blood pressure resistance. *Hypertension* **1986**, *8*, II127. [CrossRef] [PubMed]

28. Elijovich, F.; Weinberger, M.H.; Anderson, C.A.; Appel, L.J.; Bursztyn, M.; Cook, N.R.; Dart, R.A.; Newton-Cheh, C.H.; Sacks, F.M.; Laffer, C.L. Salt sensitivity of blood pressure: A scientific statement from the american heart association. *Hypertension* **2016**, *68*, e7–e46. [CrossRef] [PubMed]

29. Flack, J.M.; Ensrud, K.E.; Mascioli, S.; Launer, C.A.; Svendsen, K.; Elmer, P.J.; Grimm, R.H., Jr. Racial and ethnic modifiers of the salt-blood pressure response. *Hypertension* **1991**, *17*, I115–I121. [CrossRef]

30. Weinberger, M.H. Salt sensitivity of blood pressure in humans. *Hypertension* **1996**, *27*, 481–490. [CrossRef]

31. Newton-Cheh, C.; Johnson, T.; Gateva, V.; Tobin, M.D.; Bochud, M.; Coin, L.; Najjar, S.S.; Zhao, J.H.; Heath, S.C.; Eyheramendy, S.; et al. Genome-wide association study identifies eight loci associated with blood pressure. *Nat. Genet.* **2009**, *41*, 666–676. [CrossRef]

32. Vatner, S.F. FGF induces hypertrophy and angiogenesis in hibernating myocardium. *Circ. Res.* **2005**, *96*, 705–707. [CrossRef]

33. Xi, B.; Shen, Y.; Reilly, K.H.; Wang, X.; Mi, J. Recapitulation of four hypertension susceptibility genes (CSK, CYP17A1, MTHFR, and FGF5) in East Asians. *Metabolism* **2013**, *62*, 196–203. [CrossRef]

34. Rhee, M.Y.; Yang, S.J.; Oh, S.W.; Park, Y.; Kim, C.I.; Park, H.K.; Park, S.W.; Park, C.Y. Novel genetic variations associated with salt sensitivity in the Korean population. *Hypertens. Res.* **2011**, *34*, 606–611. [CrossRef]

35. Cheema, A.N.; Rosenthal, S.L.; Ilyas Kamboh, M. Proficiency of data interpretation: Identification of signaling SNPs/specific loci for coronary artery disease. *Database* **2017**, *2017*. [CrossRef] [PubMed]

36. Bragina, E.Y.; Goncharova, I.A.; Garaeva, A.F.; Nemerov, E.V.; Babovskaya, A.A.; Karpov, A.B.; Semenova, Y.V.; Zhalsanova, I.Z.; Gomboeva, D.E.; Saik, O.V. Molecular relationships between bronchial asthma and hypertension as comorbid diseases. *J. Integr. Bioinform.* **2018**, *15*. [CrossRef] [PubMed]

37. Tabara, Y.; Kohara, K.; Kita, Y.; Hirawa, N.; Katsuya, T.; Ohkubo, T.; Hiura, Y.; Tajima, A.; Morisaki, T.; Miyata, T.; et al. Common variants in the ATP2B1 gene are associated with susceptibility to hypertension: The Japanese Millennium Genome Project. *Hypertension* **2010**, *56*, 973–980. [CrossRef] [PubMed]

38. Kobayashi, Y.; Hirawa, N.; Tabara, Y.; Muraoka, H.; Fujita, M.; Miyazaki, N.; Fujiwara, A.; Ichikawa, Y.; Yamamoto, Y.; Ichihara, N.; et al. Mice lacking hypertension candidate gene ATP2B1 in vascular smooth muscle cells show significant blood pressure elevation. *Hypertension* **2012**, *59*, 854–860. [CrossRef]

39. Giri, A.; Hellwege, J.N.; Keaton, J.M.; Park, J.; Qiu, C.; Warren, H.R.; Torstenson, E.S.; Kovesdy, C.P.; Sun, Y.V.; Wilson, O.D.; et al. Trans-ethnic association study of blood pressure determinants in over 750,000 individuals. *Nat. Genet.* **2019**, *51*, 51–62. [CrossRef]

40. Takeuchi, F.; Akiyama, M.; Matoba, N.; Katsuya, T.; Nakatochi, M.; Tabara, Y.; Narita, A.; Saw, W.Y.; Moon, S.; Spracklen, C.N.; et al. Interethnic analyses of blood pressure loci in populations of East Asian and European descent. *Nat. Commun.* **2018**, *9*, 1–16. [CrossRef]

41. Zhu, F.; Sun, Y.; Wang, M.; Ma, S.; Chen, X.; Cao, A.; Chen, F.; Qiu, Y.; Liao, Y. Correlation between HLA-DRB1, HLA-DQB1 polymorphism and autoantibodies against angiotensin AT(1) receptors in Chinese patients with essential hypertension. *Clin. Cardiol.* **2011**, *34*, 302–308. [CrossRef]

42. Liu, W.; Morito, D.; Takashima, S.; Mineharu, Y.; Kobayashi, H.; Hitomi, T.; Hashikata, H.; Matsuura, N.; Yamazaki, S.; Toyoda, A.; et al. Identification of RNF213 as a susceptibility gene for moyamoya disease and its possible role in vascular development. *PLoS ONE* **2011**, *6*, e22542. [CrossRef]
43. Park, M.G.; Shin, J.H.; Lee, S.W.; Park, H.R.; Park, K.P. RNF213 rs112735431 polymorphism in intracranial artery steno-occlusive disease and moyamoya disease in Koreans. *J. Neurol. Sci.* **2017**, *375*, 331–334. [CrossRef]
44. Guey, S.; Kraemer, M.; Hervé, D.; Ludwig, T.; Kossorotoff, M.; Bergametti, F.; Schwitalla, J.C.; Choi, S.; Broseus, L.; Callebaut, I.; et al. Rare rnf213 variants in the c-terminal region encompassing the ring-finger domain are associated with moyamoya angiopathy in caucasians. *Eur. J. Hum. Genet.* **2017**, *25*, 995–1003. [CrossRef]
45. Liu, W.; Hitomi, T.; Kobayashi, H.; Harada, K.H.; Koizumi, A. Distribution of moyamoya disease susceptibility polymorphism p.R4810k in rnf213 in east and southeast asian populations. *Neurol Med. Chir.* **2012**, *52*, 299–303. [CrossRef]
46. Fujimura, M.; Sonobe, S.; Nishijima, Y.; Niizuma, K.; Sakata, H.; Kure, S.; Tominaga, T. Genetics and Biomarkers of Moyamoya Disease: Significance of RNF213 as a Susceptibility Gene. *J. Stroke* **2014**, *16*, 65–72. [CrossRef] [PubMed]
47. Ohkubo, K.; Sakai, Y.; Inoue, H.; Akamine, S.; Ishizaki, Y.; Matsushita, Y.; Sanefuji, M.; Torisu, H.; Ihara, K.; Sardiello, M.; et al. Moyamoya disease susceptibility gene RNF213 links inflammatory and angiogenic signals in endothelial cells. *Sci. Rep.* **2015**, *5*, 13191. [CrossRef] [PubMed]
48. Toda, N.; Arakawa, K. Salt-induced hemodynamic regulation mediated by nitric oxide. *J. Hypertens.* **2011**, *29*, 415–424. [CrossRef] [PubMed]
49. Fogarty, A.W.; Lewis, S.A.; McKeever, T.M.; Britton, J.R. Is higher sodium intake associated with elevated systemic inflammation? A population-based study. *Am. J. Clin. Nutr.* **2009**, *89*, 1901–1904. [CrossRef]
50. Koizumi, A.; Kobayashi, H.; Liu, W.; Fujii, Y.; Senevirathna, S.T.; Nanayakkara, S.; Okuda, H.; Hitomi, T.; Harada, K.H.; Takenaka, K.; et al. P.R4810K, a polymorphism of RNF213, the susceptibility gene for moyamoya disease, is associated with blood pressure. *Environ. Health Prev. Med.* **2013**, *18*, 121–129. [CrossRef]
51. Kim, Y.J.; Go, M.J.; Hu, C.; Hong, C.B.; Kim, Y.K.; Lee, J.Y.; Hwang, J.Y.; Oh, J.H.; Kim, D.J.; Kim, N.H.; et al. Large-scale genome-wide association studies in East Asians identify new genetic loci influencing metabolic traits. *Nat. Genet.* **2011**, *43*, 990–995. [CrossRef]
52. Heo, S.G.; Hwang, J.Y.; Uhmn, S.; Go, M.J.; Oh, B.; Lee, J.Y.; Park, J.W. Male-specific genetic effect on hypertension and metabolic disorders. *Hum. Genet.* **2014**, *133*, 311–319. [CrossRef]
53. Claes, G.R.; van Tienen, F.H.; Lindsey, P.; Krapels, I.P.; Helderman-van den Enden, A.T.; Hoos, M.B.; Barrois, Y.E.; Janssen, J.W.; Paulussen, A.D.; Sels, J.W.; et al. Hypertrophic remodelling in cardiac regulatory myosin light chain (MYL2) founder mutation carriers. *Eur. Heart J.* **2016**, *37*, 1815–1822. [CrossRef]
54. Lu, X.; Huang, J.; Wang, L.; Chen, S.; Yang, X.; Li, J.; Cao, J.; Chen, J.; Li, Y.; Zhao, L.; et al. Genetic predisposition to higher blood pressure increases risk of incident hypertension and cardiovascular diseases in Chinese. *Hypertension* **2015**, *66*, 786–792. [CrossRef]
55. Lu, X.; Wang, L.; Lin, X.; Huang, J.; Charles Gu, C.; He, M.; Shen, H.; He, J.; Zhu, J.; Li, H.; et al. Genome-wide association study in Chinese identifies novel loci for blood pressure and hypertension. *Hum. Mol. Genet.* **2015**, *24*, 865–874. [CrossRef] [PubMed]

Article

Increased Risk of High Body Fat and Altered Lipid Metabolism Associated to Suboptimal Consumption of Vitamin A Is Modulated by Genetic Variants rs5888 (*SCARB1*), rs1800629 (*UCP1*) and rs659366 (*UCP2*)

Sebastià Galmés [1,2,3,4], **Andreu Palou** [1,2,3,*] and **Francisca Serra** [1,2,3,4]

[1]   Laboratory of Molecular Biology, Nutrition and Biotechnology, NUO Group, Universitat de les Illes Balears, 07122 Palma, Spain; s.galmes@uib.cat (S.G.); francisca.serra@uib.es (F.S.)
[2]   CIBER de Fisiopatología de la Obesidad y Nutrición (CIBEROBN), 28029 Madrid, Spain
[3]   Institut d'Investigació Sanitària Illes Balears (IdISBa), 07120 Palma, Spain
[4]   Alimentómica S.L., Spin-off n.1 of the University of the Balearic Islands, 07121 Palma, Spain
*   Correspondence: andreu.palou@uib.es; Tel.: +34-971-173170

Received: 13 May 2020; Accepted: 24 August 2020; Published: 26 August 2020

Abstract: Obesity is characterized by an excessive body fat percentage (BF%). Animal and cell studies have shown benefits of vitamin A (VA) on BF% and lipid metabolism, but it is still controversial in humans. Furthermore, although some genetic variants may explain heterogeneity in VA plasma levels, their role in VA metabolic response is still scarcely characterized. This study was designed as a combination of an observational study involving 158 male subjects followed by a study with a well-balanced genotype–phenotype protocol, including in the design an ex vivo intervention study performed on isolated peripheral blood mononuclear cells (PBMCs) of the 41 former males. This is a strategy to accurately identify the delivery of Precision Nutrition recommendations to targeted subjects. The study assesses the influence of rs5888 (*SCARB1*), rs659366 (*UCP2*), and rs1800629 (*UCP1*) variants on higher BF% associated with suboptimal VA consumption and underlines the cellular mechanisms involved by analyzing basal and retinoic acid (RA) response on PBMC gene expression. Data show that male carriers with the major allele combinations and following suboptimal-VA diet show higher BF% (adjusted ANOVA test p-value = 0.006). Genotype–BF% interaction is observed on oxidative/inflammatory gene expression and also influences lipid related gene expression in response to RA. Data indicate that under suboptimal consumption of VA, carriers of VA responsive variants and with high-BF% show a gene expression profile consistent with an impaired basal metabolic state. The results show the relevance of consuming VA within the required amounts, its impact on metabolism and energy balance, and consequently, on men's adiposity with a clear influence of genetic variants *SCARB1*, *UCP2* and *UCP1*.

Keywords: personalized nutrition; dietary vitamin A; obesity; body fat; UCP; retinoic acid; PBMC

---

1. Introduction

Obesity is a metabolic disease characterized by excessive fat accumulation which can promote the development of associated disorders, such as diabetes or cardiovascular events, resulting in an increased risk of mortality and considerable public health costs [1]. Furthermore, its prevalence has not stopped growing over the last few decades [2]. Meanwhile, poor consumption of vitamin A (VA) has been widely related to vision problems [3] and immune system alterations [4], but also with a higher prevalence of obesity [5,6] and fat accumulation [7,8].

Therefore, scientific evidence suggests a link between dietary VA and the regulation of energy balance. Dietary intake of vegetables and fruits facilitates the availability of pro-VA in the form of β-carotene (BC), which is enzymatically converted to retinaldehyde and then irreversibly oxidized to retinoic acid (RA) [9]; whereas animal sources provide retinol, mainly as retinyl esters with fatty acids, which is metabolized in the cells to retinaldehyde at a higher efficiency than BC [10]. The main active form of VA is RA which has the capacity to influence the expression of key genes related to lipid and energy homeostasis in mammals [11,12], actively participating in the modulation of adipocyte differentiation, lipogenesis/lipolysis, thermogenesis, and fat oxidation [11–13]. Furthermore, active VA metabolites increase fatty acid oxidative metabolism [14]. The main core of studies on supplementation with BC have been based on the assessment of BC antioxidant action [15], while studies aimed at inducing weight loss or body fat reduction using carotenoid supplementation are still scarce. Nevertheless, in a recent small double-blind randomized study performed on 17 children supplemented for 6 months with a combination of carotenoids, a reduction in parameters associated with adiposity and obesity was observed [16], in accordance with a recent meta-analysis linking carotenoids and VA with the occurrence of metabolic syndrome [17]. Although higher serum levels of carotenoids are inversely associated with metabolic syndrome, large interindividual variation as regards the metabolic conversion of dietary BC is also found [17].

In particular, genetic variants may support the foundations of interindividual variability in nutritional status and handling of VA. In fact, bioavailability of BC has been shown to be dependent on genes involved in postprandial chylomicron metabolism (ATP-binding cassette, subfamily A (*ABC1*, *ABCA1*); APOB; Transcription factor 7-like 2 (*TCF7L2*); and hepatic lipase (LIPC)) as well as in the uptake, absorption, and subsequent tissue management of BC (such as scavenger receptor class B, member 1 (*SCARB1*) and ATP-binding cassette, subfamily G, (*ABCG5*)) [18]. In addition, family studies have estimated that 30% of the variation in serum retinol is heritable [19], and a single nucleotide polymorphism (SNP) has already been related to plasma retinol levels [20]. Therefore, a core of evidence suggests the existence of genetic variants that may influence the metabolism of VA, either precursors or derived metabolites; although the underlying mechanisms are not totally understood. Furthermore, to date, no studies have assessed the metabolic impact of gene variants that may be predisposed to a greater risk of obesity in the case of following suboptimal intake of VA. In this research, three genetic variants that may determine the modulation of VA efficiency and metabolism were analyzed: rs5888 located on Scavenger Receptor Class B type 1 (*SCARB1*) which encodes the protein SR-B1, a multifunctional scavenger receptor involved in dietary/blood carotenoid cell uptake and transport [21]; rs1800592 associated with the Uncoupling Protein-1 (*UCP1*) with a key role in thermogenesis and fatty acid oxidation and inducible by carotenoids and retinoids [8,22]; and rs659366 on Uncoupling Protein-2 (*UCP2*) which is involved in oxidative cell status and is induced by vitamin A [8,23]. In addition, previous studies suggest that the presence of certain alleles of these SNPs could be related to differential bioactive compound transport or responsive capacity [24–26] (see details in Table S1).

Therefore, we conducted a study to obtain further insight on the role of these specific gene variants in the susceptibility to obesity under suboptimal VA status, which was performed in two sequential parts. First, the study analyzed the role of the three genetic variants in a Spanish population (Mallorca) in order to assess whether genetics could contribute to explain the greater rate of obesity associated with a suboptimal consumption of VA described in animal models. Then, ex vivo exposure of Peripheral Blood Mononuclear Cell (PBMC) samples to RA was performed to gain further insight into the mechanisms involved by analyzing the expression of a set of lipid and oxidative metabolism key genes.

2. Materials and Methods

2.1. Subjects

Protocol was in accordance with the Declaration of Helsinki principles and approved by the ethics committee (Comitè d'Ètica de la Investig. de les Illes Balears, CEI-IB). Procedures and a flowchart are summarized in Figure 1. At the first step, information concerning specifications of the subjects, anthropometrics, and habitual intake, was collected from each participant, together with a saliva sample for genotyping (following the protocol defined at Ob-IB study (approval IB2009/13)). At the next step, we focused the study on the male cohort in order to characterize the relationship observed among adiposity and intake of retinol source foods. Subjects were selected following a "case–control" protocol and the design included an ex vivo intervention study (OptiDiet-15 study, IB2569/15) performed on isolated PBMCs of the former subjects—a selected subset of 41 men, taking into account their dietary, genetic, and anthropometric profile.

Specifically, the ninety men who did not meet nutritional requirements for vitamin A (<750 µg/day) were initially preselected for the OptiDiet-15 study and were stratified by genotype and body fat (BF). Around 60% of them were contacted again to ask for their wiliness to participate in the OptiDiet-15 study to characterize the effects of a low-VA diet associated with the presence of allele combinations of selected SNPs—rs5888, rs659366, and rs1800592. To minimize any bias, care was taken to select balanced groups concerning VA more responsive and less responsive genotypes (hereafter called Genotype A, n = 21; Genotype B, n = 20, respectively) and BF% (high BF% ≥ 25; n = 21; low BF% < 25; n = 20). Those subjects who were willing to participate were checked for diet and anthropometry to verify that the subjects still fulfilled the selection criteria, to update the information, and to further confirm their grouping. Incorporation into the different groups was managed in parallel to avoid imbalances. Eighty percent of the subjects confirmed suboptimal habitual intake of VA, and those not exceeding 25% of Population Reference Intake (PRI) (<937.5 µg/day) were considered suitable to be further analyzed. Recruitment finished when balanced groups with the desired genotype combinations and BF% were obtained.

A blood sample was taken for plasma determinations and PBMC extraction for ex vivo incubation and gene expression analysis.

2.2. Anthropometric Measures

Body fat percentage (BF%) was measured with a bio-impedance apparatus (OMRON BF306, Kyoto, Japan). The waist–hip (WHR) and the waist-to-height ratios (WtHR) were obtained by waist circumference/hip circumference and waist circumference/height, respectively. Body adiposity index (BAI) was obtained using the formula [27]: BAI = ((hip circumference/height$^{1.5}$) − 18). Bicipital, tricipital, subscapular, supraspinatus, and abdominal skinfolds were measured using a Harpenden caliper (Baty International, Burgess Hill, UK).

2.3. Estimation of Dietary Intake

Dietary intake was assessed with a 24 h recall report (24RR) during individual face-to-face interviews. We collected up to three 24RR for each participant to ensure the quality of the data recorded, wishing to diminish the influence of random measurement error and to correct for day-to-day variation within subjects. The subject mean of this set of 24RRs was used as a proxy of habitual intake of VA. In the OptiDiet-15 cohort, dietary intake was characterized with two and three 24RRs, in 71% and 27% of the subjects, respectively. Concerning the 41 subjects finally characterized at the PBMC level, we reinforced the number of subjects with three 24RRs (31%) and analyzed 66% with two 24RRs.

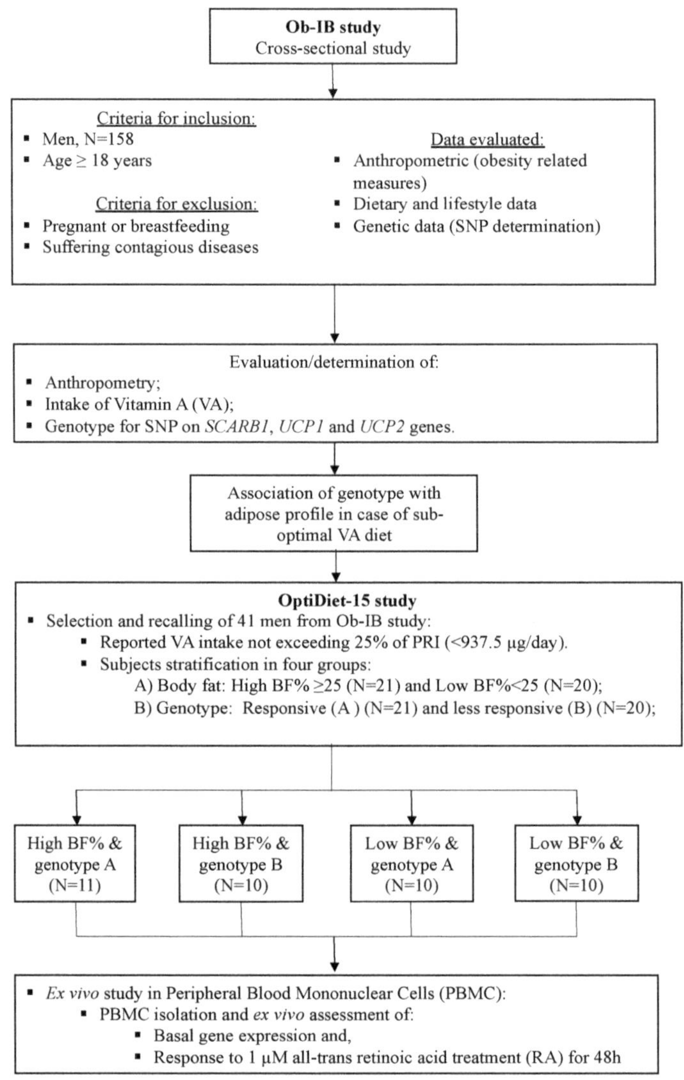

Figure 1. Flowchart diagram of the study. Flowchart diagram of the main steps of the study, including data sets recorded, criteria established for subject selection, number of subjects taken into account, and main procedures followed at each stage. PRI-Population Reference Intake, SNP-single nucleotide polymorphism.

In addition, we applied a standardized protocol aiming to minimize forgotten food items and correctly estimate portion sizes of foods. Therefore, intake data were collected during face-to face interviews. The researcher (S. Galmés) handled an image book containing habitual dishes and recipes which was used to help the individuals to form an accurate report of the last 24 h intake. In addition, the book also included different sizes either of single foods (as an apple or a slice of bread) or more complex dishes (spaghetti, paella, etc.). This was used to better define the amount of food eaten.

Then, this food information was transferred to the software DIAL that includes the composition of most Spanish foods and dishes and also allows the introduction of new food items and composition. The intake of each dietary ingredient was converted to energy and nutrient composition using the dietary software DIAL v2.0 (Alce-Ingeniería, Madrid, Spain) [28]. Total VA intake was determined (µg of retinol + (carotenoids with vitamin activity/6)). Population Reference Intake (PRI) by European Food Safety Authority (EFSA) for men was used as a cut-off point (750 µg/day) [29] to classify VA intake.

Individuals who reported taking supplements that may affect the nutritional status of vitamin A were not included in the analysis.

2.4. SNP Selection and Population Grouping According to Genotype

Genetic variants rs5888, rs659366, and rs1800592 located on *SCARB1*, Uncoupling Proteins 2 and 1 (*UCP2*, *UCP1*) genes, respectively, were selected because of their location on genes involved in VA absorption, handling, and/or metabolism and based on their response to bioactive compounds [13,22,30–37] (see Table S1 for a detailed summary).

In brief, the T variant of SNP on *SCARB1* is associated with decreased levels of its protein in vitro [38] and with lower levels of fat-soluble antioxidants in plasma [24]. Thus, it was hypothesized that the effect of the T allele (TT + TC genotypes) could be counteracted with high VA intake. Regarding rs659366 and rs1800592, T allele carriers and AA genotype, respectively, showed a greater anti-obesogenic response to bio-active compounds together with a higher expression of their respective genes [39–41]. Accordingly, a dominant model for *SCARB1* rs5888 and *UCP2* rs659366 (TT + TC vs. CC, for both of them) and a recessive model for *UCP1* rs1800592 (AA vs. AG + GG) were produced. Therefore, driving towards the hypothesis that subjects with T (rs5888), T (rs659366), and/or without G (rs1800592) would have a greater predisposition to respond more effectively to bio-active compounds, such as VA, and/or show major metabolic risk in the case of suboptimal intake.

According to this, the genotype grouping of subjects in both Ob-IB and OptiDiet-15 studies was performed according the presence of two or more "responsive" genotypes (TT or TC for rs5888; AA for rs1800592; and/or TT or TC for rs659366), tested as Genotype "Responsive" (A), in contrast with subjects having at most one of the responsive genotypes aforementioned, which were defined as Genotype B (less responsive).

2.5. DNA Extraction and Genotype Determination

Saliva samples were obtained following a standardized procedure. In brief, thirty minutes before sample collection, the subjects were requested to avoid eating, drinking, smoking, and chewing gum. Then, they were asked to spit into a collection tube until the liquid saliva reached 2 mL, without taking bubbles into account. Then, fresh saliva was immediately used to isolate genomic DNA or stored in adequate conditions (4 °C) until DNA extraction was performed. Isolation of genomic DNA was carried out using the commercial kit High Pure PCR template Preparation Kit (Roche, Basel, Switzerland). Genotyping was performed by qPCR (LightCycler®480 FastStart DNA, Roche, Basel, Switzerland), FastStart master mix (Roche, Basel, Switzerland), and specific-SNP (Tib Molbiol, Berlin, Germany) following the conditions described elsewhere [42].

2.6. Blood Sample Collection, PBMC Isolation, and Ex Vivo Treatment

Peripheral Blood Mononuclear Cells (PBMCs) constitute a source of biomarkers of metabolic status as well as a potential ex vivo system to test food bioactives' efficacy—easier to obtain than samples of other tissues, such as adipose tissue or liver [43]. Therefore, venous blood (20–25 mL) was collected from volunteers early in the morning in the fasting state using vacutainer-EDTA tubes. PBMCs were isolated by density gradient using Ficoll-Paque Plus (Healthcare Bio Science, Barcelona, Spain) following the protocol described in Cifre et al., 2016 [44]. Finally, 1×10^6 cells were activated with 0.5×10^6 of CD3/CD28 magnetic beads (Life Technologies, Madrid, Spain) and maintained in suspension in a RPMI-1640 medium, with fetal bovine serum (10%), L-glutamine (1%), penicillin (100 units/mL) streptomycin (100 µg/mL), and DMSO (control cells) or all-trans retinoic acid (RA) (1 µM) for 48 h at 37 °C and 5% CO_2. Medium components and treatments were purchased from Sigma-Aldrich (St Louis, MO, USA).

2.7. RNA Extraction and Gene Expression Analysis

After the incubation period, total RNA was isolated using Direct-zol RNA Mini-Prep (Zymo Research Corp, Irvine, CA, USA). mRNA concentration was determined by a spectrophotometric-based system (NanoDrop 1000, ThermoFisher Scientific, Waltham, MA, USA). RNA purity was assessed by the absorbance ratios at 260/280 and 260/230 nm. Randomized quality controls were introduced to check RNA integrity. This was assessed performing a 1% agarose gel electrophoresis. RNA (0.05 µg) were transcribed into cDNA using a highly sensitive first-strand cDNA synthesis kit (iScript cDNA, Bio-Rad Laboratories, Madrid, Spain). Real-time PCR was performed for each RT product to determine mRNA expression using the Power SYBR Green PCR Master Mix (Applied Biosystems, Madrid, Spain) as described in a previous publication [44]. All primers were purchased from Sigma Genosys (Sigma-Aldrich Química SA, Madrid, Spain). Two common housekeeping genes on PBMCs were analyzed—Elongation factor 1-alpha 1 (EF1a1) and Ribosomal Protein Large P0 (*RPLP0*).

The threshold cycles (Cts) were obtained using the StepOne v2.0 software (Applied Biosystems, Madrid, Spain). Then, the Cts of each analyzed gene were normalized against the housekeeping gene RPLP0 from the same sample. This housekeeping gene gave threshold cycles closer to the ones of the genes of interest. Then, relative PBMC gene expression was calculated as a percentage referring to the gene expression of control cells from genotype A and low BF% subjects (considered as 100%). Gene expression was determined for a set of genes of relevance in lipid metabolism and/or with a role in the mediation of cellular oxidative stress. Therefore, mRNA levels of *LXRA*, *SOD2*, *SLC27A2*, *SREBP1C*, *SCARB1*, *CEBPB*, *RXRA*, *CPT1A*, and *UCP2* were analyzed.

2.8. Statistical Analyses

Descriptive data are generally presented as the mean and standard deviation. Data parametricity was assessed by the Kolmogorov–Smirnov and Levene tests, otherwise variables were $\log_{10}$ transformed. ANOVA tests adjusted for confounding variables (detailed in footnotes of each table or figure) were used to compare the means between groups. Appropriate intake of macro/micronutrients was assessed by a *t* test for a single sample by comparison with the recommendations set for adult men [29,45]. Linear regression analyses adjusted for age using the level of VA intake as a dichotomous (LI and RI) independent variable and body adiposity measures (BF% and BAI) were performed to confirm the nutrigenetic relationship between dietary VA and adiposity associated with the genotype. Two-way ANOVA tests were performed to evaluate the genotype–BF% interaction effects on PBMC basal gene expression followed by a least significant difference (LSD) posthoc comparison. ANOVA tests were adjusted for the main covariates (or combinations of covariates) that could be causing biases between the specific groups that were compared. In this regard, the main confounding variable for genotype groups was total energy intake, and these comparisons were adjusted for this covariant; the main confounding variable detected relative to the BF% groups was age. Consequently, the ANOVA tests to compare these groups were adjusted for age. Therefore, interaction ANOVA tests comprising genotype–BF% groups were adjusted for both confounding variables. The RA treatment effect was evaluated by a Student's *t* test for paired data by comparing control cells with RA-treated ones. The genotype–BF% interaction on RA treatment effects was assessed by ANOVA for paired data. All statistical analyses were performed using the SPSS v25.0 (IBM, Chicago, IL, USA), and the threshold of significance was defined at $p < 0.05$ for all analyses.

3. Results

3.1. Evaluation of VA Intake Level and Genotype Impact on Adiposity in Men of the Ob-IB Study

The main characteristics of the male cohort in the Ob-IB study are summarized in Table 1. Volunteers showed an anthropometric profile of overweight: body mass index (BMI) = 26.5 ± 5.0 kg/m^2 and waist–hip (WHR) = 0.93 ± 0.07. Dietary energy intake was quite similar (2231 kcal/day) to that reported in the ANIBES study for adult men (1966 kcal/day) [46] and close to actual recommendations for moderately sedentary adult men [47]. Analysis of the dietary profile showed an imbalanced diet except for protein consumption, which was within the EFSA recommendations for the general population (15.9%; p = 0.075 vs. recommended 15–20%). Thus, fat energy contribution was higher than recommended (37.2%; p = 0.005 vs. recommended 20–35%) at the expense of carbohydrates (42.6%; p = 0.003 vs. recommended 45–60%). Accordingly, intake of fiber was also below recommendations (22.8 g/day; p < 0.001 vs. recommended > 25 g/day), but simple sugars exceeded the advisable limit of 10% (16.2%; p < 0.001). Concerning the quality of dietary fat, only monounsaturated fatty acids (MUFA) were within recommendations (15.4%; p = 0.439 vs. recommended 15–20%), whereas saturated fatty acids (SFA) were higher (12.4%; p < 0.001 vs. recommended < 10%), and polyunsaturated fatty acids (PUFA) were below the recommended intake (6.1%; p < 0.001 vs. recommended 7–10%). The dietary profile found, including total calories and energy contribution of macronutrients, was very similar to that observed in the Spanish male population [46].

Allele frequencies of the SNPs studied were in general accordance with the frequency of the European population (available at 1000Genomes (http://www.internationalgenome.org)). Genotypes including the allele identified as potentially responsive to VA showed the highest frequency in the Ob-IB population: 66.5% of TT/TC for rs5888; 51.3% AA for rs1800592; 65.2% TT/TC for rs659366 (Table 1).

Concerning intake of VA, the reported daily consumption showed an average intake that was 48% higher than the Population Reference Intake for adult men [29] (PRI: 750 µg/day; p = 0.058) although with a high range of variability (range: 110–45,239 µg/day) (Table 1). Forty three percent of volunteers (irrespective of the genotype) reported intakes comprising PRI or higher (≥750 µg/day) and were classified as the recommended intake (RI) group, whereas the rest, with VA intake <750 µg/day, were classified as the low intake (LI) group. Subsequent stratification of subjects according to the genotype (responsive, A versus less responsive, B) allowed a link between genetic background and dietary VA on adiposity to be shown. BF% of Genotype B was 25.8% and 23.8% in the LI and RI groups, respectively. Whereas in Genotype A, BF% was 25.9% in the LI and 22.1% in the RI groups (p = 0.006) (Table 2), suggesting that the allele load in Genotype A could constitute a good proxy of genetic variants driving physiological response to VA, particularly in the context of energy balance and its interaction with adiposity. Linear regression analyses using the level of VA intake (LI and RI) and BF% after adjusting by age, confirmed the nutrigenetic relationship between dietary VA and propensity to obesity associated with the genotype. Genotype A individuals were significantly associated with lower BF% (β = −4.11, p = 0.006) and BAI (β = −1.99, p = 0.029) when fulfilling dietary VA recommendations (RI group) taking low VA intake as the reference group, whereas this was not the case for Genotype B subjects (BF% (β = −0.10, p = 0.972) and BAI (β = −0.28, p = 0.988)) in comparison with the low VA intake group (Table 3). In agreement with the hypothesis, data pointed out that Genotype A would be specifically susceptible to improving body composition by increasing VA consumption to current recommendations.

3.2. Assessment of the Influence of Suboptimal VA Intake, Genotype, and Adiposity on PBMC Metabolism

Aiming to further characterize the metabolic relevance of genotype on BF%, particularly under the influence of suboptimal VA intake, the OptiDiet-15 study was undertaken in the following steps. Subjects who had initially reported a VA intake below recommendations (<750 µg/day) were contacted again and rechecked. Eighty percent of them confirmed suboptimal habitual intake of VA, and subjects not exceeding 25% of PRI (<937.5 µg/day) were considered suitable to be further analyzed. Therefore, a case–control design, including balanced groups in terms of genotype (A or B) and BF% (cut off at 25%), was accomplished ($n = 41$) to study gene expression in isolated PBMC as well as to analyze PBMC ex vivo response to RA.

Anthropometric and dietary characteristics are shown in Table 4. No major anthropometric differences were found between genotypes within the same group of body fat. However, concerning dietary characteristics in High-BF% groups, Genotype A subjects reported higher energy intake than those belonging to Genotype B ($p = 0.007$), involving greater energy from MUFA (19.1%; $p = 0.003$) and higher protein intake (94.1 g/day; $p = 0.014$) in comparison to subjects with Genotype B (13.7% of MUFA; 72.1 g protein/day). In contrast, in Low-BF% groups, Genotype A subjects showed lower PUFA intake (4.9%) than Genotype B subjects (7.0%, $p = 0.027$).

Table 1. General anthropometric dietary and genetic characteristics of male Ob-IB study participants [1].

Male Subjects (Ob-IB Study) ($n = 158$)	Mean	SD	
Age (years)	37	17	
Anthropometric measures			
Height (cm)	175	7.59	
Weight (kg)	80.9	15.5	
Hip (cm)	98.6	10.1	
Waist (cm)	92.0	14.8	
WHR	0.93	0.07	
BAI	24.8	5.03	
BMI (kg/m^2)	26.5	5.04	
BF%	24.5	8.13	
Skinfolds (mm)			
Bicipital	7.27	4.52	
Tricipital	11.1	5.62	
Subscapular	14.1	6.59	
Supraspinatus	17.2	8.98	
Abdominal	21.7	9.84	
Dietary parameters			Recommendation
Energy intake (kcal/day)	2231	521	2000–2600
Carbohydrate (g/day)	237 (42.6%)	80.5	45–60% *
Fat (g/day)	92.1 (37.2%)	31.5	20–35% *
Proteins (g/day)	88.4 (15.9%)	26.1	15–20%
Fiber (g/day)	22.8	9.08	>25 g/day *
Vitamin A (µg/day)	1113	3718	750 [#]
Genetic features		Ob-IB (%)	1000 genomes (%)
rs5888 (*SCARB1*)	TT + TC	66.5	70.6
	CC	33.5	29.4
rs659366 (*UCP2*)	TT + TC	65.2	61.0
	CC	34.8	39.0
rs1800592 (*UCP1*)	AA	51.3	58.1
	AG + GG	48.7	41.9

[1] All values are the mean and SD (standard deviation). Dietary data are based on the analysis of a set of up to three dietary recalls of 24 h, and reported means are compared with current European Food Safety Authority (EFSA) recommendations. Statistical differences between Ob-IB study intakes and recommendations were assessed by a single sample t-test. * p-value < 0.05; [#] $0.05 < p$ value < 0.06. Genetic frequencies of the Ob-Ib study and 1000Genomes (for European populations) database are expressed as % for each genotype. Abbreviations: WHR (waist–hip ratio); BAI (body adiposity index); BMI (body mass index); BF% (body fat percentage).

Table 2. Anthropometric and dietary parameters by genotype group (A, Responsive to VA; B, Less responsive to VA) and Vitamin A (VA) intake level (LI, Low Intake <750 µg/day; RI, recommended intake ≥750 µg/day) [1].

| Variables | Genotype VA Responsive (A) (n = 106) | | | | | | | | | Genotype Less VA Responsive (B) (n = 52) | | | | | | | | | GxVA Interaction (p-Value) |
| | Low VA Intake (LI) (n = 60) | | | | Recommended VA Intake (RI) (n = 46) | | | | | Low VA Intake (LI) (n = 30) | | | | Recommended VA Intake (RI) (n = 22) | | | | | |
	Mean	SD	Min	Max	Mean	SD	Min	Max	p-Value	Mean	SD	Min	Max	Mean	SD	Min	Max	p-Value	
Weight (kg)	81.8	17.3	49.5	140	79.3	14.1	47.9	122	0.352	81.2	15.3	60.0	133	81.1	14.1	61.9	127	0.611	0.383
WHR	0.93	0.08	0.70	1.12	0.92	0.08	0.72	1.11	0.430	0.95	0.08	0.81	1.12	0.91	0.05	0.85	1.01	0.466	0.905
BMI (kg/m^2)	27.0	5.51	19.2	45.1	25.5	4.47	18.7	36.8	0.076	27.2	5.39	19.8	42.9	26.3	4.24	20.5	38.3	0.737	0.249
BF%	25.9	8.12	5.30	43.7	22.1	8.44	6.70	38.8	0.006	25.8	8.24	11.3	47.6	23.8	6.39	13.0	36.6	0.972	0.114
BAI	25.5	5.30	17.2	44.8	23.7	4.66	15.1	34.6	0.033	25.8	5.67	18.9	39.6	24.1	3.59	16.7	30.5	0.799	0.271
Bicipital SF	8.41	5.59	2.60	30.0	5.83	2.85	2.20	13.0	0.002	7.97	4.67	2.80	19.6	6.33	2.87	2.90	13.5	0.509	0.187
Energy (kcal/day)	2229	538	1253	374	2276	536	1384	4256	0.551	2058	456	1061	3132	2377	497	1615	3666	0.040	0.167
Carbohydrate (% EC)	43.7	10.5	21.8	73.7	41.1	9.64	20.5	57.6	0.194	43.2	9.18	30.0	64.3	42.1	10.2	19.9	63.3	0.508	0.739
Fat (% EC)	35.9	9.60	16.8	61.5	38.6	8.33	24.3	57.9	0.135	36.4	10.2	12.9	57.8	38.7	10.6	13.9	63.5	0.479	0.853
Proteins (g/day)	86.9	25.7	37.0	180	94.5	29.3	55.8	182	0.122	76.2	18.5	30.0	104	96.6	23.7	57.9	154	0.002	0.224
Fiber (g/day)	22.7	11.0	4.50	63.8	23.2	7.24	13.0	39.4	0.324	20.9	5.86	10.4	36.7	24.7	10.5	11.0	56.8	0.104	0.570
Vitamin A (µg/day)	458	148	110	744	2368	6762	751	45,239	0.000	498	131	196	722	1118	342	764	1887	0.000	0.138
Retinol (µg/day)	238	115	2.10	527	1703	6829	22.3	45,116	0.000	258	126	19.0	518	479	320	35.2	1625	0.020	0.558
ß-carotene (µg/day)	951	681	10.4	2999	3090	1930	15.6	9264	0.000	1140	788	120	3153	2795	1481	318	5166	0.000	0.682

[1] All values are expressed as the mean, SD (standard deviation), min (minimum), and max (maximum). Genotype A (responsive): TT or TC for rs5888, AA for rs1800592, and/or TT or TC for rs659366; Genotype B (less responsive), having at most one of the responsive genotypes aforementioned. Abbreviations: VA (vitamin A); G (genotype); WHR (waist–hip ratio); BMI (body mass index); BF% (body fat percentage); BAI (Body Adiposity Index); SF (skinfold); and EC (energy contribution). Statistical assessment of the differences between LI and RI groups within each genotype was performed by an ANOVA test adjusted for age.

Table 3. Linear regression table of the effect of low vs. high vitamin A intake on adiposity depending on the Genotype A/B [1].

Genotype A	Beta	SE	*p*-Value
BF%	−4.11	1.47	0.006
BAI	−1.99	0.92	0.033
Genotype B	**Beta**	**SE**	***p*-Value**
BF%	−0.10	1.75	0.972
BAI	−0.28	1.10	0.799

[1] Linear regression analyses were performed using low VA intake as the reference group within each genotype group. Genotype A (responsive): TT or TC for rs5888, AA for rs1800592, and/or TT or TC for rs659366; Genotype B (less responsive), having at most one of the responsive genotypes aforementioned. Regressions were adjusted for age. Abbreviations: BF% (body fat percentage), BAI (Body Adiposity index), SE (Standard Error).

To assess the impact of genotype and BF% on cell metabolism, the expression of genes involved in lipid homeostasis was analyzed in PBMCs under basal conditions. In High-BF% subjects, the influence of Genotype A was reflected by the increased expression of Liver X receptor alpha (*LXRA*) ($p = 0.010$) (Figure 2A), Superoxide Dismutase-2 (*SOD2*) ($p = 0.047$) (Figure 2B), and Acyl-CoA synthase (*SLC27A2*) ($p = 0.037$) (Figure 2C), in comparison with Low-BF% subjects with the same genotype. In contrast, High-BF% subjects with Genotype B showed increased mRNA levels of Sterol Response Element Binding Protein 1c (*SREBP1C*) ($p = 0.014$) (Figure 2D) and Retinoid X Receptor alpha (*RXRA*) ($p = 0.025$) (Figure 2G) in comparison with Low-BF% and the same genotype. Furthermore, the specific influence of genotype was observed on *SREBP1C* ($p = 0.027$) and *RXRA* ($p = 0.023$) expression only in Low-BF%, in which Genotype A presented a higher expression than Genotype B. Furthermore, gene expression showed interaction effects between genotype and adiposity. Specifically, interactions were shown on gene expression of *LXRA* ($p = 0.032$) (Figure 2A), *SREBP1C* ($p = 0.004$), Scavenger Receptor class B member 1 (*SCARB1*) ($p = 0.035$), CCAAT/enhancer-binding protein beta (*CEBPB*) ($p = 0.011$), and *RXRA* ($p = 0.033$) (Figure 2D–G).

To obtain further insight into the molecular mechanisms involved, cellular response was analyzed in PBMCs incubated with RA. Interestingly, the interactive effects of genotype–BF% were found on gene expression of *SREBP1C* ($p = 0.001$), *SCARB1* ($p = 0.011$), *CEBPB* ($p = 0.010$) (Figure 3A–C), and *RXRA* ($p = 0.022$) (Figure 3G). In addition, RA treatment caused specific increases in *CPT1A* ($p < 0.001$) and *UCP2* ($p = 0.030$) (Figure 3E,F) in Genotype A subjects, with Low-BF% and High-BF%, respectively. Meanwhile, in Genotype B individuals, *SCARB1* (High-BF%, $p = 0.007$; Low-BF%, $p = 0.005$) and *RVLDL* (High-BF%, $p = 0.001$; Low-BF%, $p = 0.021$) expression was decreased by RA regardless of BF% (Figure 3B,D).

Concerning the gene expression of transcription factors, RA incubation decreased mRNA levels of *RXRA* ($p = 0.024$) (Figure 3G) in Genotype A and Low-BF% subjects and increased *SREBP1C* gene expression in Low-BF% (Figure 3A), regardless of the genotype (Genotype A, $p = 0.007$; Genotype B, $p = 0.004$). Altogether, data on gene expression showed the functional impact of the allelic load considered in handling VA and disclosed the cellular adaptations to RA delivery depending on BF%.

Table 4. Anthropometric and dietary parameters by body fat % (High BF $\geq$ 25% or Low BF < 25%) and genotype groups (A, Responsive to VA; B, Less responsive to VA) of the OptiDiet-15 cohort [1].

Variables	High Body Fat % (n = 21)									Low Body Fat% (n = 20)								
	Genotype A (n = 11)				Genotype B (n = 10)				p-Value	Genotype A (n = 10)				Genotype B (n = 10)				p-Value
	Mean	SD	Min	Max	Mean	SD	Min	Max		Mean	SD	Min	Max	Mean	SD	Min	Max	
Weight (kg)	93.4	23.0	63.6	152	82.0	10.8	68.8	106	0.179	68.7	9.92	52.7	85.0	74.0	8.14	63.3	86.7	0.176
WHR	0.93	0.10	0.78	1.11	0.85	0.09	0.66	1.01	0.077	0.87	0.09	0.78	1.11	0.83	0.06	0.73	0.90	0.249
BMI (kg/m^2)	31.1	6.17	24.9	46.6	28.0	4.88	21.7	39.0	0.199	22.6	2.52	19.7	27.1	23.8	2.42	19.5	27.1	0.246
BF%	32.2	4.92	26.5	43.6	29.6	2.66	26.2	35.1	0.152	19.0	4.20	10.1	24.0	18.8	4.50	12.6	24.8	0.794
Bicipital SF	11.2	9.95	3.90	40.0	11.1	3.31	6.00	15.8	0.915	4.77	1.54	2.70	8.00	6.38	3.85	3.60	15.8	0.123
Energy (kcal/day)	2311	415	1880	3052	1860	212	1530	2130	0.007	2238	424	1253	2853	2047	276	1695	2500	0.226
Carbohydrate (% EC)	38.9	5.92	30.9	47.4	40.9	7.43	30.0	51.2	0.472	46.2	8.48	38.4	60.4	42.5	8.02	32.8	56.4	0.326
Fat (% EC)	40.7	4.72	33.8	47.8	35.6	9.81	12.9	46.4	0.190	33.9	7.41	20.9	46.0	39.0	8.33	25.1	48.3	0.208
Proteins (g/day)	94.1	21.4	54.1	122	72.1	11.6	44.5	85.4	0.014	80.2	20.9	53.6	112	82.0	10.3	66.1	94.30	0.922
Fiber (g/day)	19.9	5.78	14.7	35.5	19.4	5.53	8.10	25.7	0.599	21.6	7.51	12.0	32.0	18.6	5.75	10.1	25.50	0.511
Vitamin A (µg/day)	583	165	372	794	543	148	386	819	0.711	532	141	287	754	618	207	365	934	0.218
Retinol (µg/day)	231	140	59.9	429	254	121	28.3	397	0.227	250	86.4	132	436	280	106	111	497	0.553
ß-carotene (µg/day)	1612	1065	480	4300	1540	993	303	3450	0.791	1295	912	241	2999	1741	1160	308	4333	0.217

[1] All values are expressed as the mean, SD (standard deviation), min (minimum), and max (maximum). Abbreviations: VA (vitamin A); WHR (waist–hip ratio); BMI (body mass index); BF% (body fat percentage); SF (skinfold); and EC (energy contribution). Statistical assessment of the differences between each genotype group were assessed by an ANOVA test adjusted for age.

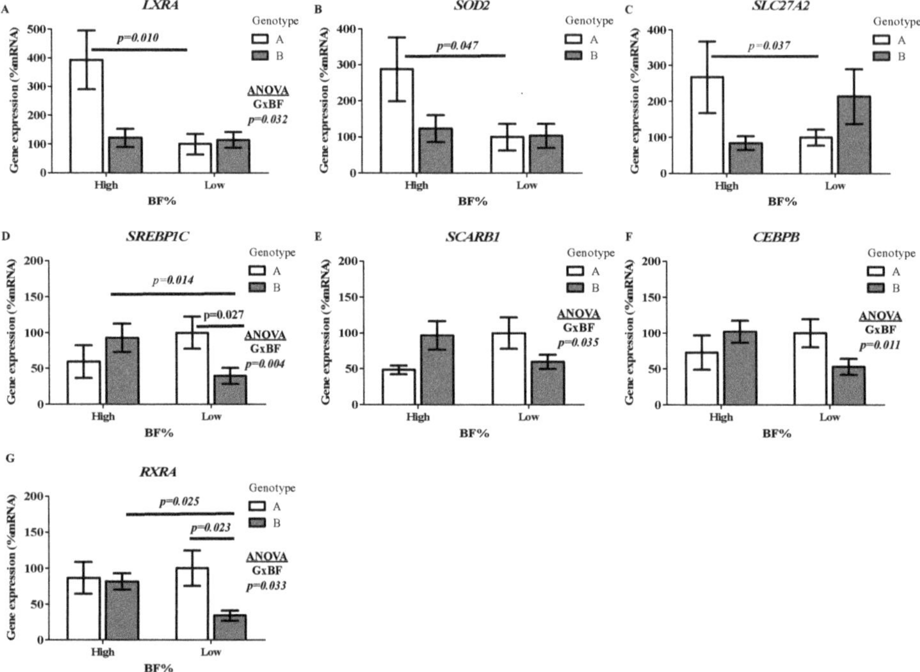

Figure 2. Gene expression under basal conditions in Peripheral Blood Mononuclear Cells. Gene expression under basal conditions in Peripheral Blood Mononuclear Cells (PBMCs) of subjects according to genotype (A/B) and BF% level (High/Low) groups. Data correspond to mRNA levels of genes encoding transcription factors involved in cellular response to retinoic acid, lipid, and energy metabolic homeostasis: (**A**) Liver X Receptor alpha (LXRA), (**B**) Superoxide dismutase 2 (SOD2), (**C**) Very long-chain acyl-CoA synthetase (SLC27A2), (**D**) Sterol Response Element Binding Protein 1c (SREBP1C), (**E**) Scavenger Receptor Class B Member 1 (*SCARB1*), (**F**) CCAAT/enhancer-binding protein beta (CEBPB), and (**G**) Retinol-X Receptor α (RXRA). The mRNA levels were normalized to RPLP0 and expressed in relation to the expression found in Genotype A and Low BF% which was set at 100%. The statistical analysis of genotype–BF% interaction was carried out by ANOVA adjusted for total energy intake and age; GxBF indicates $p < 0.05$. Differences between genotype groups were established by an ANOVA test adjusted for total energy intake. Differences between BF% groups were analyzed by an ANOVA test adjusted for age.

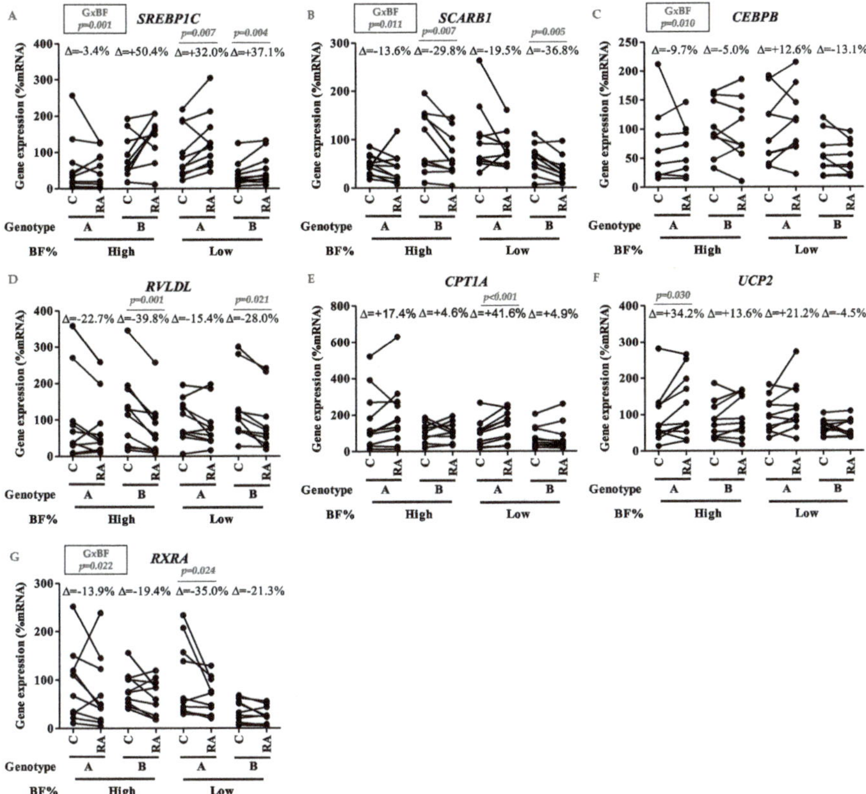

Figure 3. Effect of all-trans retinoic acid treatment on mRNA expression in Peripheral Blood Mononuclear Cells of Genotype (A/B) and BF% level (High/Low). Effect of all-trans retinoic acid (RA, 1 µM) treatment on mRNA expression in Peripheral Blood Mononuclear Cells (PBMCs) of subjects according to genotype (A/B) and BF% level (High/Low) groups. Data correspond to mRNA levels of (**A**) Sterol Response Element Binding Protein 1c (SREBP1C), (**B**) Scavenger Receptor Class B Member 1 (SCARB1), (**C**) CCAAT/enhancer-binding protein beta (CEBPB), (**D**) Very Low Density Lipoprotein Receptor (RVLDL), (**E**) Carnitine palmitoyltransferase I alpha (CPT1A), (**F**) Uncoupling Protein 2 (UCP2), and (**G**) Retinol-X Receptor α (RXRA). The % increase (Δ) was obtained by the difference between the mean of the gene expression in RA-treated cells and the mean of gene expression under baseline conditions (C) in each experimental group. Statistical assessment of the effect of treatment on each experimental group was estimated using a paired t-test. Genotype–BF% interaction regarding the effect of treatment was assessed using repeated measures ANOVA (GxBF, $p < 0.05$).

4. Discussion

Although a recent study underlined that different carotenoids in serum and adipose tissue in humans are associated with metabolic benefits, such as improvement in insulin sensitivity in both liver and adipose tissue [8,48], direct evidence in humans concerning the effects of dietary VA or carotenoid supplementation on body adiposity is limited [15]. Furthermore, it has been shown that variants in genes involved in carotenoid absorption could be responsible for high interindividual variability in their bioavailability [18,25,49].

Thereby, our findings show that three genetic variants, which have not been previously related to fat accumulation, might be predisposed to greater adiposity depending on the amount of VA intake; although we cannot rule out other genes or genetic variants also being relevant, particularly depending on population characteristics (e.g., sex, age, or lifestyle). However, the role of SNPs and the functional assessment carried out may contribute to providing further insight into the mechanisms involved in this approach. Specifically, allele combinations performing the responsive Genotype (A), involving T allele for rs5888 (*SCARB1*) and rs659366 (*UCP2*), respectively, and/or the absence of G for rs1800592 (*UCP1*), were associated with a higher BF% in individuals reporting suboptimal VA intake. In contrast, optimal consumption of VA was associated with normal BF%, suggesting that Genotype A may contribute to a more efficient adipose metabolism in a way that depends on dietary VA.

The rs5888 is located on exon 8 of the *SCARB1* gene and entails a synonym amino acid exchange (A350A) in SR-B1 protein, causing splicing activity alterations [50]. Interestingly, T allele related to a better adipose profile under optimal VA consumption in our study has been previously associated with a lower risk of coronary heart disease [51] and lower levels of plasma triglycerides, specifically in men [52]. Moreover, SR-B1 is a multifunctional scavenger involved in carotenoid uptake from diet [21], and the presence of genetic variants may change its effectiveness.

The rs659366 is located on the *UCP2* gene promoter and has a crucial role in ROS dissipation [37]. The T allele has been associated with higher *UCP2* expression and energy expenditure [41], lower risk of obesity [53], and lower Homeostatic Model Assessment (HOMA) index [54]. However, to our knowledge, this is the first report showing its functional relationship with VA intake.

The rs1800592 is located on the *UCP1* gene promotor, whose protein plays a role in mitochondrial thermogenesis by uncoupling oxidative phosphorylation and, therefore, diminishing the synthesis of ATP from nutrients [13]. The VA and UCP1 activity relationship is well documented in cell cultures [34] and rodents [22]: VA enhances fatty acid oxidation and promotes UCP1-associated thermogenesis. Major allele of rs1800592 (A) is associated with higher *UCP1* expression and greater thermogenic capacity [39,40] which confers differential outcomes to bio-active compounds with anti-obesogenic effects [25,55]. This would fit with our findings of the genotypes of subjects, who under optimal VA intake show leaner phenotype.

Therefore, aiming to characterize in further detail the metabolic consequences of adiposity associated with suboptimal VA intake and the role played by the genotype, a novel methodological approach was carried out following an ex vivo study in blood cells from selected individuals. Peripheral Blood Mononuclear Cells (PBMCs) are in a continuous interplay with the tissues, including adipose tissue depots, express key hormones involved in energy homeostasis and body weight control (leptin, visfatin, ghrelin), and respond to metabolic challenges. In addition, PBMCs are readily accessible, the sample collection is less invasive than adipose specimens, and similarities in gene expression between PBMCs and white adipose tissue have been shown under different nutritional status and metabolic challenges [43]. Despite the fact that previous results from our research group do not show differences between PBMC populations between normal weight and obese subjects [44], some studies indicate that the lymphocyte fraction may be altered in obesity [56,57]. Therefore, there is the possibility that the differences observed that the baseline expression of the analyzed genes between Low-BF% vs. High-BF% experimental groups could be influenced by differential lymphocyte proportions. For this reason, each genotype group includes its respective group with both Low and High-BF% representation. Furthermore, PBMC fraction is a tool increasingly used in nutrition and obesity research as a source of transcriptomic biomarkers, because its capacity to reflect the homeostatic state of key tissues and different blood cell count proportions in obesity is assumed to be one of the causes of transcriptomic alterations [58].

Furthermore, the possibility that obesity may change the proportion of immune cell types can be discarded as the proportion of lymphocytes and monocytes in PBMC samples do not differ between normal-weight and overweight/obese men. Therefore, we may assume that the baseline gene expression differences associated to fat content are not due to differences in the immune cell type proportion. Therefore, PBMCs have been widely used and are considered a very useful tool in testing the efficacy of bioactive compounds in humans [43,44] as well as in nutrigenomics and transcriptomic based studies [42,59]

The overexpression of the genes *LXRA*, *SLC27A2*, and *SOD2* specifically found in High-BF% individuals with Genotype A would be indicative of altered metabolic status, reflected in their PBMCs, and resulting from the interaction of the genetic baggage with the low intake of VA. The *LXRA* gene codes for a transcription factor that exerts an enhancing role of the reverse transport of cholesterol and mediates inflammatory process in macrophages [60]. *SLC27A2* encodes a key enzyme that plays an important role in both lipid biosynthesis and fatty acid degradation [61]. Furthermore, its overexpression in rat PBMCs has been proposed as an early marker of overweight development, particularly related to inadequate diets [62]. *SOD2* is a member of the superoxide dismutase family, involved in the dissipation of ROS, which would confer protection against the cellular damage caused by oxidative stress and pro-inflammatory cytokines [63], raising its expression in elevated oxidative stress environments [64] in an attempt to prevent cell death [63]. Subsequently, increased expression of *LXRA* and *SLC27A2* in Genotype A individuals with High-BF% could be an indicator of lipotoxicity [65,66]. Altogether, the data reflect a compromised metabolic status that is in accordance with the findings from the observational study, in which men with Genotype A and low vitamin A intake showed greater risk of being overfat.

Additionally, genotype–BF% interactive effects were found regarding the expression of three genes involved in fat metabolism and lipid homeostasis. *SREBP1C* is a key regulator of cholesterol and fatty acid metabolism [67], *SCARB1* is a gene involved in the cellular transport of cholesterol [30], and *CEBPB* is a transcription factor that coordinates regulatory networks for inflammation and lipid metabolism in macrophages and adipocytes [68]. Thus, their gene expression would be conditioned by the adiposity of the subjects, their genetic load, and would affect cell lipid homeostasis.

Aiming to characterize the metabolic flexibility, PBMCs of the subjects were incubated with retinoic acid (RA). PBMCs from Genotype A subjects (which showed lower adiposity in the case of optimal VA consumption) treated with RA presented greater increases in *CPT1A* and *UCP2* expression than in Genotype B subjects (whose VA intake level was not related to BF%), suggesting greater the induction of mitochondrial fatty acid oxidation [14,33] (Figure 3). The increased mitochondrial activity would not enhance the *NFKB* mediated inflammatory response [69], inasmuch as *NFKB* expression decreased in these subjects (data not shown).

Further, genotype seems to influence lipid metabolism and cell fat management, since RA was associated with a decreased expression of *SCARB1* and *RVLDL*—both genes involved in the uptake of fat-soluble compounds—which were only significant in Genotype B regardless of BF% (Figure 3). In addition, the expression of lipid key genes (*SREBP1C*, *SCARB1*, and *CEBPB*) reflected the interactive effect of genotype and BF%. Thus, the combined effect of these two factors (body fat and genotype) conditions lipid homeostasis, as well as exerts a modulating effect of RA on the expression of key genes in PBMCs.

The influence of BF% on the response to RA was also noted, since RA caused an increase in the expression of *SREBP1C* in cells from individuals with Low-BF% of both genotypes (Figure 3A) and a decreased *RXRA* expression in Genotype A with Low-BF% but not in Genotype B individuals (Figure 3G). The differential activation/inhibition of response elements and transcription factors related to lipid and VA metabolism could be key to explaining the differential effects of RA, depending on genotype and BF%, by regulating a large network of genes involved in cell activity and metabolism [60,70]. In this sense, high adiposity could short-circuit some pathways related to the bioactive function of RA. The attenuation of the beneficial effects of certain bioactives on health due to the presence of obesity has been previously reported in prior studies using PBMCs. In particular, overweight/obesity alters the anti-inflammatory response to polyunsaturated fatty acids [44]. In addition, our previous results also indicate a synergistic effect of excess body weight with the presence of specific genetic variants, modulating the inflammatory status [42].

In summary, subjects following suboptimal consumption of dietary VA with a responsive Genotype (A) were related to a higher risk of adiposity, in contrast with Genotype B subjects that did not show such a relationship between VA intake and body fat content. Furthermore, body adiposity degree, influenced by genotype characteristics, was a conditioning factor of gene expression under both baseline conditions and in response to RA. In PBMCs from suboptimal VA consumers, Genotype A showed increased *LXRA*, *SLC27A2*, and *SOD2* expression, especially in individuals with High-BF%, which was consistent with an impaired basal metabolic state. Interestingly, gene expression after incubation with RA was found to drive an improvement in the cellular status, particularly in Low-BF% (by increasing *CPT1A* and *UCP2*). In contrast, gene expression in Genotype B appeared less sensitive to BF% in basal conditions and also after incubation with RA. In unstimulated cells, gene expression was in accordance with a better metabolic state in comparison with Genotype A subjects. Hence, RA treatment led to decreasing the expression of key genes for the uptake/management of fat-soluble substances, such as *SCARB1* and *RVLDL*, in comparison with Genotype A.

However, there may be some limitations in this study that should be taken into account, such as the relative small sample size; the multiple testing performed on gene expression; and the potential effect of other environmental confounders, socio-economic factors, or genetic variants not considered.

5. Conclusions

Obesity is a multifactorial disorder; thus, its treatment and prevention are masked by several factors, including genetics and lifestyle factors [71]. In the present study, we show that suboptimal consumption of vitamin A is related to an increased risk of body fat accumulation in genetically predisposed individuals, who would represent approximately 38% of the individuals in the studied population. The results show the relevance of consuming VA within the required amounts, its impact on metabolism and energy balance, and consequently, on men's adiposity with a modulating influence of genetic variants of *SCARB1*, *UCP2*, and *UCP1*.

Finally, the combination of an observational study together with a genotype–phenotype ex vivo intervention on PBMCs seems to be an effective strategy to contribute to Precision Nutrition. The study of genetics together with the characterization of gene–nutrient–phenotype interactions in a well-balanced study design facilitates personalized dietary and lifestyle advice and promotes healthier status through a focused and integrated approach.

Supplementary Materials: The following are available online at http://www.mdpi.com/2072-6643/12/9/2588/s1. Table S1. Rs code, gene and chromosome (Chr) location of SNPs studied for their potential modulating effect on vitamin A metabolism.

Author Contributions: Conceptualization: F.S. and A.P.; formal analysis: S.G., F.S. and A.P.; funding acquisition, A.P.; investigation and methodology: S.G.; supervision: F.S. and A.P.; writing—original draft: S.G.; writing—review and editing: F.S., S.G., and A.P. All authors have read and agreed to the published version of the manuscript.

Funding: SG was supported by a fellowship (FPI/1680/2014) from the Conselleria d'Educació, Cultura i Universitats, del Govern de les Illes Balears. This work was supported by the Spanish Government: INTERBIOBES-AGL2015–67019-P (AEI, MINECO/FEDER, EU).

Acknowledgments: We thank Balaguer from the Health Center of Vilafranca de Bonany and the staff for facilitating volunteer enrolment. We acknowledge the technical support of Miguel García-Gabilondo and María Alomar in blood sampling.

Conflicts of Interest: The authors declare no conflict of interest.

References

1. Seidell, J.C.; Halberstadt, J. The Global Burden of Obesity and the Challenges of Prevention. *Ann. Nutr. Metab.* **2015**, *66*, 7–12. [CrossRef]
2. Arroyo-Johnson, C.; Mincey, K.D. Obesity Epidemiology Worldwide. *Gastroenterol. Clin. N. Am.* **2016**, *45*, 571–579. [CrossRef] [PubMed]
3. Wiseman, E.M.; Bar-El Dadon, S.; Reifen, R. The vicious cycle of vitamin a deficiency: A review. *Crit. Rev. Food Sci. Nutr.* **2017**, *57*, 3703–3714. [CrossRef] [PubMed]

4.   García, O.P. Effect of vitamin A deficiency on the immune response in obesity. *Proc. Nutr. Soc.* **2012**, *71*, 290–297. [CrossRef] [PubMed]

5.   Vaughan, L.A.; Benyshek, D.C.; Martin, J.F. Food acquisition habits, nutrient intakes, and anthropometric data of Havasupai adults. *J. Am. Diet. Assoc.* **1997**, *97*, 1275–1282. [CrossRef]

6.   Viroonudomphol, D.; Pongpaew, P.; Tungtrongchitr, R.; Changbumrung, S.; Tungtrongchitr, A.; Phonrat, B.; Vudhivai, N.; Schelp, F.P. The relationships between anthropometric measurements, serum vitamin A and E concentrations and lipid profiles in overweight and obese subjects. *Asia Pac. J. Clin. Nutr.* **2003**, *12*, 73–79. [PubMed]

7.   García, O.; Ronquillo, D.; del Carmen Caamaño, M.; Martínez, G.; Camacho, M.; López, V.; Rosado, J. Zinc, Iron and Vitamins A, C and E Are Associated with Obesity, Inflammation, Lipid Profile and Insulin Resistance in Mexican School-Aged Children. *Nutrients* **2013**, *5*, 5012–5030. [CrossRef]

8.   Bonet, M.L.; Ribot, J.; Galmés, S.; Serra, F.; Palou, A. Carotenoids and carotenoid conversion products in adipose tissue biology and obesity: Pre-clinical and human studies. *Biochim. Biophys. Acta Mol. Cell Biol. Lipids* **2020**, *1865*, 158676. [CrossRef]

9.   Blomhoff, R.; Blomhoff, H.K. Overview of retinoid metabolism and function. *J. Neurobiol.* **2006**, *66*, 606–630. [CrossRef]

10.   Bonet, M.L.; Canas, J.A.; Ribot, J.; Palou, A. Carotenoids and their conversion products in the control of adipocyte function, adiposity and obesity. *Arch. Biochem. Biophys.* **2015**, *572*, 112–125. [CrossRef]

11.   Bonet, M.L.; Ribot, J.; Felipe, F.; Palou, A. Vitamin A and the regulation of fat reserves. *Cell. Mol. Life Sci.* **2003**, *60*, 1311–1321. [CrossRef] [PubMed]

12.   Bonet, M.L.; Ribot, J.; Palou, A. Lipid metabolism in mammalian tissues and its control by retinoic acid. *Biochim. Biophys. Acta Mol. Cell Biol. Lipids* **2012**, *1821*, 177–189. [CrossRef] [PubMed]

13.   Bonet, M.L.; Mercader, J.; Palou, A. A nutritional perspective on UCP1-dependent thermogenesis. *Biochimie* **2017**, *134*, 99–117. [CrossRef] [PubMed]

14.   Amengual, J.; Petrov, P.; Bonet, M.L.; Ribot, J.; Palou, A. Induction of carnitine palmitoyl transferase 1 and fatty acid oxidation by retinoic acid in HepG2 cells. *Int. J. Biochem. Cell Biol.* **2012**, *44*, 2019–2027. [CrossRef]

15.   Eggersdorfer, M.; Wyss, A. Carotenoids in human nutrition and health. *Arch. Biochem. Biophys.* **2018**, *652*, 18–26. [CrossRef]

16.   Canas, J.A.; Lochrie, A.; McGowan, A.G.; Hossain, J.; Schettino, C.; Balagopal, P.B. Effects of mixed carotenoids on adipokines and abdominal adiposity in children: A pilot study. *J. Clin. Endocrinol. Metab.* **2017**, *102*, 1983–1990. [CrossRef]

17.   Beydoun, M.A.; Chen, X.; Jha, K.; Beydoun, H.A.; Zonderman, A.B.; Canas, J.A. Carotenoids, vitamin A, and their association with the metabolic syndrome: A systematic review and meta-analysis. *Nutr. Rev.* **2019**, *77*, 32–45. [CrossRef]

18.   Borel, P.; Desmarchelier, C.; Nowicki, M.; Bott, R. A Combination of Single-Nucleotide Polymorphisms Is Associated with Interindividual Variability in Dietary -Carotene Bioavailability in Healthy Men. *J. Nutr.* **2015**, *145*, 1740–1747. [CrossRef]

19.   Gueguen, S.; Leroy, P.; Gueguen, R.; Siest, G.; Visvikis, S.; Herbeth, B. Genetic and environmental contributions to serum retinol and alpha-tocopherol concentrations: The Stanislas Family Study. *Am. J. Clin. Nutr.* **2005**, *81*, 1034–1044. [CrossRef]

20.   Mondul, A.M.; Yu, K.; Wheeler, W.; Zhang, H.; Weinstein, S.J.; Major, J.M.; Cornelis, M.C.; Männistö, S.; Hazra, A.; Hsing, A.W.; et al. Genome-wide association study of circulating retinol levels. *Hum. Mol. Genet.* **2011**, *20*, 4724–4731. [CrossRef]

21.   Valacchi, G.; Sticozzi, C.; Lim, Y.; Pecorelli, A. Scavenger receptor class B type I: A multifunctional receptor. *Ann. N. Y. Acad. Sci.* **2011**, *1229*, E1–E7. [CrossRef] [PubMed]

22.   Mercader, J.; Ribot, J.; Murano, I.; Felipe, F.; Cinti, S.; Bonet, M.L.; Palou, A. Remodeling of White Adipose Tissue after Retinoic Acid Administration in Mice. *Endocrinology* **2006**, *147*, 5325–5332. [CrossRef] [PubMed]

23.   Felipe, F.; Bonet, M.L.; Ribot, J.; Palou, A. Up-regulation of muscle uncoupling protein 3 gene expression in mice following high fat diet, dietary vitamin A supplementation and acute retinoic acid-treatment. *Int. J. Obes.* **2003**, *27*, 60–69. [CrossRef] [PubMed]

24.   Borel, P.; Moussa, M.; Reboul, E.; Lyan, B.; Defoort, C.; Vincent-Baudry, S.; Maillot, M.; Gastaldi, M.; Darmon, M.; Portugal, H.; et al. Human Plasma Levels of Vitamin E and Carotenoids Are Associated with Genetic Polymorphisms in Genes Involved in Lipid Metabolism. *J. Nutr.* **2007**, *137*, 2653–2659. [CrossRef]

25. Ebrahimzadeh Attari, V.; Asghari Jafarabadi, M.; Zemestani, M.; Ostadrahimi, A. Effect of Zingiber officinale Supplementation on Obesity Management with Respect to the Uncoupling Protein 1 -3826A > G and β3-adrenergic Receptor Trp64Arg Polymorphism. *Phyther. Res.* **2015**, *29*, 1032–1039. [CrossRef]

26. Snitker, S.; Fujishima, Y.; Shen, H.; Ott, S.; Pi-Sunyer, X.; Furuhata, Y.; Sato, H.; Takahashi, M. Effects of novel capsinoid treatment on fatness and energy metabolism in humans: Possible pharmacogenetic implications. *Am. J. Clin. Nutr.* **2009**, *89*, 45–50. [CrossRef] [PubMed]

27. Bergman, R.N.; Stefanovski, D.; Buchanan, T.A.; Sumner, A.E.; Reynolds, J.C.; Sebring, N.G.; Xiang, A.H.; Watanabe, R.M. A Better Index of Body Adiposity. *Obesity* **2011**, *19*, 1083–1089. [CrossRef]

28. Ortega, R.M.; Andrés, P.; Requejo, A.M.; Aparicio, A.; Molinero, L.M.; López-Sobaler, A.M. DIAL software for assessing diets and food calculations (for Windows, version 2.0). Department of Nutrition (UCM) & Alce Ingeniería, S.L. Madrid, Spain. Available online: https://www.alceingenieria.net/infodial.htm (accessed on 20 July 2020).

29. European Food Safety Authority (EFSA). Dietary Reference Values: Vitamin A Advice Published|European Food Safety Authority. Available online: https://www.efsa.europa.eu/en/press/news/150305 (accessed on 12 June 2017).

30. Shen, W.-J.; Azhar, S.; Kraemer, F.B. SR-B1: A Unique Multifunctional Receptor for Cholesterol Influx and Efflux. *Annu. Rev. Physiol.* **2018**, *80*, 95–116. [CrossRef]

31. Borel, P.; Lietz, G.; Goncalves, A.; Szabo de Edelenyi, F.; Lecompte, S.; Curtis, P.; Goumidi, L.; Caslake, M.J.; Miles, E.A.; Packard, C.; et al. CD36 and SR-BI Are Involved in Cellular Uptake of Provitamin A Carotenoids by Caco-2 and HEK Cells, and Some of Their Genetic Variants Are Associated with Plasma Concentrations of These Micronutrients in Humans. *J. Nutr.* **2013**, *143*, 448–456. [CrossRef]

32. Murholm, M.; Isidor, M.S.; Basse, A.L.; Winther, S.; Sørensen, C.; Skovgaard-Petersen, J.; Nielsen, M.M.; Hansen, A.S.; Quistorff, B.; Hansen, J.B. Retinoic acid has different effects on UCP1 expression in mouse and human adipocytes. *BMC Cell Biol.* **2013**, *14*, 41. [CrossRef]

33. Pierelli, G.; Stanzione, R.; Forte, M.; Migliarino, S.; Perelli, M.; Volpe, M.; Rubattu, S. Uncoupling protein 2: A key player and a potential therapeutic target in vascular diseases. *Oxid. Med. Cell. Longev.* **2017**, *2017*, 7348372. [CrossRef] [PubMed]

34. Serra, F.; Bonet, M.L.; Puigserver, P.; Oliver, J.; Palou, A. Stimulation of uncoupling protein 1 expression in brown adipocytes by naturally occurring carotenoids. *Int. J. Obes. Relat. Metab. Disord.* **1999**, *23*, 650–655. [CrossRef] [PubMed]

35. Larose, M.; Cassard-Doulcier, A.M.; Fleury, C.; Serra, F.; Champigny, O.; Bouillaud, F.; Ricquier, D. Essential cis-acting elements in rat uncoupling protein gene are in an enhancer containing a complex retinoic acid response domain. *J. Biol. Chem.* **1996**, *271*, 31533–31542. [CrossRef] [PubMed]

36. Bonet, M.L.; Puigserver, P.; Serra, F.; Ribot, J.; Vázquez, F.; Pico, C.; Palou, A. Retinoic acid modulates retinoid X receptor alpha and retinoic acid receptor alpha levels of cultured brown adipocytes. *FEBS Lett.* **1997**, *406*, 196–200. [CrossRef]

37. Cadenas, S. Mitochondrial uncoupling, ROS generation and cardioprotection. *Biochim. Biophys. Acta Bioenerg.* **2018**, *1859*, 940–950. [CrossRef]

38. Constantineau, J.; Greason, E.; West, M.; Filbin, M.; Kieft, J.S.; Carletti, M.Z.; Christenson, L.K.; Rodriguez, A. A synonymous variant in scavenger receptor, class B, type I gene is associated with lower SR-BI protein expression and function. *Atherosclerosis* **2010**, *210*, 177–182. [CrossRef]

39. Nagai, N.; Sakane, N.; Tsuzaki, K.; Moritani, T. UCP1 genetic polymorphism (−3826 A/G) diminishes resting energy expenditure and thermoregulatory sympathetic nervous system activity in young females. *Int. J. Obes.* **2011**, *35*, 1050–1055. [CrossRef]

40. Yoneshiro, T.; Ogawa, T.; Okamoto, N.; Matsushita, M.; Aita, S.; Kameya, T.; Kawai, Y.; Iwanaga, T.; Saito, M. Impact of UCP1 and β3AR gene polymorphisms on age-related changes in brown adipose tissue and adiposity in humans. *Int. J. Obes.* **2013**, *37*, 993–998. [CrossRef]

41. Kovacs, P.; Ma, L.; Hanson, R.L.; Franks, P.; Stumvoll, M.; Bogardus, C.; Baier, L.J. Genetic variation in UCP2 (uncoupling protein-2) is associated with energy metabolism in Pima Indians. *Diabetologia* **2005**, *48*, 2292–2295. [CrossRef]

42. Galmés, S.; Cifre, M.; Palou, A.; Oliver, P.; Serra, F. A Genetic Score of Predisposition to Low-Grade Inflammation Associated with Obesity May Contribute to Discern Population at Risk for Metabolic Syndrome. *Nutrients* **2019**, *11*, 298. [CrossRef]

43. Reynés, B.; Díaz-Rúa, R.; Cifre, M.; Oliver, P.; Palou, A. Peripheral blood mononuclear cells as a potential source of biomarkers to test the efficacy of weight-loss strategies. *Obesity* **2014**, *23*, 28–31. [CrossRef] [PubMed]

44. Cifre, M.; Díaz-Rua, R.; Varela Calviño, R.; Reynés, B.; Pericás-Beltrán, J.; Palou, A.; Oliver, P. Human peripheral blood mononuclear cell in vitro system to test the efficacy of food bioactive compounds: Effects of polyunsaturated fatty acids and their relation with BMI. *Mol. Nutr. Food Res.* **2016**, *61*, 1600353. [CrossRef] [PubMed]

45. European Food Safety Authority (EFSA). Dietary Reference Values for nutrients Summary report. *EFSA Support. Publ.* **2017**, *14*, e15121. [CrossRef]

46. Ruiz, E.; Ávila, J.M.; Valero, T.; del Pozo, S.; Rodriguez, P.; Aranceta-Bartrina, J.; Gil, Á.; González-Gross, M.; Ortega, R.M.; Serra-Majem, L.; et al. Energy Intake, Profile, and Dietary Sources in the Spanish Population: Findings of the ANIBES Study. *Nutrients* **2015**, *7*, 4739–4762. [CrossRef] [PubMed]

47. European Food Safety Authority (EFSA). Scientific Opinion on Dietary Reference Values for energy. *EFSA J.* **2013**, *11*, 3333. [CrossRef]

48. Harari, A.; Coster, A.C.; Jenkins, A.; Xu, A.; Greenfield, J.R.; Harats, D.; Shaish, A.; Samocha-Bonet, D. Obesity and Insulin Resistance Are Inversely Associated with Serum and Adipose Tissue Carotenoid Concentrations in Adults. *J. Nutr.* **2020**, *150*, 38–46. [CrossRef]

49. Borel, P.; Desmarchelier, C. Genetic Variations Associated with Vitamin A Status and Vitamin A Bioavailability. *Nutrients* **2017**, *9*, 246. [CrossRef]

50. Morabia, A.; Ross, B.M.; Costanza, M.C.; Cayanis, E.; Flaherty, M.S.; Alvin, G.B.; Das, K.; James, R.; Yang, A.-S.; Evagrafov, O.; et al. Population-based study of SR-BI genetic variation and lipid profile. *Atherosclerosis* **2004**, *175*, 159–168. [CrossRef]

51. Ma, R.; Zhu, X.; Yan, B. SCARB1 rs5888 gene polymorphisms in coronary heart disease: A systematic review and a meta-analysis. *Gene* **2018**, *678*, 280–287. [CrossRef]

52. Ye, L.-F.; Zheng, Y.-R.; Zhang, Q.-G.; Yu, J.-W.; Wang, L.-H. Meta-analysis of the association between SCARB1 polymorphism and fasting blood lipid levels. *Oncotarget* **2017**, *8*, 81145–81153. [CrossRef]

53. Qian, L.; Xu, K.; Xu, X.; Gu, R.; Liu, X.; Shan, S.; Yang, T. UCP2 -866G/A, Ala55Val and UCP3 -55C/T polymorphisms in association with obesity susceptibility—A meta-analysis study. *PLoS ONE* **2013**, *8*, e58939. [CrossRef]

54. Andersen, G.; Dalgaard, L.T.; Justesen, J.M.; Anthonsen, S.; Nielsen, T.; Thørner, L.W.; Witte, D.; Jørgensen, T.; Clausen, J.O.; Lauritzen, T.; et al. The frequent UCP2 −866G > A polymorphism protects against insulin resistance and is associated with obesity: A study of obesity and related metabolic traits among 17 636 Danes. *Int. J. Obes.* **2013**, *37*, 175–181. [CrossRef] [PubMed]

55. Lee, M.; Chae, S.; Cha, Y.; Park, Y. Supplementation of Korean fermented soy paste doenjang reduces visceral fat in overweight subjects with mutant uncoupling protein-1 allele. *Nutr. Res.* **2012**, *32*, 8–14. [CrossRef] [PubMed]

56. Dixon, J.; O'Brien, P. Obesity and the White Blood Cell Count: Changes with Sustained Weight Loss. *Obes. Surg.* **2006**, *16*, 251–257. [CrossRef]

57. Koca, T.T. Does obesity cause chronic inflammation? The association between complete blood parameters with body mass index and fasting glucose. *Pak. J. Med. Sci.* **2017**, *33*, 65–69. [CrossRef] [PubMed]

58. Reynés, B.; Priego, T.; Cifre, M.; Oliver, P.; Palou, A. Peripheral Blood Cells, a Transcriptomic Tool in Nutrigenomic and Obesity Studies: Current State of the Art. *Compr. Rev. Food Sci. Food Saf.* **2018**, *17*, 1006–1020. [CrossRef]

59. Picó, C.; Serra, F.; Rodríguez, A.M.; Keijer, J.; Palou, A. Biomarkers of Nutrition and Health: New Tools for New Approaches. *Nutrients* **2019**, *11*, 1092. [CrossRef] [PubMed]

60. Laurencikiene, J.; Rydé, M. Liver X receptors and fat cell metabolism. *Int. J. Obes.* **2012**, *3621*, 1494–1502. [CrossRef]

61. Anderson, C.M.; Stahl, A. SLC27 fatty acid transport proteins. *Mol. Asp. Med.* **2013**, *34*, 516–528. [CrossRef]

62. Caimari, A.; Oliver, P.; Rodenburg, W.; Keijer, J.; Palou, A. Slc27a2 expression in peripheral blood mononuclear cells as a molecular marker for overweight development. *Int. J. Obes.* **2010**, *34*, 831–839. [CrossRef]

63. Becuwe, P.; Ennen, M.; Klotz, R.; Barbieux, C.; Grandemange, S. Manganese superoxide dismutase in breast cancer: From molecular mechanisms of gene regulation to biological and clinical significance. *Free Radic. Biol. Med.* **2014**, *77*, 139–151. [CrossRef] [PubMed]

64. Grdina, D.J.; Murley, J.S.; Miller, R.C.; Mauceri, H.J.; Sutton, H.G.; Thirman, M.J.; Li, J.J.; Woloschak, G.E.; Weichselbaum, R.R. A Manganese Superoxide Dismutase (SOD2)-Mediated Adaptive Response. *Radiat. Res.* **2013**, *179*, 115–124. [CrossRef] [PubMed]

65. Ulven, S.M.; Dalen, K.T.; Gustafsson, J.-Å.; Nebb, H.I. LXR is crucial in lipid metabolism. *Prostaglandins Leukot. Essent. Fat. Acids* **2005**, *73*, 59–63. [CrossRef] [PubMed]

66. Liu, K.; Yeh, I.; Chou, S.; Yen, M.; Kuo, P. Regulatory mechanism of fatty acid-CoA metabolic enzymes under endoplasmic reticulum stress in lung cancer. *Oncol. Rep.* **2018**, *40*, 2674–2682. [CrossRef] [PubMed]

67. Oishi, Y.; Spann, N.J.; Link, V.M.; Muse, E.D.; Strid, T.; Edillor, C.; Kolar, M.J.; Matsuzaka, T.; Hayakawa, S.; Tao, J.; et al. SREBP1 Contributes to Resolution of Pro-inflammatory TLR4 Signaling by Reprogramming Fatty Acid Metabolism. *Cell Metab.* **2017**, *25*, 412–427. [CrossRef]

68. Rahman, S.M.; Janssen, R.C.; Choudhury, M.; Baquero, K.C.; Aikens, R.M.; De La Houssaye, B.A.; Friedman, J.E. CCAAT/enhancer-binding protein β (C/EBPβ) expression regulates dietary-induced inflammation in macrophages and adipose tissue in mice. *J. Biol. Chem.* **2012**, *287*, 34349–34360. [CrossRef]

69. Baker, R.G.; Hayden, M.S.; Ghosh, S. NF-κB, inflammation, and metabolic disease. *Cell Metab.* **2011**, *13*, 11–22. [CrossRef]

70. Zhang, R.; Wang, Y.; Li, R.; Chen, G. Transcriptional Factors Mediating Retinoic Acid Signals in the Control of Energy Metabolism. *Int. J. Mol. Sci.* **2015**, *16*, 14210–14244. [CrossRef]

71. Goodarzi, M.O. Genetics of obesity: What genetic association studies have taught us about the biology of obesity and its complications. *Lancet Diabetes Endocrinol.* **2018**, *6*, 223–236. [CrossRef]

Review

Associations between Genotype–Diet Interactions and Weight Loss—A Systematic Review

Sandra Bayer [1,†]**, Vincent Winkler** [1,†]**, Hans Hauner** [1,2] **and Christina Holzapfel** [1,*]

[1]   Institute for Nutritional Medicine, University Hospital Klinikum rechts der Isar, School of Medicine, Technical University of Munich, 80992 Munich, Germany; sandra.bayer@tum.de (S.B.); vincent.winkler@tum.de (V.W.); hans.hauner@tum.de (H.H.)

[2]   Else Kröner-Fresenius-Center for Nutritional Medicine, School of Life Sciences, Technical University of Munich, 85354 Freising, Germany

*   Correspondence: christina.holzapfel@tum.de; Tel.: +49-89-2892-4923

†   These authors contributed equally to this work.

Received: 21 August 2020; Accepted: 18 September 2020; Published: 22 September 2020

Abstract: Studies on the interactions between single nucleotide polymorphisms (SNPs) and macronutrient consumption on weight loss are rare and heterogeneous. This review aimed to conduct a systematic literature search to investigate genotype–diet interactions on weight loss. Four databases were searched with keywords on genetics, nutrition, and weight loss (PROSPERO: CRD42019139571). Articles in languages other than English and trials investigating special groups (e.g., pregnant women, people with severe diseases) were excluded. In total, 20,542 articles were identified, and, after removal of duplicates and further screening steps, 27 articles were included. Eligible articles were based on eight trials with 91 SNPs in 63 genetic loci. All articles examined the interaction between genotype and macronutrients (carbohydrates, fat, protein) on the extent of weight loss. However, in most cases, the interaction results were not significant and represented single findings that lack replication. The publications most frequently analyzed genotype–fat intake interaction on weight loss. Since the majority of interactions were not significant and not replicated, a final evaluation of the genotype–diet interactions on weight loss was not possible. In conclusion, no evidence was found that genotype–diet interaction is a main determinant of obesity treatment success, but this needs to be addressed in future studies.

Keywords: genetic variant; single nucleotide polymorphism; weight loss; nutrigenomics; dietary intervention; personalized nutrition

---

1. Introduction

In the last four decades, obesity has been identified as one of the major health risks worldwide and has reached pandemic extents [1]. According to the World Health Organization (WHO), over one-third of the world's population is overweight and 13% are described as obese [2]. Obesity adversely affects almost all physiological functions of the body and increases the risk of developing multiple diseases such as type 2 diabetes, cardiovascular diseases, and certain cancers [3,4]. Overweight and obesity are mainly caused by a long-term positive energy balance as a result of the modern lifestyle which is characterized by low physical activity and high consumption of energy-dense food [5]. To tackle obesity, multiple lifestyle intervention strategies have been developed with limited average success rates. Different diets varying in macronutrient content (e.g., low-fat/low-carb) have been investigated and compared to identify dietary regimes for successful weight loss [6]. The "one size fits all" approach for weight reduction is critically discussed. As a consequence, customized, personalized dietary recommendations are gaining more attention to fit individual needs.

In general, people lose weight to a varying extent under specific diets and this heterogeneity may depend on various factors, e.g., adherence to treatment or genetic factors [7]. The identification of multiple genetic loci associated with body mass index (BMI) and body fat distribution in genome-wide association studies (GWAS) supports the hypothesis of strong genetic interference [8–11]. A recent GWAS identified 941 BMI-associated genetic loci, which account for approximately 6% of BMI variation [11], and some specific single nucleotide polymorphisms (SNP) have been discussed as being involved in the pathogenesis of obesity [11–13]. Frayling et al. demonstrated an additive association between the risk allele of SNP rs9939609 of the fat mass and obesity-associated (*FTO*) gene and higher body weight [14]. Furthermore, the A allele of the SNP rs571312 of the melanocortin-4 receptor (*MC4R*) gene is associated with an increased BMI by 0.23 kg/m^2 [10].

Previous findings from genetic association studies led to the investigation of the relationship between certain genotypes and the effect of diets on body weight. In the Diet Intervention Examining The Factors Interacting with Treatment Success (DIETFITS) randomized clinical trial, Gardner et al. found that there was no significant difference in weight change between the low-carb and low-fat diet group after 12 months and no diet-genotype interaction for weight loss was found [15]. The Nutrient-Gene Interactions in Human Obesity: Implications for Dietary Guidelines (NUGENOB); Diet, Obesity, and Genes (DiOGenes); and Food4Me trials showed similar results [16–18]. The Food4Me trial investigated various stages of tailored nutrition. In comparison to the control group, a personalized dietary recommendation led to significantly greater weight loss [16]. In contrast, personalization based on specific SNPs had no further benefit in this study compared to other strategies of personalization [16]. However, Xiang et al. concluded in their meta-analysis with 6951 participants that *FTO* risk allele carriers (SNP rs9939609) show significantly greater weight loss than non-carriers [19]. This supports the hypothesis that specific genotypes may play a role in weight management [6]. In addition to that, studies have shown substantial inter-individual differences in metabolic response to certain meal challenges [20,21]. These results can be partly explained by genetic variations between the participants. Therefore, there is a growing interest to investigate and to understand genotype-diet interactions. A review by Livingstone et al. investigated the association between certain risk alleles and macronutrient intake [22]. They concluded that these risk alleles play a role in altering the dietary consumption of fat and protein and thus influencing weight loss [22]. The same research group also investigated the relationship between *FTO* minor alleles and weight loss in a meta-analysis [23]. The result indicated that individuals carrying *FTO* minor alleles do not show any significant differences regarding body weight response to a dietary intervention compared to non-carriers [23]. Taken together, the results were inconclusive and substantiate the need for further analyses regarding the association between genotype, diet, and weight loss.

This systematic literature search aimed to investigate whether there are genotype–diet interactions on weight loss. By including interaction terms, this analysis provides new information on the combined effect of SNPs and macronutrients on weight loss. The results may contribute to substantiate genotype-based dietary recommendations for the prevention and treatment of overweight and obesity.

2. Materials and Methods

This review was registered in the International Prospective Register for Systematic Reviews (PROSPERO, registration number CRD42019139571) and followed the Preferred Reporting Items for Systematic Review and Meta-Analyses (PRISMA) [24].

2.1. Search Strategy

The four electronic databases PubMed, Embase, Web of Science, and Cochrane Library were searched on 2 July 2019 by one person (S.B.). To identify articles examining the research question of this review, the search items were subdivided into three blocks: genetics, nutrition, and weight. For the genetic block, the following search items were used: "single nucleotide polymorphism", "SNP", "genotype", "genetic variant", and "gene variant". To include nutritional aspects, we applied the following search items: "energy", "caloric", "calorie", "fat", "carbohydrate", "carb", "diet", "dietary", "nutrition", and "nutritional". The nutritional item "protein" was not included in the search strategy as it plays a minor role in the treatment of obesity. Search items for the block weight included the following terms: "weight", "weight loss", "weight reduction", "BMI", and "body mass index". The search items in each block were combined with the Boolean operator "OR". The three blocks were then combined with the Boolean operator "AND". Depending on the database, plural forms of the search items, quotation marks, and/or asterisk were used and filters for language ("English") and species ("human") were applied. Reference lists of eligible articles were checked by hand to identify additional articles.

2.2. Study Selection

The study selection adhered to the PICO (population, intervention, control, and outcomes) criteria [25]. Studies with the following criteria were included: (a) intervention study, (b) diet described, (c) availability of SNP data, (d) outcome: weight loss, (e) interaction term of genotype x diet. Literature not in English, animal studies, and studies with participants having a severe disease (e.g., cancer) or impaired mobility were excluded. Furthermore, studies in children, pregnant and breastfeeding women, and transplant patients were excluded. Studies with no statistical application term of a genotype–diet interaction on weight loss were excluded as well. Studies with dietary interventions not focused on weight loss or on macronutrients were not considered. Two reviewers (S.B., V.W.) independently screened titles, abstracts, and full texts for eligibility. In cases of discrepant evaluations, a third reviewer (C.H.) assessed the article for eligibility. Reasons for exclusion were documented. If the full text was not available, we contacted the authors. For the screening process, we used the reference management software EndNote X9 (Thomsen Reuters, New York, NY, USA) and Microsoft Excel 2016 (Microsoft Corp, Redmond, WA, USA).

2.3. Data Extraction

The data extraction was performed independently by two reviewers (S.B, V.W.) with the software program Microsoft Excel 2016 (Microsoft Corp, Redmond, WA, USA). A third reviewer (C.H.) was consulted if inconsistencies emerged. The following data were extracted: authors, publication year, study name, study design, description of the study population, sample size, measurement of weight, intervention time, description of dietary intervention, assessment of dietary intake, genes of interest, SNP, statistical results, and statistical adjustment procedures.

2.4. Reporting Strategy

Due to the expected heterogeneity of the eligible studies, we did not perform a meta-analysis of the data. Studies with statistically significant and not significant genotype–diet interactions were treated equally in this review. Data are presented narratively.

2.5. Risk of Bias and Quality Assessment

The risk of bias of randomized controlled trials (RCTs) was assessed by the Cochrane Collaboration's risk-of-bias tool for randomized trials (RoB 2) [26]. The studies were evaluated for the randomization process, deviations from the intended intervention, missing outcome data, measurement of the outcome, and selection of the reported results. The risk of bias was judged either as low risk, some concerns, or high risk.

Non-randomized intervention studies were examined by the Cochrane Collaboration's Risk of Bias in Non-Randomized Studies—of Intervention (ROBINS-I) assessment tool [27]. Assessment was performed for confounding, selection of participants into the study, classification of intervention, deviations from intended interventions, missing data, measurement of outcomes, and selection of the reported results. Here, the risk of bias was also judged as low risk, some concerns, or high risk.

For the genotype–diet interaction term, we further applied an assessment tool for the quality of genetic association studies [28]. The validity of associations was assessed by 11 questions focusing on chance, risk, and confounding. Points ranging from −11 to +11 were given to rate quality as follows: rather high (+4 to +11 points), intermediate (−3 to +3 points), or low (−4 to −11 points).

3. Results

In total, 20,542 articles were identified, of whom 6993 articles were removed as duplicates (Figure 1). During title and abstract screening, a further 13,249 articles were excluded. Twenty-seven articles (26 publications based on eight RCTs and one publication based on one non-randomized trial) met the PICO criteria and were included in this systematic review.

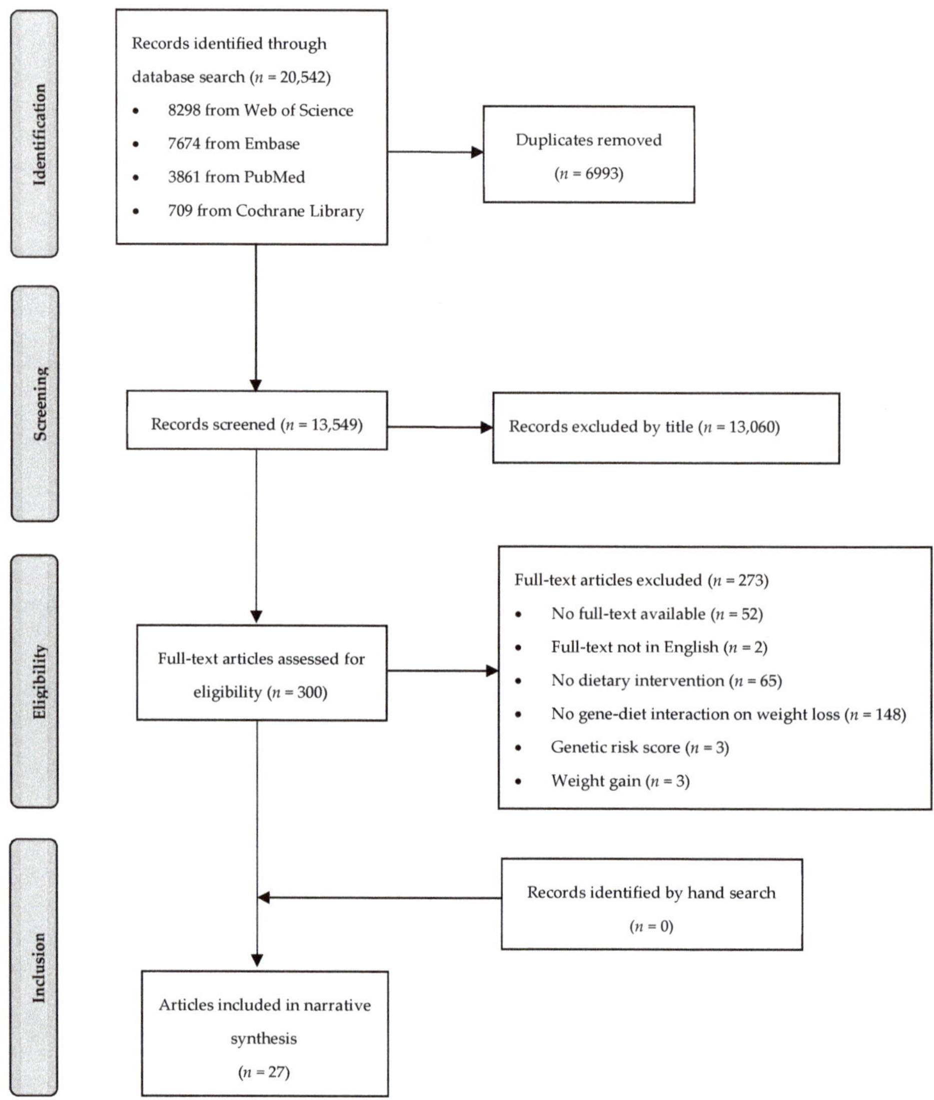

Figure 1. Flow chart of the systematic literature search according to Moher et al. [24].

3.1. Characteristics of Studies Included

The eight human intervention trials included in this systematic review are described in Table 1: NUGENOB [17,29–34], Development of Nutrigenetic Test for Personalized Prescription of Body Weight Loss Diets (Obekit) [35,36], Preventing Overweight Using Novel Dietary Strategies (POUNDS Lost) [37–51], Dietary Intervention Randomized Controlled Trial (DIRECT) [41,45], Prevención con Diet Mediterránea (PREDIMED) [52], DiOGenes [33], one trial from Italy [53], and one trial from Spain [54] (Table 1). The publication time ranged from 2006 to 2019. The sample sizes were from 147 to 7447 participants. The duration of the interventions ranged from 4 weeks to 4 years. For the collection of dietary intake, most articles used 24 h recalls [37–51], followed by dietary records [17,29–36], a combination of 24 h recalls and food frequency questionnaires (FFQ) [41,45], as well as other questionnaires [52], or no further information was given [53,54]. The studies differed in characteristics of participants such as ethnicity, age, BMI, and disease status, as well as the kind of dietary intervention (Table 1).

Table 1. Description of the trials.

Study Name	Country, Ethnicity	Study Population	Intervention	Duration of Intervention	Weight Loss in kg (Mean ± SD)	Collection of Dietary Data	Reference
NUGENOB	United Kingdom, Netherlands, France, Spain, Czech Republic, Sweden, Denmark Caucasian	771 participants inclusion: BMI ≥ 30 kg/m^2; age 20–50 years exclusion: weight change > 3 kg in last 3 months; drug treated hypertension; diabetes mellitus; hyperlipidaemia; untreated thyroid disease; surgically/drug-treated obesity; pregnancy; alcohol/drug abuse; participation in other study.	600 kcal/day less (1) low-fat diet: 20–25 E% fat, 15 E% protein, 60–65 E% carbs (2) high-fat diet: 40–45 E% fat, 15 E% protein, 40–45 E% carbs	10 weeks	(1) −6.9 ± 3.4 (2) −6.6 ± 3.5	Dietary record	[55]
Obekit	Spain: Caucasian Hispanics	147 participants inclusion: BMI 25–40 kg/m^2; unrelated exclusion: cardiovascular disease; diabetes mellitus treated with insulin; pregnant and lactating women; use of medications that affect body weight; weight change > 3 kg in last 3 months; unstable use of medication for hyperlipidaemia, type 2 diabetes and treatment of hypertension.	30% energy restriction (1) low-fat diet: 22 E% fat, 18 E% protein, 60 E% carbs (2) moderately high protein diet: 30 E% fat, 30 E% protein, 40 E% carbs	16 weeks	(1) −8.1 ± 4.1 (2) −7.6 ± 4.0	Dietary record	[35]
POUNDS Lost	United States: 80% Whites, 15% African Americans, 3% Hispanics, 2% Asians or other	811 participants inclusion: BMI 25–40 kg/m^2; age 30–70 years exclusion: diabetes mellitus; cardiovascular disease; medications that affect body weight; insufficient motivation.	750 kcal/day less (1) low fat/low-protein diet: 20 E% fat, 15 E% protein, 65 E% carbs (2) low-fat/high-protein diet: 20 E% fat, 25 E% protein, 55 E% carbs (3) high-fat/low-protein diet: 40 E% fat, 15 E% protein, 45 E% carbs (4) high-fat/high-protein diet: 40 E% fat, 25 E% protein, 35 E% carbs	2 years	6 months (1) −6.54 ± 0.42 (2) −6.80 ± 0.42 (3) −6.37 ± 0.42 (4) −6.42 ± 0.42 2 years (1) −3.26 ± 0.56 (2) −5.03 ± 0.58 (3) −3.87 ± 0.59 (4) −3.98 ± 0.42	24 h recall	[56,57]
DIRECT	Israel	322 participants inclusion: BMI ≥ 27 kg/m^2; age 40–65 years; presence of type 2 diabetes or coronary heart disease exclusion: pregnant or lactating women; serum creatinine level ≥ 2 mg/dl; liver dysfunction; gastrointestinal problems; active cancer; participating in another diet trial.	(1) low-fat diet: 1500 kcal women, 1800 kcal men, 30 E% fat, 10 E% saturated fats, 300 mg cholesterol intake (2) Mediterranean diet: 1500 kcal women, 1800 kcal men, no more than 35 E% fat, 30 to 45 g of added olive oil and a handful of nuts (3) low-carbohydrate diet: non-restricted calorie diet, 120 g carbohydrates, based on Atkins diet	2 years	(1) −2.9 ± 4.2 (2) −4.4 ± 6.0 (3) −4.7 ± 6.5	FFQ and 24 h recall	[58]

Table 1. *Cont.*

Study Name	Country, Ethnicity	Study Population	Intervention	Duration of Intervention	Weight Loss in kg (Mean ± SD)	Collection of Dietary Data	Reference
PREDIMED	Spain: European	7447 participants inclusion: age 55–80 (men)/60–80 (women) years; diabetes or three or more major cardiovascular risk factors exclusion: history of cardiovascular disease; severe chronic illness; drug or alcohol addiction; history of allergy or intolerance to olive oil or nuts; low predicted likelihood of changing dietary habits.	(1) low-fat diet (2) Mediterranean diet + olive oil (3) Mediterranean diet + nuts	4 years	(1) −0.10 ± 0.3 (2) −0.21 ± 0.2 (3) −0.07 ± 3.8	Questionnaire	[52,59]
DiOGenes	Netherlands, Denmark, United Kingdom, Greece, Germany, Spain, Bulgaria, Czech Republic	938 participants inclusion: BMI 27–45 kg/m^2; age < 65 years exclusion: > 3 kg weight change within 2 months prior to the study; medication; certain disease.	low-calorie diet: Modifast diet, four items per day, one item between 202–218 kcal, 880 kcal, fat 20 E%, carbs 54 E%, protein 26 E%	8 weeks	−11.1 ± 3.5	Dietary record	[18]
No acronym	Italy: Caucasian	300 participants inclusion: Caucasian; Italian; age > 16 years.	(1) control group: general recommendations (2) Mediterranean diet: isocaloric, <25 E% fat, 20 E% protein, 55 E% carbs	4 weeks	(1) TT genotype: −1.27 ± 3.89 A carriers: −0.62 ± 1.26 (2) TT genotype: −3.41 ± 6.47 A carriers: −2.25 ± 11.79	n. a.	[53]
No acronym	Spain	1465 participants inclusion: BMI 25–39.99 kg/m^2; age 20–65 years exclusion: medication for blood pressure, lowering glucose or lipids; diabetes mellitus; chronic renal failure; hepatic disease; cancer.	600 kcal less; women: 1200–1800 kcal/day; men: 1500–2000 kcal/day; Mediterranean diet: 35 E% fat (<10 E% saturated fats + 20 E% monounsaturated fats), 15–20 E% protein, 50 E% carbs	different between participants	G carriers: −6.84 ± 5.54 CC genotype: −7.35 ± 5.68	n. a.	[54]

BMI, body mass index; carbs, carbohydrates; DiOGenes, Diet, Obesity, and Genes; DIRECT, Dietary Intervention Randomized Controlled Trial; E%, energy%; FFQ, Food Frequency Questionnaire; h, hour; kcal, kilocalories; kg, kilograms; n.a., not available; NUGENOB, Nutrient-Gene Interactions in Human Obesity: Implications for Dietary Guidelines; Obekit, Development of Nutrigenetic Test for Personalized Prescription of Body Weight Loss Diets; POUNDS Lost, Preventing Overweight Using Novel Dietary Strategies; PREDIMED, Prevención con Diet Mediterránea; SD, standard deviation.

3.2. Study Quality and Risk of Bias

The risk of bias assessment of the selected studies is shown in Figure 2. In summary, two out of 26 articles from the RCTs were judged to be at low risk of bias for all domains. Thirteen articles were judged to raise some concerns. This was due to high drop-out rates during intervention. Because of missing data on sample size in the intervention groups at the end of intervention, we judged 13 articles to be at high risk of bias. The non-randomized trial was judged to be at moderate risk of bias for all domains. The latter was due to missing information about exclusion criteria of participants during intervention (Figure 2).

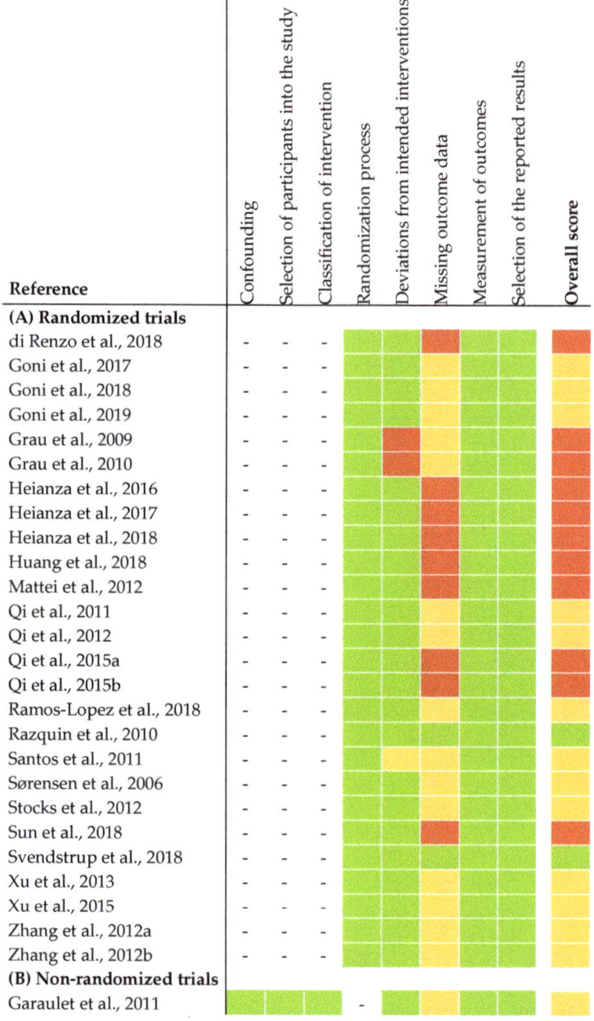

Figure 2. Risk of bias assessment of articles included in narrative synthesis. (**A**) Risk of bias assessment of randomized trials [26]. (**B**) Risk of bias assessment of non-randomized trials [27]. Overall score: the risk of bias was judged as low risk (green), some concerns (yellow), high risk (red).

The quality of analyses concerning genetics was judged to be rather high in 21 articles (Figure 3). The quality of six articles was judged as being intermediate. This was due to missing statistical analysis concerning confounding parameters (e.g., adjustment for ethnicity).

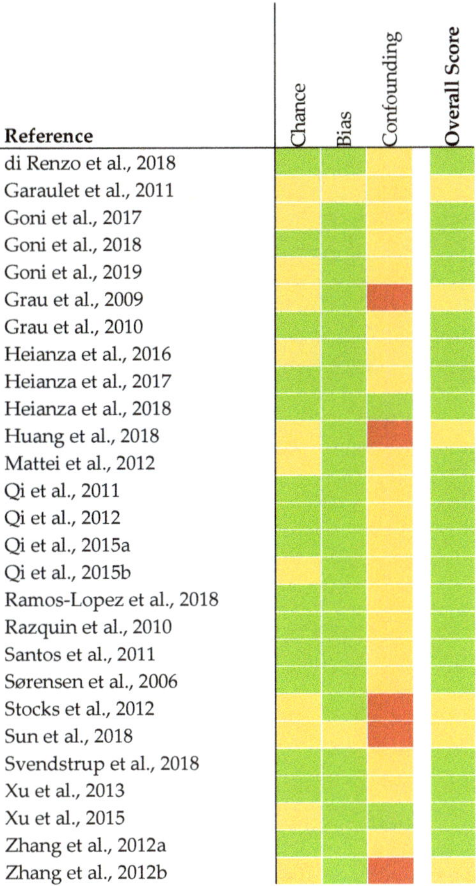

Figure 3. Quality assessment of genetic association studies [28]. The quality was judged as rather high (green), intermediate (yellow), or low (red).

3.3. Main Findings

3.3.1. Interaction of Genotype and Fat Intake on Weight Loss

In total, an interaction of genotype and fat intake on weight loss was assessed for 60 different genetic loci and 88 different SNPs (Table 2).

Table 2. Interaction of genotype and fat intake on weight loss.

Gene	SNP	Study Name	Sample Size	Study Population of Interaction Term	Time Point of Weight Measurement	Results (p-Value)	Reference
ADAMTS9	rs6795735	NUGENOB	559–580		10 weeks	0.2 [1,5,21]	[33]
ADCY3	rs10182181	Obekit	101		16 weeks	$p = 0.02$ [2,6,21], $p = 0.04$ [2,6,22] Additive model: carriers of the GG (minor allele G) genotype greater weight loss with low-fat diet than carriers of the AG or AA genotypes Co-dominant model: carriers of the GG and AA genotype less weight loss with low-fat diet than carriers of the AG genotype	[35]
ADIPOQ	rs266729	NUGENOB	642	All participants	10 weeks	0.029 [2,7,23] Carriers of the GG and GC (minor allele G) genotype greater weight loss on high-fat diet than carriers of the CC genotype; Carriers of the GC (minor allele G) genotype greater weight loss on low-fat diet than carriers of the CC genotype	[32]
	rs2241766					0.18 [2,7,23]	
	rs1501299					0.14 [2,7,23]	
	rs17300539					0.07 [2,7,23]	
ADRB2	rs1042713	Obekit	107		4 months	0.71 [3,8,23]	[36]
	rs1042714					0.86 [3,8,23]	
AMY1-AMY2	rs11185098	POUNDS Lost	692	Whites + Blacks	2 years	n. s. [2,9,21]	[40]
APOA5	rs964184		734			n. s. [2,10,21]	[50]
CART	rs7379701 *	NUGENOB	642	All participants	10 weeks	0.36 [2,7,23]	[32]
	rs6453132 *					0.10 [2,7,23]	
	rs17358216					0.72 [2,7,23]	
	rs5868607					0.68 [2,7,23]	
CD36	rs2232169					0.42 [2,7,23]	
CETP	rs3764261	POUNDS Lost + DIRECT	723 + 171	Pooled	2 years	n. s. [2,11,23]	[45]

Table 2. *Cont.*

Gene	SNP	Study Name	Sample Size	Study Population of Interaction Term	Time Point of Weight Measurement	Results (*p*-Value)	Reference
CTNNBL1	rs9939609	NUGENOB	559–580		10 weeks	0.7 [1,5,21]	[33]
CYP2R1	rs10741657		732	All participants	6 months	0.22 [2,12,21]	[46]
					2 years	0.02 [2,12,21] Carriers of the AA (minor allele A) genotype less weight loss with low-fat diet than carriers of the AG or GG genotype	
		POUNDS Lost	576	Whites	6 months	n. s. [2,11,21]	
					2 years	< 0.05 [2,11,21] Carriers of the AA (minor allele A) genotype less weight loss with low-fat diet than carriers of the AG or GG genotype	
DHCR7	rs12785878		732	All participants	6 months	0.80 [2,12,21]	
					2 years	0.22 [2,12,21]	
			584	Whites	6 months	n. s. [2,10,21]	
					2 years	n. s. [2,10,21]	
DNM3-PIGC	rs1011731		559–580			0.2 [1,5,21]	[33]
ENPP1	rs1799774		642			0.62 [2,7,23]	[32]
	rs1044498	NUGENOB			10 weeks	0.85 [2,7,23]	
	rs7754561					0.13 [2,7,23]	
FANCL	rs887912		559–580			0.8 [1,5,21]	[33]
FOXC2	rs34221221		642	All participants		0.75 [2,7,23]	[32]
FTO	rs9939609	No acronym	188		4 weeks	0.87 [3,13,23]	[53]
			734			0.55 [2,7,24]	[30]
GAD2	rs928197 *	NUGENOB	642		10 weeks	0.56 [2,7,23]	[32]
	rs992990					0.78 [2,7,23]	
	rs2236418 *					0.88 [2,7,23]	
GC	rs2282679	POUNDS Lost	732		6 months	0.17 [2,12,21]	[46]
					2 years	0.08 [2,12,21]	
			576	Whites	6 months	n. s. [2,10,21]	
					2 years	n. s. [2,10,21]	

Table 2. *Cont.*

Gene	SNP	Study Name	Sample Size	Study Population of Interaction Term	Time Point of Weight Measurement	Results (*p*-Value)	Reference
GHRL	rs696217	NUGENOB	642		10 weeks	0.33 [2,7,23]	[32]
GHSR	rs2232169			All participants	10 weeks	0.35 [2,7,23]	
GIPR	rs2287019	POUNDS Lost	737	All participants	6 months	0.08 [2,14,21]	[43]
					2 years	n. s. [2,14,21]	
			590	Whites	6 months	0.18[2,13,21]	
					2 years	n. s.[2,10,21]	
GPRC5B	rs12444979	NUGENOB	559–580		10 weeks	0.046[1,5,21]	[33]
HNF1A	rs7957197	POUNDS Lost	722	All participants	6 months	0.006 [2,12,23] Carriers of the TT and AT (minor allele T) genotype greater weight loss with high-fat diet than carriers of the AA genotype	[41]
			575	Whites		0.001 [2,15,23] Carriers of the TT and AT (minor allele T) genotype greater weight loss with high-fat diet than carriers of the AA genotype	
		DIRECT	171	All participants		0.03 [2,16,23] Carriers of the TT and AT (minor allele T) genotype greater weight loss with high-fat diet than carriers of the AA genotype	
		POUNDS Lost + DIRECT	722 + 171	Pooled		0.001 [2,12,23] Carriers of the TT and AT (minor allele T) genotype greater weight loss with high-fat diet than carriers of the AA genotype	
		POUNDS Lost	722	All participants	2 years	n. s. [2,10,23]	
		DIRECT	171			n. s. [2,10,23]	
		POUNDS Lost + DIRECT		Pooled		n. s. [2,10,23]	

Table 2. *Cont.*

Gene	SNP	Study Name	Sample Size	Study Population of Interaction Term	Time Point of Weight Measurement	Results (*p*-Value)	Reference
HSD11B1	rs846919	NUGENOB	642	All participants	10 weeks	0.49 [2,7,23]	[32]
IGF2	rs3168310 *					0.34 [2,7,23]	
	rs680 *					0.59 [2,7,23]	
	rs3842759					0.77 [2,7,23]	
IL6	rs1800795	PREDIMED	737		3 years	0.028 [3,17,25] Carriers of the CC (minor allele C) genotype greater weight loss on Mediterranean diet with olive oil than carriers of the CG and GG genotype	[52]
			480	Non-diabetics		0.007 [3,10,25] Carriers of the CC (minor allele C) genotype greater weight loss on Mediterranean diet with olive oil than carriers of the CG and GG genotype	
			257	Diabetics		n. s. [3,10,25]	
KCNJ11	rs5219	NUGENOB	642		10 weeks	0.60 [2,7,23]	[32]
						0.10 [2,7,23]	
LEPROTL1	−2625 C > T					0.12 [2,7,23]	
	rs6082					0.42 [2,7,23]	
LIPC	rs1800588 *					0.67 [2,7,23]	
	rs2070895 *	POUNDS Lost	734	All participants	2 years	n. s. [2,15,21]	[48]
LRRN6C	rs10968576		559–580			0.1 [1,5,21]	[33]
LY86	rs1294421					0.6 [1,5,21]	
MAF	rs1424233					0.1 [1,5,21]	
MAP2K5	rs2241423					0.3 [1,5,21]	
MC3R	rs6024728	NUGENOB	760		10 weeks	0.89 [2,7,21]	[31]
	rs6014646					0.48 [2,7,21]	
	rs6024730					0.57 [2,7,21]	
	rs6024731					0.72 [2,7,21]	
	rs11697509					0.20 [2,7,21]	
	rs6127698					0.81 [2,7,21]	

Table 2. *Cont.*

Gene	SNP	Study Name	Sample Size	Study Population of Interaction Term	Time Point of Weight Measurement	Results (*p*-Value)	Reference
	rs3746619 *					0.81 [2,7,21]	
	rs3827103 *					0.90 [2,7,21]	
	rs1543873					0.42 [2,7,21]	
	rs6099058					0.80 [2,7,21]	
MC4R	rs12970134		559–580			0.4 [1,5,21]	[33]
MKKS	rs1547		642			0.47 [2,7,23]	[32]
MTIF3	rs4771122		559–580			0.3 [1,5,21]	[33]
MTNR1B	rs10830963		575	Whites	6 months	< 0.05 [2,10,21] Carriers of the GG (minor allele G) genotype greater weight loss with low-fat diet than carriers of the GC or CC genotype	[37]
		POUNDS Lost	722			0.01 [2,12,21] Carriers of the GG (minor allele G) genotype greater weight loss with low-fat diet than carriers of the GC or CC genotype	
				All participants	2 years	0.19 [2,12,21]	
NFATC2IP	rs11150675		692			0.005 [2,18,25] Carriers of the AA (minor allele A) genotype less weight loss with low-fat diet than carriers of the AG and GG genotype	[47]
NPC1	rs1805081	NUGENOB	559–580		10 weeks	0.2 [1,5,21]	[33]
NPY	rs16147	POUNDS Lost	723		2 years	0.464 [2,15,21]	[51]
			575	Whites		n. s. [2,10,21]	
			264	Hypertensive		0.688 [2,15,21]	
			459	Non-hypertensive		0.547 [2,15,21]	

Table 2. *Cont.*

Gene	SNP	Study Name	Sample Size	Study Population of Interaction Term	Time Point of Weight Measurement	Results (*p*-Value)	Reference
PCSK1	rs6235	NUGENOB	642		10 weeks	0.76 [2,7,23]	[32]
PPARG2	rs1801282 *	No acronym	1236		Different between participants	0.001 [4,19,23] Carriers of the GG and GC (minor allele G) genotype less weight loss (% of baseline weight) on high-fat diet than carriers of the CC genotype	[54]
		NUGENOB	642	All participants	10 weeks	0.88 [2,7,23]	[32]
	rs3856806					0.45 [2,7,23]	
	rs7649970 *					0.87 [2,7,23]	
PPARG3	rs10865710					0.76 [2,7,23]	
PPARGC1A	rs8192678					0.39 [2,7,23]	
	rs2932963					0.36 [2,7,23]	
			757			0.94 [2,7,21]	[29]
PPM1K	rs1440581	POUNDS Lost	734		6 months	0.002 [2,15,21] Carriers of the CC (minor allele C) genotype less weight loss with high-fat diet than carriers of the CT or TT genotype	[49]
					2 years	0.008 [2,15,21] Carriers of the CC (minor allele C) genotype less weight loss with high-fat diet than carriers of the CT or TT genotype	
			587	Whites	6 months	0.02 [2,16,21] Carriers of the CC (minor allele C) genotype less weight loss with high-fat diet than carriers of the CT or TT genotype	
					2 years	0.01 [2,16,21] Carriers of the CC (minor allele C) genotype less weight loss with high-fat diet than carriers of the CT or TT genotype	

Table 2. *Cont.*

Gene	SNP	Study Name	Sample Size	Study Population of Interaction Term	Time Point of Weight Measurement	Results (*p*-Value)	Reference
RSPO3	rs9491696	NUGENOB	559–580	All participants	10 weeks	0.5 [1,5,21]	[33]
SERPINE1	rs1799889		642			0.29 [2,7,23]	[32]
SLC39A8	rs13107325		559–580			0.8 [1,5,21]	[33]
SLC6A14	rs2011162		481	Women		0.06 [2,7,23]	[32]
			161	Men		0.78 [2,7,23]	
SREBF1	17 C > G		642	All participants		0.20 [2,7,23]	
TCF7L2	rs7903146 *		739			0.023 [2,7,25] Carriers of the TT (minor allele T) genotype less weight loss on high-fat diet than carriers of the TC and CC genotype	[17]
		POUNDS Lost	588	Whites	6 months	0.28 [2,7,21]	[42]
					2 years	0.692 [2,7,21]	
	rs12255372 *		591		6 months	0.057 [2,7,21]	
					2 years	0.517 [2,7,21]	
TFAP2B	rs987237	DiOGenes	640		8 weeks	0.4 [1,10,21]	[33]
			559–580			0.03 [1,5,21]	
TNFα	rs1800629	NUGENOB	642	All participants	10 weeks	0.04 [2,7,23] Carriers of the AA (minor allele A) genotype greater weight loss on a high-fat diet than carriers of the GG genotype; Carriers of the AG genotype greater weight loss on low-fat diet than carriers of the GG genotype	[32]
UCP2	rs6593669					0.25 [2,7,23]	

Table 2. *Cont.*

Gene	SNP	Study Name	Sample Size	Study Population of Interaction Term	Time Point of Weight Measurement	Results (*p*-Value)	Reference
UCP3	rs1900849					0.86 [2,7,23]	
	rs6905288		559–580			0.4 [1,7,23]	[33]
VEGFA	rs1358980		707			0.26 [2,13,26]	[34]
			174	Men		0.58 [2,20,26]	
			533	Women		0.06 [2,20,26]	
WAC	rs2807761		642	All participants		0.17 [2,7,23]	[32]
ZNF608	rs4836133		559–580			0.9 [1,5,21]	[33]

ADAMTS9, a disintegrin-like and metallopeptidase (reprolysin type) with thrombospondin type 1 motif, 9; ADCY3, adenylate cyclase 3; ADIPOQ, adiponectin, C1Q and collagen domain-containing; ADRB2, adrenoceptor beta 2; AMY1-AMY2, amylase alpha 1A-amylase alpha 2A/B; APOA5, apolipoprotein A5; CART, cocaine- and amphetamine-regulated transcript; CD36, CD36 antigen; CETP, cholesteryl ester transfer protein; CTNNBL1, catenin beta like 1; CYP2R1, cytochrome P450 family 2 subfamily R member; DHCR7, 7-dehydrocholesterol reductase; DiOGenes, Diet, Obesity, and Genes; DIRECT, Dietary Intervention Randomized Controlled Trial; DNM3-PIGC, dynamin 3-phosphatidylinositol glycan anchor biosynthesis class C; ENPP1, ectonucleotide pyrophosphatase/phosphodiesterase 1; FANCL, Fanconi anemia complementation group L; FOXC2, forkhead box C2; FTO, fat mass and obesity-associated; GAD2, glutamate decarboxylase 2; GC, GC vitamin D-binding protein; GHRL, ghrelin; GHSR, growth hormone secretagogue receptor; GIPR, gastric inhibitory polypeptide receptor; GPRC5B, G protein-coupled receptor class C group 5 member B; HNF1A, HNF1 homeobox A; HSD11B1, hydroxysteroid 11-beta dehydrogenase 1; IGF2, insulin-like growth factor 2; IL6, interleukin 6; KCNJ11, potassium inwardly rectifying channel subfamily J member 11; LEPROTL1, leptin receptor overlapping transcript-like 1; LIPC, hepatic lipase C; LRRN6C, leucine rich repeat and immunoglobulin domain-containing 2; LY86, lymphocyte antigen 86; MAF, MAF basic leucine zipper domain transcription factor; MAP2K5, mitogen-activated protein kinase kinase 5; MC3R, melanocortin 3 receptor; MC4R, melanocortin 4 receptor; MKKS, McKusick–Kaufman syndrome; MTIF3, mitochondrial translational initiation factor 3; MTNR1B, melatonin receptor 1B; NFATC2IP, nuclear factor of activated T cells 2-interacting protein; NPC1, Niemann-Pick disease, type C1 intracellular cholesterol transporter 1; NPY, neuropeptide Y; n. s., not significant; NUGENOB, Nutrient-Gene Interactions in Human Obesity: Implications for Dietary Guidelines; Obekit, Development of Nutrigenetic Test for Personalized Prescription of Body Weight Loss Diets; PCSK1, proprotein convertase subtilisin/kexin type 1; POUNDS Lost, Preventing Overweight Using Novel Dietary Strategies; PPARG2, peroxisome proliferator-activated receptor gamma isoform 2; PPARG3, peroxisome proliferator-activated receptor gamma isoform 3; PPARGC1A, peroxisome proliferator-activated receptor gamma coactivator 1 alpha; PPM1K, protein phosphatase, Mg2+/Mn2+-dependent 1K; PREDIMED, Prevención con Diet Mediterránea; RSPO3, R-spondin 3; SERPINE1, serine proteinase inhibitor 1; SLC39A8, solute carrier family 39 member 8; SLC6A14, solute carrier family 6 member 14; SNP, single nucleotide polymorphism; SREBF1, sterol regulatory element-binding transcription factor 1; TCF7L2, transcription factor 7-like 2; TFAP2B, transcription factor activating enhancer binding protein 2 beta; TNFα, tumor necrosis factor alpha; UCP2, uncoupling protein 2; UCP3, uncoupling protein 3; VEGFA, vascular endothelial growth factor A; WAC, WW domain-containing adaptor with coiled-coil; ZNF608, zinc finger protein 608. * SNPs within this gene locus are in high linkage disequilibrium ($r^2 > 0.8$). [1] Interaction term on weight loss in kg for genotype and fat intake in dietetic change (difference in initial and end-of-intervention percentage intake of fat). [2] Interaction term on weight loss in kg for genotype and fat intake in energy %. [3] Interaction term on weight loss in kg for genotype and Mediterranean diet. [4] Interaction term on weight loss in kg for genotype and fat intake according to the intake of the population (below or above the median). [5] Adjusted for age, sex, baseline weight, baseline weight x sex, center, genotype, fat intake. [6] Adjusted for age, sex, baseline weight. [7] Adjusted for age, sex, baseline weight, center. [8] Adjusted for age, sex, lipid-lowering medications. [9] Adjusted for age, sex, ethnicity, body mass index, weight at baseline, fasting glucose concentration, medication use at baseline. [10] No information about adjustment. [11] Adjusted for age, sex, ethnicity, baseline body mass index, lipid-lowering medication use. [12] Adjusted for age, sex, ethnicity, baseline weight. [13] Adjusted for age, sex. [14] Adjusted for age, sex, ethnicity. [15] Adjusted for age, sex, ethnicity, baseline body mass index. [16] Adjusted for age, sex, baseline body mass index. [17] Adjusted for age, sex, baseline body mass index, diabetes. [18] Adjusted for age, sex, ethnicity, baseline body mass index, smoking status, intervention time. [19] Adjusted for age, sex, clinic. [20] Adjusted for age. [21] Additive genetic model. [22] Co-dominant genetic model. [23] Dominant genetic model. [24] No assumption of genetic model. [25] Recessive genetic model. [26] No information about genetic model.

Most of the SNPs (*n* = 80) were analyzed once and six of them (adenylate cyclase 3 (*ADCY3*) SNP rs10182181 [35]; adiponectin, C1Q and collagen domain containing (*ADIPOQ*) SNP rs266729 [32]; tumor necrosis factor alpha (*TNFα*) SNP rs1800629 [32]; cytochrome P450 family 2 subfamily R member (*CYP2R1*) SNP rs10741657 [46]; melatonin receptor 1B (*MTNR1B*) SNP rs10830963 [37]; nuclear factor of activated T cells 2-interacting protein (*NFATC2IP*) SNP rs11150675 [47]) showed a statistically significant interaction with fat intake on weight loss.

SNPs within eight genetic loci (*FTO* SNP rs9939609; HNF1 homeobox A (*HNF1A*) SNP rs7957197; interleukin 6 (*IL6*) SNP rs1800795; hepatic lipase C (*LIPC*) SNPs rs6082, rs1800588, rs2070895; peroxisome proliferator-activated receptor gamma isoform 2 (*PPARG2*) SNP rs1801282; protein phosphatase, Mg2+/Mn2+-dependent 1K (*PPM1K*) SNP rs1440581; transcription factor 7-like 2 (*TCF7L2*) SNPs rs7903146, rs12255372; transcription factor AP-2 beta (*TFAP2B*) SNP rs987237) were examined for the genotype x fat intake interaction on weight loss twice in 12 articles [17,29,30,32,33,41,42,48,49,52–54]. No statistically significant interaction between the *FTO* SNP rs9939609 and fat intake on weight loss was seen [30,53]. After 6 months of intervention, a statistically significant interaction between the *HNF1A* SNP rs7957197 and fat intake on weight loss was seen in the POUNDS Lost and the DIRECT trial, as well as in the pooled data [41]. A greater weight loss was observed in participants with the T allele with a high-fat diet compared to those without the T allele. However, after 2 years of intervention, no statistically significant interaction could be found between the *HNF1A* SNP rs7957197 and fat intake on weight loss [41].

The study participants of the PREDIMED [52] and the NUGENOB [32] trials were analyzed to investigate the *IL6* SNP rs1800795 x fat intake interaction on weight loss. In the PREDIMED trial, homozygous carriers of the risk allele showed a greater weight loss with the Mediterranean diet with olive oil supplementation compared to a low-fat diet than heterozygous carriers and non-carriers after 3 years of intervention [52]. This result could not be replicated in a similar analysis from the NUGENOB trial [32]. Here, no statistically significant interaction between the *IL6* SNP rs1800795 and fat intake on weight loss was found after 10 weeks of dietary intervention. Furthermore, there was no statistically significant interaction between the *LIPC* SNPs rs6082, rs1800588 in the NUGENOB trial, and SNP rs2070895 in the POUNDS Lost trial and fat intake on weight loss. The SNPs rs1800588 and rs2070895 are in high linkage disequilibrium (r^2 > 0.8) [32,48]. The results on the *PPARG2* SNP rs1801282 x fat intake on weight loss were controversial. While a study from Spain with 1465 participants found a lower percental weight loss compared to baseline weight in homozygous and heterozygous carriers compared to the non-carriers on the high-fat diet [54], no statistically significant interaction was seen in an analysis on the NUGENOB dataset [32].

While the NUGENOB trial did not reveal a statistically significant interaction between the *PPM1K* SNP rs1440581 and fat intake on weight loss after 10 weeks of intervention [29], the homozygous carriers of this gene variant in the POUNDS Lost trial showed a lower weight loss on a high-fat diet after 6 months as well as after 2 years of intervention compared to the heterozygous and the non-carriers [49]. Contrary results were also found for the *TCF7L2* SNP rs7903146 x fat intake interaction on weight loss. After 10 weeks of intervention, the NUGENOB trial identified a lower weight loss on a high-fat diet in homozygous carriers compared to heterozygous carriers and non-carriers combined [17]. In the POUNDS Lost trial, no statistically significant interaction between the *TCF7L2* SNP rs7903146 x fat intake on weight loss was found [42]. Additionally, no statistically significant interaction was seen between the *TCF7L2* SNP rs12255372 and fat intake on weight loss. The *TCF7L2* SNPs rs7903146 and rs12255372 were in high LD (r^2 > 0.8) [42].

One article examined the interaction of *TFAP2B* SNP rs987237 and fat intake on weight loss in the DiOGenes and NUGENOB trial, respectively [33]. An additive genotype–diet interaction model of the NUGENOB trial showed that homozygotes for the A allele lost more weight with the low-fat than the high-fat diet, whereas homozygotes for the G allele lost more weight on the high-fat diet compared to the low-fat diet. These findings were not confirmed by the DiOGenes trial [33].

3.3.2. Interaction of Genotype and Carbohydrate Intake on Weight Loss

Three articles investigated the interaction of genotype and carbohydrate intake on weight loss (Table 3) [38,42,44]. No significant interactions could be found between the consumption of carbohydrates and the fibroblast growth factor 21 (*FGF21*) SNP rs838147 [38] and the *TCF7L2* SNPs rs7903146 and rs12255372 (LD r^2 > 0.8) on weight loss [42]. Among the participants of the POUNDS Lost trial, a statistically significant interaction between the insulin receptor substrate 1 (*IRS1*) SNP rs2943641 and the highest-carbohydrate diet was found after 6 months of intervention [44]. Homozygous carriers of the risk allele showed greater weight loss on the highest-carbohydrate diet (65 energy%) than heterozygous carriers or non-carriers after 6 months of intervention. After 2 years of intervention, no statistically significant effect of the interaction between *IRS1* SNP rs2943641 and carbohydrate diet on weight loss could be found (Table 3).

3.3.3. Interaction of Genotype and Protein Intake on Weight Loss

In total, seven publications assessed the interaction of genotype and protein intake on weight loss [39,40,42,43,46,49,51] (Table 4). All of them analyzed data from the POUNDS Lost trial and used an additive model for genetic analysis. No significant interactions between the amylase alpha 1A- amylase alpha 2A/B (*AMY1-AMY2*) SNP rs11185098 [40]; *CYR2R1* SNP rs10741657 [46]; 7-dehydrocholesterol reductase (*DHCR7*) SNP rs12785878 [46]; GC vitamin D-binding protein (*GC*) SNP rs2282679 [46]; gastric inhibitory polypeptide receptor (*GIPR*) SNP rs2287019 [43]; lactase (*LCT*) SNP rs4988235 [39]; neuropeptide Y (*NPY*) SNP rs16147 [51]; *PPM1K* SNP rs1440581 [49]; and *TCF7L2* SNPS rs7903146, rs12255372 (LD r^2 > 0.8) [42] and the protein intake on weight loss was found (Table 4).

Table 3. Interaction of genotype and carbohydrate intake on weight loss.

Gene	SNP	Sample Size	Study Population of Interaction Term	Time Point of Weight Measurement	Results (p-Value) [1]	Reference
FGF21	rs838147	715	All participants	2 years	0.07 [2,5]	[38]
		573	Whites		n. s. [3,5]	
IRS1	rs2943641	738	All participants	6 months	$p = 0.037$ [2,5]; $p = 0.058$ [2,6] Additive model: carriers of the TT (minor allele T) genotype greater weight loss with the highest-carbohydrate diet than carriers of the TC or CC genotype.	[44]
				2 years	$p = 0.84$ [2,5]; $p = 0.59$ [2,6]	
		591	Whites	6 months	$p < 0.05$ [3,5]; $p < 0.05$ [3,6] Additive model: carriers of the TT (minor allele T) genotype greater weight loss with the highest-carbohydrate diet than carriers of the TC or CC genotype.	
				2 years	n. s. [3,5], n. s. [3,6]	
TCF7L2	rs7903146 *	588		6 months	0.811 [4,5]	[42]
				2 years	0.948 [4,5]	
	rs12255372 *	591		6 months	0.21 [4,5]	
				2 years	0.403 [4,5]	

FGF21, fibroblast growth factor 21; IRS1, insulin receptor substrate 1; n. s., not significant; SNP, single nucleotide polymorphism; TCF7L2, transcription factor 7-like 2. * SNPs are in high linkage disequilibrium ($r^2 > 0.8$). [1] interaction term on weight loss in kg for genotype and carbohydrate intake in energy %. All articles investigated data from the Preventing Overweight Using Novel Dietary Strategies (POUNDS Lost) trial. [2] Adjusted for age, sex, baseline weight, ethnicity. [3] No information about adjustment. [4] Adjusted for age, sex, center, baseline weight. [5] Additive genetic model. [6] Dominant genetic model.

Table 4. Interaction of genotype and protein intake on weight loss.

Gene	SNP	Sample Size	Study Population of Interaction Term	Time Point of Weight Measurement	Results (p-Value) [1]	Reference
AMY1-AMY2	rs11185098	692		2 years	n. s. [2]	[40]
CYP2R1	rs10741657	732	All participants	6 months	0.48 [3]	[46]
				2 years	0.19 [3]	
		576	Whites	6 months	n. s. [8]	
				2 years	n. s. [8]	
DHCR7	rs12785878	732	All participants	6 months	0.25 [3]	
				2 years	0.52 [3]	
		584	Whites	6 months	n. s. [8]	
				2 years	n. s. [8]	
GC	rs2282679	732	All participants	6 months	0.68 [3]	
				2 years	0.41 [3]	
		576	Whites	6 months	n. s. [8]	
				2 years	n. s. [8]	
GIPR	rs2287019	737	All participants	6 months	> 0.35 [4]	[43]
				2 years		
		590	Whites	6 months	> 0.35 [9]	
LCT	rs4988235	583		2 years	n. s. [5]	[39]
NPY	rs16147	723	All participants		n. s. [6]	[51]
		575	Whites		n. s. [8]	

Table 4. *Cont.*

Gene	SNP	Sample Size	Study Population of Interaction Term	Time Point of Weight Measurement	Results (p-Value)[1]	Reference
PPM1K	rs1440581	734	All participants	6 months	> 0.05 [6]	[49]
				2 years		
		587		6 months	> 0.05 [10]	
				2 years		
TCF7L2	rs7903146 *	588	Whites	6 months	0.906 [7]	[42]
				2 years	0.515 [7]	
	rs12255372 *	591		6 months	0.746 [7]	
				2 years	0.328 [7]	

AMY1-AMY2, amylase alpha 1A-amylase alpha 2A/B; CYP2R1, cytochrome P450 family 2 subfamily R member; DHCR7, 7-dehydrocholesterol reductase; GC, GC vitamin D-binding protein; GIPR, gastric inhibitory polypeptide receptor; LCT, lactase; NPY, neuropeptide Y; n. s., not significant; PPM1K, protein phosphatase, Mg2+/Mn2+-dependent 1K; SNP, single nucleotide polymorphism; TCF7L2, transcription factor 7-like 2. * SNPs are in high linkage disequilibrium ($r^2 > 0.8$). [1] Interaction term on weight loss in kg for genotype and protein intake in energy %. All interactions were analyzed with an additive genetic model and all articles investigated data from the Preventing Overweight Using Novel Dietary Strategies (POUNDS Lost) trial. [2] Adjusted for age, sex, ethnicity, body mass index, baseline weight, fasting glucose concentration, medication use at baseline. [3] Adjusted for age, sex, ethnicity, baseline weight. [4] Adjusted for age, sex, ethnicity. [5] Adjusted for age, sex, body mass index, baseline weight. [6] Adjusted for age, sex, ethnicity, baseline body mass index. [7] Adjusted for age, sex, center, baseline weight. [8] No information about adjustment. [9] Adjusted for age, sex. [10] Adjusted for age, sex, baseline body mass index.

4. Discussion

This systematic review gives an overview of genotype x diet interactions and their association with weight loss. The literature search identified 27 articles in which the interaction of 91 SNPs within 63 genetic loci and fat [17,29–37,40–43,45–54], carbohydrate [38,42,44], or protein intake [39,40,42,43,46,49,51] on weight loss was investigated.

Most publications (*n* = 24) focused their interaction term on the macronutrient fat. This may be due to the fact that fat is the most energy-dense macronutrient and, therefore, most weight-loss studies have a focus on reducing fat intake. Furthermore, it might be assumed that obesity-associated SNPs play a role in fat intake [60]. However, the results of our systematic review present an inconsistent picture for genotype–fat intake interaction and weight loss. Most findings were not significant and not replicated in other trials. The statistically significant findings were related to 12 SNPs in 12 distinct genetic loci. However, the *ADCY3* SNP rs10182181 [35], *ADIPOQ* SNP rs266729 [32], *CYP2R1* SNP rs10741657 [46], *MTNR1B* SNP rs10830963 [37], *NFATC2IP* SNP rs11150675 [47], and *TNFα* SNP rs1800629 [32] were examined in only one trial each. This lack of replication excludes a robust interpretation of the results.

The general findings of this systematic review are in line with a publication about the association between the genetic variant *FTO* rs9939609, and dietary intake and BMI, indicating no significant interaction between the *FTO* variant and dietary intake on BMI [60].

A genotype–fat intake interaction on weight loss was examined twice with eight SNPs in 12 articles, but the findings showed inconsistent results as in one study a significant interaction between the SNP and fat intake on weight loss was found, whereas this result could not be replicated in the other study [17,29,32,33,42,49,52,54]. This might be explained by the different sample sizes, study durations, or dietary interventions. Furthermore, a statistically significant interaction between the *HNF1A* SNP rs7957197 and a high-fat diet on weight loss could be seen in the POUNDS Lost trial as well as in the DIRECT trial [41]. Carriers of the T allele (minor allele) showed a greater weight loss on a high-fat diet than non-carriers after 6 months of intervention. Variants of the *HNF1A* gene are known to be associated with diabetes [61,62]. Thereby, the risk of developing diabetes might be influenced by the lifestyle and weight status of the individual [41,63]. One possible explanation of the underlying mechanism might be that a high-fat diet downregulates *HNF1A* gene expression in the pancreas [64,65] and thereby might cause weight loss and improvement of insulin resistance [64,65]. However, after 2 years of intervention, no statistically significant interaction was found between the *HNF1A* SNP rs7957197 and fat intake on weight loss in both studies as well as in the pooled data [41]. This result may also be explained by a loss of adherence to the diet and high drop-out rates [56]. Moreover, after 12 months of intervention, the participants in both trials regained body weight [56,58].

Similar inconclusive results were found for the potential interactions of SNPs and both carbohydrate and protein intake and their effect on weight loss. All long-term results reached no statistical significance and were investigated only within the POUNDS Lost trial. Due to the fact that the sample size of the POUNDS Lost trial with 811 participants is low compared to most genetic association studies [10,14], the statistical power to reach significant results was rather limited. Moreover, the high drop-out rates (*n* = 179) and the regain of body weight, as described earlier [56], may further decrease statistical power.

Irrespective of such inconsistencies, this systematic review provides three main findings on the topic of genotype–diet interactions and weight loss. First, there are many "significant" findings observed in single and mostly small studies without replication in others. Second, the number and size of studies to examine genotype–diet interactions on weight loss are rather limited. The 27 publications identified refer to only eight weight loss trials of which 15 publications were based on data from the POUNDS Lost trial and another seven papers analyzed potential interactions in the database of the NUGENOB trial. The third message from this analysis is the considerable heterogeneity of the identified studies. There were substantial differences among the trials not only in the selection of SNPs, but also in study design, dietary interventions, intervention duration, and sample size. In addition, dietary intake data—highly relevant for the research question—were collected using different methods, all of those self-reported and with a high risk of recall and reporting bias, which may further complicate such analyses. It is noteworthy that many intervention studies were excluded because they investigated the association between single SNPs and weight loss without considering any interaction term for genotype x diet on weight loss. This might be explained by the fact that weight loss was not the primary outcome [66] or was due to publication bias, as negative results concerning genetics are commonly not published [67]. To promote the field of genotype–diet interactions on weight loss it is, therefore, crucial to collect and analyze genetic material more frequently or regularly in dietary intervention studies. However, as the treatment of overweight and obesity is a complex process with many factors involved, this request may be a great challenge.

All studies selected for this review are based on a hypothesis-driven approach investigating defined candidate genes. This means, that no hypothesis-free GWAS are available for weight loss intervention trials. It appears likely that other genetic loci rather than the obesity-associated loci may play a role in weight loss and macronutrient intake. Therefore, broader approaches may be needed to overcome the limitations of current studies, e.g., by pooling of RCTs and broader genotyping. The identification of SNPs associated with thinness might be an innovative approach [68].

Strengths and Limitations

The strength of this systematic review is the inclusion of all SNPs for which a genotype–diet interaction on weight loss was available. The narrative synthesis was based on any nutritional intervention differing in the macronutrient distribution as well as hypocaloric diets, as others mainly focus on other types of trials [19,23]. We assessed the risk of bias as well as the methodological quality of the included articles. A limitation of many publications was that studying the interaction between genetic loci and macronutrient composition of a diet was not the primary aim of the respective study and was usually a post hoc analysis. Due to the high heterogeneity of the SNPs, a confirmation of the findings was not possible in most cases and it is also not possible to perform meta-analyses of the extracted studies. Our review could not identify studies investigating the additive effect of common SNPs in the form of genetic risk scores and specific diets on weight loss. Furthermore, we did not include studies focusing on copy number variants; mutation analysis; haplotypes; and studies investigating the association of genotype–diet interaction and BMI, body fat, or other obesity-related variables.

5. Conclusions

This systematic review summarized the results of genotype–diet interactions on weight loss. Independent of the kind of dietary intervention, most of the genotype–diet interactions on weight loss were not significant. The high heterogeneity of the SNPs and the lack of replications does not allow us to draw a final conclusion, as robust data on possible genotype–diet interactions on weight loss are missing. Most findings were based on the POUNDS Lost and NUGENOB trials. Therefore, more studies with larger sample sizes are needed to adequately address this highly relevant question in obesity research.

Author Contributions: Conceptualization, S.B. and C.H.; validation, S.B., V.W., and C.H.; writing—original draft preparation, S.B. and V.W.; writing—review and editing, C.H. and H.H.; visualization, S.B. and V.W.; supervision, C.H. and H.H.; project administration, C.H. All authors have read and agreed to the published version of the manuscript.

Funding: This research was funded by the German Federal Ministry of Education and Research (Bundesministerium für Bildung und Forschung, BMBF), grant number 01EA1709, *enable* publication number 062. This article was written by the Junior Research Group "Personalized Nutrition & eHealth" within the Nutrition Cluster *enable*.

Acknowledgments: The authors thank the German Federal Ministry of Education and Research (Bundesministerium für Bildung und Forschung, BMBF) for funding. Furthermore, we want to thank Peter von Philipsborn and Roxana Raab for helping with methodical questions.

Conflicts of Interest: S.B. and V.W. declare no conflict of interest. C.H. is a member of the scientific advisory board of 4sigma GmbH. H.H. is a member of the scientific advisory board of Oviva AG, Zurich, Switzerland.

The funders had no role in the design of the study; in the collection, analysis, or interpretation of data; in the writing of the manuscript; or in the decision to publish the results.

References

1. Bluher, M. Obesity: Global epidemiology and pathogenesis. *Nat. Rev. Endocrinol.* **2019**, *15*, 288–298. [CrossRef]
2. World Health Organization. Fact Sheet about Obesity and Overweigt. Available online: https://www.who.int/en/news-room/fact-sheets/detail/obesity-and-overweight (accessed on 27 April 2020).
3. Sanghera, D.K.; Bejar, C.; Sharma, S.; Gupta, R.; Blackett, P.R. Obesity genetics and cardiometabolic health: Potential for risk prediction. *Diabetes Obes. Metab.* **2019**, *21*, 1088–1100. [CrossRef]
4. Al-Goblan, A.S.; Al-Alfi, M.A.; Khan, M.Z. Mechanism linking diabetes mellitus and obesity. *Diabetes Metab Syndr. Obes.* **2014**, *7*, 587–591. [CrossRef] [PubMed]
5. Hill, J.O.; Wyatt, H.R.; Reed, G.W.; Peters, J.C. Obesity and the environment: Where do we go from here? *Science* **2003**, *299*, 853–855. [CrossRef]
6. Seid, H.; Rosenbaum, M. Low Carbohydrate and Low-Fat Diets: What We Don't Know and Why we Should Know It. *Nutrients* **2019**, *11*, 2749. [CrossRef]
7. Drabsch, T.; Holzapfel, C. A Scientific Perspective of Personalised Gene-Based Dietary Recommendations for Weight Management. *Nutrients* **2019**, *11*, 617. [CrossRef]
8. Lindgren, C.M.; Heid, I.M.; Randall, J.C.; Lamina, C.; Steinthorsdottir, V.; Qi, L.; Speliotes, E.K.; Thorleifsson, G.; Willer, C.J.; Herrera, B.M.; et al. Genome-wide association scan meta-analysis identifies three Loci influencing adiposity and fat distribution. *PLoS Genet.* **2009**, *5*, e1000508. [CrossRef]
9. Locke, A.E.; Kahali, B.; Berndt, S.I.; Justice, A.E.; Pers, T.H.; Day, F.R.; Powell, C.; Vedantam, S.; Buchkovich, M.L.; Yang, J.; et al. Genetic studies of body mass index yield new insights for obesity biology. *Nature* **2015**, *518*, 197–206. [CrossRef]
10. Speliotes, E.K.; Willer, C.J.; Berndt, S.I.; Monda, K.L.; Thorleifsson, G.; Jackson, A.U.; Lango Allen, H.; Lindgren, C.M.; Luan, J.; Magi, R.; et al. Association analyses of 249,796 individuals reveal 18 new loci associated with body mass index. *Nat. Genet.* **2010**, *42*, 937–948. [CrossRef]
11. Yengo, L.; Sidorenko, J.; Kemper, K.E.; Zheng, Z.; Wood, A.R.; Weedon, M.N.; Frayling, T.M.; Hirschhorn, J.; Yang, J.; Visscher, P.M.; et al. Meta-analysis of genome-wide association studies for height and body mass index in approximately 700000 individuals of European ancestry. *Hum. Mol. Genet.* **2018**, *27*, 3641–3649. [CrossRef]
12. Choquet, H.; Meyre, D. Genetics of Obesity: What have we Learned? *Curr. Genom.* **2011**, *12*, 169–179. [CrossRef] [PubMed]
13. Willyard, C. Heritability: The family roots of obesity. *Nature* **2014**, *508*, S58–S60. [CrossRef] [PubMed]
14. Frayling, T.M.; Timpson, N.J.; Weedon, M.N.; Zeggini, E.; Freathy, R.M.; Lindgren, C.M.; Perry, J.R.; Elliott, K.S.; Lango, H.; Rayner, N.W.; et al. A common variant in the FTO gene is associated with body mass index and predisposes to childhood and adult obesity. *Science* **2007**, *316*, 889–894. [CrossRef] [PubMed]
15. Gardner, C.D.; Trepanowski, J.F.; Del Gobbo, L.C.; Hauser, M.E.; Rigdon, J.; Ioannidis, J.P.A.; Desai, M.; King, A.C. Effect of Low-Fat vs Low-Carbohydrate Diet on 12-Month Weight Loss in Overweight Adults and the Association with Genotype Pattern or Insulin Secretion: The DIETFITS Randomized Clinical Trial. *JAMA* **2018**, *319*, 667–679. [CrossRef] [PubMed]
16. Celis-Morales, C.; Livingstone, K.M.; Marsaux, C.F.; Macready, A.L.; Fallaize, R.; O'Donovan, C.B.; Woolhead, C.; Forster, H.; Walsh, M.C.; Navas-Carretero, S.; et al. Effect of personalized nutrition on health-related behaviour change: Evidence from the Food4Me European randomized controlled trial. *Int. J. Epidemiol.* **2017**, *46*, 578–588. [CrossRef] [PubMed]
17. Grau, K.; Cauchi, S.; Holst, C.; Astrup, A.; Martinez, J.A.; Saris, W.H.; Blaak, E.E.; Oppert, J.M.; Arner, P.; Rossner, S.; et al. TCF7L2 rs7903146-macronutrient interaction in obese individuals' responses to a 10-wk randomized hypoenergetic diet. *Am. J. Clin. Nutr.* **2010**, *91*, 472–479. [CrossRef]
18. Larsen, T.M.; Dalskov, S.M.; van Baak, M.; Jebb, S.A.; Papadaki, A.; Pfeiffer, A.F.; Martinez, J.A.; Handjieva-Darlenska, T.; Kunesova, M.; Pihlsgard, M.; et al. Diets with high or low protein content and glycemic index for weight-loss maintenance. *N. Engl. J. Med.* **2010**, *363*, 2102–2113. [CrossRef]

19. Xiang, L.; Wu, H.; Pan, A.; Patel, B.; Xiang, G.; Qi, L.; Kaplan, R.C.; Hu, F.; Wylie-Rosett, J.; Qi, Q. FTO genotype and weight loss in diet and lifestyle interventions: A systematic review and meta-analysis. *Am. J. Clin. Nutr.* **2016**, *103*, 1162–1170. [CrossRef]

20. Krug, S.; Kastenmuller, G.; Stuckler, F.; Rist, M.J.; Skurk, T.; Sailer, M.; Raffler, J.; Romisch-Margl, W.; Adamski, J.; Prehn, C.; et al. The dynamic range of the human metabolome revealed by challenges. *FASEB J.* **2012**, *26*, 2607–2619. [CrossRef]

21. Zeevi, D.; Korem, T.; Zmora, N.; Israeli, D.; Rothschild, D.; Weinberger, A.; Ben-Yacov, O.; Lador, D.; Avnit-Sagi, T.; Lotan-Pompan, M.; et al. Personalized Nutrition by Prediction of Glycemic Responses. *Cell* **2015**, *163*, 1079–1094. [CrossRef]

22. Livingstone, K.M.; Celis-Morales, C.; Lara, J.; Ashor, A.W.; Lovegrove, J.A.; Martinez, J.A.; Saris, W.H.; Gibney, M.; Manios, Y.; Traczyk, I.; et al. Associations between FTO genotype and total energy and macronutrient intake in adults: A systematic review and meta-analysis. *Obes. Rev.* **2015**, *16*, 666–678. [CrossRef] [PubMed]

23. Livingstone, K.M.; Celis-Morales, C.; Papandonatos, G.D.; Erar, B.; Florez, J.C.; Jablonski, K.A.; Razquin, C.; Marti, A.; Heianza, Y.; Huang, T.; et al. FTO genotype and weight loss: Systematic review and meta-analysis of 9563 individual participant data from eight randomised controlled trials. *BMJ* **2016**, *354*, i4707. [CrossRef] [PubMed]

24. Moher, D.; Liberati, A.; Tetzlaff, J.; Altman, D.G.; Group, P. Preferred reporting items for systematic reviews and meta-analyses: The PRISMA statement. *PLoS Med.* **2009**, *6*, e1000097. [CrossRef] [PubMed]

25. Booth, A.; Brice, A. *Evidence-Based Practice for Information Professionals: A Handbook*; Facet Pub.: London, UK, 2004; 304p.

26. Sterne, J.A.C.; Savovic, J.; Page, M.J.; Elbers, R.G.; Blencowe, N.S.; Boutron, I.; Cates, C.J.; Cheng, H.Y.; Corbett, M.S.; Eldridge, S.M.; et al. RoB 2: A revised tool for assessing risk of bias in randomised trials. *BMJ* **2019**, *366*, l4898. [CrossRef]

27. Sterne, J.A.C.; Hernan, M.A.; Reeves, B.C.; Savovic, J.; Berkman, N.D.; Viswanathan, M.; Henry, D.; Altman, D.G.; Ansari, M.T.; Boutron, I.; et al. ROBINS-I: A tool for assessing risk of bias in non-randomised studies of interventions. *BMJ Brit. Med. J.* **2016**, *355*. [CrossRef]

28. Campbell, H.; Rudan, I. Interpretation of genetic association studies in complex disease. *Pharm. J.* **2002**, *2*, 349–360. [CrossRef]

29. Goni, L.; Qi, L.; Cuervo, M.; Milagro, F.I.; Saris, W.H.; MacDonald, I.A.; Langin, D.; Astrup, A.; Arner, P.; Oppert, J.M.; et al. Effect of the interaction between diet composition and the PPM1K genetic variant on insulin resistance and beta cell function markers during weight loss: Results from the Nutrient Gene Interactions in Human Obesity: Implications for dietary guidelines (NUGENOB) randomized trial. *Am. J. Clin. Nutr.* **2017**, *106*, 902–908. [CrossRef]

30. Grau, K.; Hansen, T.; Holst, C.; Astrup, A.; Saris, W.H.; Arner, P.; Rossner, S.; Macdonald, I.; Polak, J.; Oppert, J.M.; et al. Macronutrient-specific effect of FTO rs9939609 in response to a 10-week randomized hypo-energetic diet among obese Europeans. *Int. J. Obes. (Lond.)* **2009**, *33*, 1227–1234. [CrossRef]

31. Santos, J.L.; De la Cruz, R.; Holst, C.; Grau, K.; Naranjo, C.; Maiz, A.; Astrup, A.; Saris, W.H.; MacDonald, I.; Oppert, J.M.; et al. Allelic variants of melanocortin 3 receptor gene (MC3R) and weight loss in obesity: A randomised trial of hypo-energetic high- versus low-fat diets. *PLoS ONE* **2011**, *6*, e19934. [CrossRef]

32. Sorensen, T.I.; Boutin, P.; Taylor, M.A.; Larsen, L.H.; Verdich, C.; Petersen, L.; Holst, C.; Echwald, S.M.; Dina, C.; Toubro, S.; et al. Genetic polymorphisms and weight loss in obesity: A randomised trial of hypo-energetic high- versus low-fat diets. *PLoS Clin. Trials* **2006**, *1*, e12. [CrossRef]

33. Stocks, T.; Angquist, L.; Banasik, K.; Harder, M.N.; Taylor, M.A.; Hager, J.; Arner, P.; Oppert, J.M.; Martinez, J.A.; Polak, J.; et al. TFAP2B influences the effect of dietary fat on weight loss under energy restriction. *PLoS ONE* **2012**, *7*, e43212. [CrossRef] [PubMed]

34. Svendstrup, M.; Allin, K.H.; Sorensen, T.I.A.; Hansen, T.H.; Grarup, N.; Hansen, T.; Vestergaard, H. Genetic risk scores for body fat distribution attenuate weight loss in women during dietary intervention. *Int. J. Obes. (Lond.)* **2018**, *42*, 370–375. [CrossRef] [PubMed]

35. Goni, L.; Riezu-Boj, J.I.; Milagro, F.I.; Corrales, F.J.; Ortiz, L.; Cuervo, M.; Martinez, J.A. Interaction between an ADCY3 Genetic Variant and Two Weight-Lowering Diets Affecting Body Fatness and Body Composition Outcomes Depending on Macronutrient Distribution: A Randomized Trial. *Nutrients* **2018**, *10*, 789. [CrossRef] [PubMed]

36. Ramos-Lopez, O.; Riezu-Boj, J.I.; Milagro, F.I.; Goni, L.; Cuervo, M.; Martinez, J.A. Differential lipid metabolism outcomes associated with ADRB2 gene polymorphisms in response to two dietary interventions in overweight/obese subjects. *Nutr. Metab. Cardiovasc. Dis.* **2018**, *28*, 165–172. [CrossRef] [PubMed]

37. Goni, L.; Sun, D.; Heianza, Y.; Wang, T.; Huang, T.; Martinez, J.A.; Shang, X.; Bray, G.A.; Smith, S.R.; Sacks, F.M.; et al. A circadian rhythm-related MTNR1B genetic variant modulates the effect of weight-loss diets on changes in adiposity and body composition: The POUNDS Lost trial. *Eur. J. Nutr.* **2019**, *58*, 1381–1389. [CrossRef]

38. Heianza, Y.; Ma, W.; Huang, T.; Wang, T.; Zheng, Y.; Smith, S.R.; Bray, G.A.; Sacks, F.M.; Qi, L. Macronutrient Intake-Associated FGF21 Genotype Modifies Effects of Weight-Loss Diets on 2-Year Changes of Central Adiposity and Body Composition: The POUNDS Lost Trial. *Diabetes Care* **2016**, *39*, 1909–1914. [CrossRef]

39. Heianza, Y.; Sun, D.; Ma, W.; Zheng, Y.; Champagne, C.M.; Bray, G.A.; Sacks, F.M.; Qi, L. Gut-microbiome-related LCT genotype and 2-year changes in body composition and fat distribution: The POUNDS Lost Trial. *Int. J. Obes. (Lond.)* **2018**, *42*, 1565–1573. [CrossRef]

40. Heianza, Y.; Sun, D.; Wang, T.; Huang, T.; Bray, G.A.; Sacks, F.M.; Qi, L. Starch Digestion-Related Amylase Genetic Variant Affects 2-Year Changes in Adiposity in Response to Weight-Loss Diets: The POUNDS Lost Trial. *Diabetes* **2017**, *66*, 2416–2423. [CrossRef]

41. Huang, T.; Wang, T.; Heianza, Y.; Sun, D.; Ivey, K.; Durst, R.; Schwarzfuchs, D.; Stampfer, M.J.; Bray, G.A.; Sacks, F.M.; et al. HNF1A variant, energy-reduced diets and insulin resistance improvement during weight loss: The POUNDS Lost trial and DIRECT. *Diabetes Obes. Metab.* **2018**, *20*, 1445–1452. [CrossRef]

42. Mattei, J.; Qi, Q.; Hu, F.B.; Sacks, F.M.; Qi, L. TCF7L2 genetic variants modulate the effect of dietary fat intake on changes in body composition during a weight-loss intervention. *Am. J. Clin. Nutr.* **2012**, *96*, 1129–1136. [CrossRef]

43. Qi, Q.; Bray, G.A.; Hu, F.B.; Sacks, F.M.; Qi, L. Weight-loss diets modify glucose-dependent insulinotropic polypeptide receptor rs2287019 genotype effects on changes in body weight, fasting glucose, and insulin resistance: The Preventing Overweight Using Novel Dietary Strategies trial. *Am. J. Clin. Nutr.* **2012**, *95*, 506–513. [CrossRef] [PubMed]

44. Qi, Q.; Bray, G.A.; Smith, S.R.; Hu, F.B.; Sacks, F.M.; Qi, L. Insulin receptor substrate 1 gene variation modifies insulin resistance response to weight-loss diets in a 2-year randomized trial: The Preventing Overweight Using Novel Dietary Strategies (POUNDS LOST) trial. *Circulation* **2011**, *124*, 563–571. [CrossRef]

45. Qi, Q.; Durst, R.; Schwarzfuchs, D.; Leitersdorf, E.; Shpitzen, S.; Li, Y.; Wu, H.; Champagne, C.M.; Hu, F.B.; Stampfer, M.J.; et al. CETP genotype and changes in lipid levels in response to weight-loss diet intervention in the POUNDS LOST and DIRECT randomized trials. *J. Lipid Res.* **2015**, *56*, 713–721. [CrossRef] [PubMed]

46. Qi, Q.; Zheng, Y.; Huang, T.; Rood, J.; Bray, G.A.; Sacks, F.M.; Qi, L. Vitamin D metabolism-related genetic variants, dietary protein intake and improvement of insulin resistance in a 2 year weight-loss trial: POUNDS Lost. *Diabetologia* **2015**, *58*, 2791–2799. [CrossRef] [PubMed]

47. Sun, D.; Heianza, Y.; Li, X.; Shang, X.; Smith, S.R.; Bray, G.A.; Sacks, F.M.; Qi, L. Genetic, epigenetic and transcriptional variations at NFATC2IP locus with weight loss in response to diet interventions: The POUNDS Lost Trial. *Diabetes Obes. Metab.* **2018**, *20*, 2298–2303. [CrossRef] [PubMed]

48. Xu, M.; Ng, S.S.; Bray, G.A.; Ryan, D.H.; Sacks, F.M.; Ning, G.; Qi, L. Dietary Fat Intake Modifies the Effect of a Common Variant in the LIPC Gene on Changes in Serum Lipid Concentrations during a Long-Term Weight-Loss Intervention Trial. *J. Nutr.* **2015**, *145*, 1289–1294. [CrossRef]

49. Xu, M.; Qi, Q.; Liang, J.; Bray, G.A.; Hu, F.B.; Sacks, F.M.; Qi, L. Genetic determinant for amino acid metabolites and changes in body weight and insulin resistance in response to weight-loss diets: The Preventing Overweight Using Novel Dietary Strategies (POUNDS LOST) trial. *Circulation* **2013**, *127*, 1283–1289. [CrossRef]

50. Zhang, X.; Qi, Q.; Bray, G.A.; Hu, F.B.; Sacks, F.M.; Qi, L. APOA5 genotype modulates 2-y changes in lipid profile in response to weight-loss diet intervention: The Pounds Lost Trial. *Am. J. Clin. Nutr.* **2012**, *96*, 917–922. [CrossRef]

51. Zhang, X.; Qi, Q.; Liang, J.; Hu, F.B.; Sacks, F.M.; Qi, L. Neuropeptide Y promoter polymorphism modifies effects of a weight-loss diet on 2-year changes of blood pressure: The preventing overweight using novel dietary strategies trial. *Hypertension* **2012**, *60*, 1169–1175. [CrossRef]

52. Razquin, C.; Martinez, J.A.; Martinez-Gonzalez, M.A.; Fernandez-Crehuet, J.; Santos, J.M.; Marti, A. A Mediterranean diet rich in virgin olive oil may reverse the effects of the -174G/C IL6 gene variant on 3-year body weight change. *Mol. Nutr. Food Res.* **2010**, *54* (Suppl. 1), S75–S82. [CrossRef]

53. Di Renzo, L.; Cioccoloni, G.; Falco, S.; Abenavoli, L.; Moia, A.; Sinibaldi Salimei, P.; De Lorenzo, A. Influence of FTO rs9939609 and Mediterranean diet on body composition and weight loss: A randomized clinical trial. *J. Transl. Med.* **2018**, *16*, 308. [CrossRef] [PubMed]

54. Garaulet, M.; Smith, C.E.; Hernandez-Gonzalez, T.; Lee, Y.C.; Ordovas, J.M. PPARgamma Pro12Ala interacts with fat intake for obesity and weight loss in a behavioural treatment based on the Mediterranean diet. *Mol. Nutr. Food Res.* **2011**, *55*, 1771–1779. [CrossRef]

55. Petersen, M.; Taylor, M.A.; Saris, W.H.; Verdich, C.; Toubro, S.; Macdonald, I.; Rossner, S.; Stich, V.; Guy-Grand, B.; Langin, D.; et al. Randomized, multi-center trial of two hypo-energetic diets in obese subjects: High- versus low-fat content. *Int. J. Obes. (Lond.)* **2006**, *30*, 552–560. [CrossRef] [PubMed]

56. Sacks, F.M.; Bray, G.A.; Carey, V.J.; Smith, S.R.; Ryan, D.H.; Anton, S.D.; McManus, K.; Champagne, C.M.; Bishop, L.M.; Laranjo, N.; et al. Comparison of weight-loss diets with different compositions of fat, protein, and carbohydrates. *N. Engl. J. Med.* **2009**, *360*, 859–873. [CrossRef]

57. de Jonge, L.; Bray, G.A.; Smith, S.R.; Ryan, D.H.; de Souza, R.J.; Loria, C.M.; Champagne, C.M.; Williamson, D.A.; Sacks, F.M. Effect of diet composition and weight loss on resting energy expenditure in the POUNDS LOST study. *Obesity (Silver Spring)* **2012**, *20*, 2384–2389. [CrossRef]

58. Shai, I.; Schwarzfuchs, D.; Henkin, Y.; Shahar, D.R.; Witkow, S.; Greenberg, I.; Golan, R.; Fraser, D.; Bolotin, A.; Vardi, H.; et al. Weight loss with a low-carbohydrate, Mediterranean, or low-fat diet. *N. Engl. J. Med.* **2008**, *359*, 229–241. [CrossRef] [PubMed]

59. Estruch, R.; Ros, E.; Salas-Salvado, J.; Covas, M.I.; Corella, D.; Aros, F.; Gomez-Gracia, E.; Ruiz-Gutierrez, V.; Fiol, M.; Lapetra, J.; et al. Primary prevention of cardiovascular disease with a Mediterranean diet. *N. Engl. J. Med.* **2013**, *368*, 1279–1290. [CrossRef]

60. Qi, Q.; Kilpelainen, T.O.; Downer, M.K.; Tanaka, T.; Smith, C.E.; Sluijs, I.; Sonestedt, E.; Chu, A.Y.; Renstrom, F.; Lin, X.; et al. FTO genetic variants, dietary intake and body mass index: Insights from 177,330 individuals. *Hum. Mol. Genet.* **2014**, *23*, 6961–6972. [CrossRef]

61. Consortium, S.T.D.; Estrada, K.; Aukrust, I.; Bjorkhaug, L.; Burtt, N.P.; Mercader, J.M.; Garcia-Ortiz, H.; Huerta-Chagoya, A.; Moreno-Macias, H.; Walford, G.; et al. Association of a low-frequency variant in HNF1A with type 2 diabetes in a Latino population. *JAMA* **2014**, *311*, 2305–2314. [CrossRef]

62. Najmi, L.A.; Aukrust, I.; Flannick, J.; Molnes, J.; Burtt, N.; Molven, A.; Groop, L.; Altshuler, D.; Johansson, S.; Bjorkhaug, L.; et al. Functional Investigations of HNF1A Identify Rare Variants as Risk Factors for Type 2 Diabetes in the General Population. *Diabetes* **2017**, *66*, 335–346. [CrossRef]

63. Morita, K.; Saruwatari, J.; Tanaka, T.; Oniki, K.; Kajiwara, A.; Otake, K.; Ogata, Y.; Nakagawa, K. Associations between the common HNF1A gene variant p.I27L (rs1169288) and risk of type 2 diabetes mellitus are influenced by weight. *Diabetes Metab.* **2015**, *41*, 91–94. [CrossRef] [PubMed]

64. Columbus, J.; Chiang, Y.; Shao, W.; Zhang, N.; Wang, D.; Gaisano, H.Y.; Wang, Q.; Irwin, D.M.; Jin, T. Insulin treatment and high-fat diet feeding reduces the expression of three Tcf genes in rodent pancreas. *J. Endocrinol.* **2010**, *207*, 77–86. [CrossRef] [PubMed]

65. Harries, L.W.; Brown, J.E.; Gloyn, A.L. Species-specific differences in the expression of the HNF1A, HNF1B and HNF4A genes. *PLoS ONE* **2009**, *4*, e7855. [CrossRef] [PubMed]

66. de Luis, D.A.; Izaola, O.; Primo, D.; Aller, R. Effect of the rs1862513 variant of resistin gene on insulin resistance and resistin levels after two hypocaloric diets with different fat distribution in subjects with obesity. *Eur. Rev. Med. Pharmacol. Sci.* **2018**, *22*, 3865–3872. [CrossRef] [PubMed]

67. Munafo, M.R.; Clark, T.G.; Flint, J. Assessing publication bias in genetic association studies: Evidence from a recent meta-analysis. *Psychiatry Res.* **2004**, *129*, 39–44. [CrossRef] [PubMed]

68. Orthofer, M.; Valsesia, A.; Magi, R.; Wang, Q.P.; Kaczanowska, J.; Kozieradzki, I.; Leopoldi, A.; Cikes, D.; Zopf, L.M.; Tretiakov, E.O.; et al. Identification of ALK in Thinness. *Cell* **2020**, *181*, 1246–1262.e22. [CrossRef]

 *nutrients*

Article

The Impact of *FTO* Genetic Variants on Obesity and Its Metabolic Consequences is Dependent on Daily Macronutrient Intake

Przemyslaw Czajkowski [1,†], Edyta Adamska-Patruno [1,*,†], Witold Bauer [1], Joanna Fiedorczuk [2], Urszula Krasowska [1], Monika Moroz [2], Maria Gorska [3] and Adam Kretowski [1,2,3]

[1] Clinical Research Centre, Medical University of Bialystok, Jana Kilinskiego 1, 15-089 Bialystok, Poland; przemyslaw.czajkowski@umb.edu.pl (P.C.); witold.bauer@umb.edu.pl (W.B.); urszula.krasowska@umb.edu.pl (U.K.); adamkretowski@wp.pl (A.K.)

[2] Clinical Research Centre, Medical University of Bialystok Clinical Hospital, Marii Sklodowskiej-Curie 24A, 15-276 Bialystok, Poland; j.fiedorczuk@wp.pl (J.F.); monika_bakun@wp.pl (M.M.)

[3] Department of Endocrinology, Diabetology and Internal Medicine, Medical University of Bialystok, Jana Kilinskiego 1, 15-089 Bialystok, Poland; mgorska25@wp.pl

* Correspondence: edyta.adamska@umb.edu.pl

† These authors contributed equally to this work.

Received: 14 September 2020; Accepted: 22 October 2020; Published: 23 October 2020

Abstract: Numerous studies have identified the various fat mass and obesity-associated (*FTO*) genetic variants associated with obesity and its metabolic consequences; however, the impact of dietary factors on these associations remains unclear. The aim of this study was to evaluate the association between *FTO* single nucleotide polymorphisms (SNPs), daily macronutrient intake, and obesity and its metabolic consequences. From 1549 Caucasian subjects of Polish origin, genotyped for the *FTO* SNPs (rs3751812, rs8044769, rs8050136, and rs9939609), 819 subjects were selected for gene–diet interaction analysis. Anthropometric measurements were performed and total body fat content and distribution, blood glucose and insulin concentration during oral glucose tolerance test (OGTT), and lipid profile were determined. Macronutrient intake was analyzed based on three-day food records, and daily physical activity levels were evaluated using the International Physical Activity Questionnaire Long Form (IPAQ-LF). Our study shows that carriers of the GG genotype of rs3751812 presented lower body weight, body mass index (BMI), total body fat content, and hip and waist circumference and presented lower obesity-related markers if more than 48% of daily energy intake was derived from carbohydrates and lower subcutaneous and visceral fat content when energy intake derived from dietary fat did not exceed 30%. Similar results were observed for rs8050136 CC genotype carriers. We did not notice any significant differences in obesity markers between genotypes of rs8044769, but we did observe a significant impact of diet-gene associations. Body weight and BMI were significantly higher in TT and CT genotype carriers if daily energy intake derived from carbohydrates was less than 48%. Moreover, in TT genotype carriers, we observed higher blood glucose concentration while fasting and during the OGTT test if more than 18% of total energy intake was derived from proteins. In conclusion, our results indicate that daily macronutrient intake may modulate the impact of *FTO* genetic SNPs on obesity and obesity-related metabolic consequences.

Keywords: *FTO* gene; obesity; dietary protein; dietary carbohydrates; dietary fat; macronutrients; gene-diet interaction; glucose homeostasis

---

1. Introduction

Obesity is a major public health problem worldwide [1] and a leading risk factor for type 2 diabetes mellitus in adolescents [2,3] and children [3,4]. It has already been established that the predictions for diabetes prevalence are not optimistic [5]. Moreover, the increasing prevalence of obesity is associated with lipid metabolism disturbances, such as high concentrations of total cholesterol and low-density lipoprotein (LDL) and low concentrations of high-density lipoprotein (HDL) [6]. Considering the above, obesity also increases the risk of cardiovascular disease [7].

In general, obesity is a result of imbalanced energy homeostasis, but genome-wide association studies have identified many single nucleotide polymorphisms (SNPs) in the fat mass and obesity-associated (*FTO*) gene, melanocortin-4 receptor (*MC4R*) gene, and other genes [8,9] that are associated with the risk of developing obesity. Among these genes, *FTO* has been reported as the gene with the strongest significant correlation with obesity [10]. The *FTO* gene is profoundly expressed in the hypothalamus region, which is involved in appetite regulation [11]. The associations between *FTO* genetic variants, dietary factors, and body weight gain are still under investigation, although it has been postulated that some *FTO* genetic variants may influence the risk of weight gain through larger amounts of consumed food [12] or appetite and satiety regulation [13]. *FTO* rs9939609 SNPs have been associated with increased macronutrient consumption, especially fat and carbohydrates, as well as total energy intake [11,14,15], but these genetic variants do not seem to influence energy expenditure [11,16]. Moreover, environmental factors such as diet may influence the associations between genetic risk and obesity development. Over the last decade, the study of dietary patterns and their relation to genetic risk of obesity has received more attention [17,18]; nevertheless, the associations between *FTO* single nucleotide polymorphisms and dietary patterns need further investigation [19]. Therefore, the aim of our study was to evaluate whether daily macronutrient intake could modify the association between genetic variations of the *FTO* gene and obesity and obesity-related metabolic consequences among the Polish population.

2. Materials and Methods

2.1. Participants

The study was conducted among 1549 Caucasian volunteers of Polish origin (18–79 years old) enrolled in the 1000PLUS Cohort Study (registered at www.clinicaltrials.gov as NCT03792685) from 2007 to 2019, described previously [20–22], which were seeking personalized nutrition for prevention of obesity and treatment of type 2 diabetes mellitus. Individuals who used to take medicines (weight loss, anti-diabetic, lipid-lowering, or any other medication that could have an impact on body weight, body fat content, blood glucose, and other investigated parameters) or diet supplements, which could affect the results, were not enrolled in this study. Subjects who reported endocrine, gastrointestinal, hepatic, renal, metabolic, immunological, or psychiatric disorders or who had bariatric surgery, which could have an impact on investigated parameters, were excluded from the study analysis as well. We excluded all subjects who used to take anti-diabetic (56 subjects, 6.8%) or lipid-lowering medications (47 subjects, 5.7%) and 109 (13.3%) with a previous history of prediabetes or diabetes, as potential cofounders, and others who met the exclusion criteria mentioned above. Moreover, individuals who followed any special diet or dietary pattern (vegetarian, vegan, Atkins, etc.) were not included in the analysis.

2.2. Anthropometric and Body Composition Measurements

The following anthropometric data were collected: body weight, height, and waist and hip circumference. Body mass index (BMI) was calculated using the following formula: body weight (kg) divided by height squared (m). Waist–hip ratio (WHR) was estimated by dividing waist circumference by hip circumference. Total body composition (including fat mass, fat-free mass, and skeletal muscle mass) was evaluated by the bioelectrical impedance method (InBody 220, Biospace, Korea). Body fat distribution analysis, including visceral adipose tissue (VAT) and subcutaneous adipose tissue (SAT) content, was performed by the multi-frequency bioimpedance method (Maltron BioScan 920-2, Maltron International Ltd., United Kingdom). The VAT/SAT ratio was calculated by dividing visceral adipose tissue by subcutaneous adipose tissue content.

2.3. Blood Collection, Biochemical Analysis, and Calculations

Oral glucose tolerance tests (OGTTs) were performed in non-diabetic participants according to the World Health Organization (WHO) recommendations with a dose of 75 g oral glucose. The subjects were instructed to fast for 8–12 h prior to the test and to not restrict carbohydrate intake in the 3 days before the test. Blood was collected at 0, 30, 60, and 120 min after glucose administration. Blood samples were obtained and collected to evaluate the concentrations of plasma glucose, insulin, LDL, HDL, total cholesterol and triglyceride (TG), and hemoglobin A1c (HbA1c). The samples were prepared for assessment according to the laboratory kit instructions. Serum insulin concentrations were evaluated by immunoradiometric assay (INS-Irma, DIASource S.A., Belgium; Wallac Wizard 1470 Automatic Gamma Counter, PerkinElmer Life Sciences, Turku, Finland). Plasma glucose concentration was measured using the hexokinase enzymatic method (Cobas c111, Roche Diagnostics Ltd., Switzerland), and lipid profile was evaluated by enzymatic colorimetric assay using commercially available kits (Cobas c111, Roche Diagnostic Ltd., Switzerland). HbA1c levels were measured using high-performance liquid chromatography (HPLC) (D-10 Hemoglobin Testing System, Bio-Rad Laboratories Inc., Hercules, CA, USA; Bio-Rad, Marnes-la-Coquette, France).

The homeostatic model assessment of insulin resistance (HOMA-IR) was calculated following the standard formula: (fasting plasma glucose concentration (mmol/L)) × (fasting insulin concentration (μU/mL))/22.5.

2.4. Daily Physical Activity and Dietary Intake Analyses

To evaluate daily physical activity, the International Physical Activity Questionnaire-Long Form (IPAQ-LF) was used. Metabolic equivalent (MET, min per week) was determined using the following formula: (MET level) × (minutes of activity) × (events per week) [23]. Individuals were stratified as having a low, moderate, or high level of physical activity.

Subjects were asked to record 3-day food intake diaries and were instructed on how to estimate portion sizes of foods based on the provided color photograph albums of portion sizes. Moreover, the subjects were instructed on how to weigh the food, if possible. Daily carbohydrate, protein, fat, and total energy intake were estimated using Dieta 6.0 software (National Food and Nutrition Institute, Warsaw, Poland). Dieta software was developed and is continuously updated by the National Food and Nutrition Institute (Warsaw, Poland), and it is used to calculate the nutritional value of food and diets based on tables of the nutritional value of local food products and dishes. In order to study the interactions between genetic factors and diet, study participants were divided into 2 quantiles based on average daily protein, fat, and carbohydrate intake: lower and higher than median dietary protein intake (≤18% and >18% of total energy intake, respectively), lower and higher than median dietary carbohydrate intake (≤48% and >48% of total energy intake, respectively), and lower and higher than median dietary fat intake (≤30% and >30% of total energy intake, respectively).

2.5. Genetic Analyses

We genotyped 4 common *FTO* SNPs in rs3751812 (G > T), rs8044769 (C > T), rs8050136 (A > C), and rs9939609 (T > A). DNA was extracted from peripheral blood leukocytes using a classical salting-out method. The SNPs were genotyped with TaqMan SNP technology from a ready-to-use human assay library (Applied Biosystems, MA, USA) using a high-throughput genotyping system, OpenArray (Life Technologies, CA, USA). SNP analysis was performed in duplicate, following the manufacturer's instructions. We used a sample without template as a negative control to detect possible false positive signals caused by contamination.

2.6. Ethics Statement

The study methods were carried out in accordance with the ethical standards on human experimentation and with the Helsinki Declaration of 1975 as revised in 1983. Written informed consent was obtained from all participants before inclusion in the study. The study protocol was approved by the local Ethics Committee of the Medical University of Bialystok, Poland (R-I-002/35/2009).

2.7. Statistical Analysis

Numerical data were summarized with number of observations (N), arithmetic mean, and standard deviation (SD). For categorical data, number of observations and frequency (percentage) were presented. Study participants were divided into quantiles based on average daily protein, carbohydrate, and fat intake, with the thresholds set as the median value of each parameter. Risk genotypes of the 4 common *FTO* SNPs were predefined based on the literature and our previous findings. Because of the relatively small sample size, we did not include a comparison of the allelic and genotypic frequencies and odds ratio calculations in this study. Continuous parameters were tested for normality with Shapiro-Wilk's test as well as visual inspection. Homogeneity of variance across groups was studied using Levene's test. Nonparametric tests were used for response variables that failed the mentioned statistical tests. Differences between selected parameters and dietary groups were then compared using analysis of variance (ANOVA) or Kruskal-Wallis test for numerical variables, with either Tukey's or Dunn's post-hoc test with Holm *p*-value adjustment (in case multiple pairwise tests were performed, or when there were multiple grouping variables, as presented in tables and figures), and chi-squared test for categorical variables. In order to study the hypothesis that the relationship between *FTO* genotypes and continuous responses varies in average daily protein, fat, and carbohydrate intake groups, we added (dietary macronutrient quantile) x (genotype) interaction terms to the multivariate linear regression models. These models were adjusted for age, sex, BMI (when applicable), total average energy intake (kcal/day), and physical activity. The Huber-White robust standard errors (HC1) were calculated. Model fit was estimated using R-squared values plus adjusted R-squared values. Some of the models were optimized by a stepwise backward elimination based on the Akaike information criterion (AIC). The statistical significance level was set at <0.05 for all 2-sided tests and multivariate comparisons. All calculations were prepared in R (version 4.0.2) [24].

3. Results

Our analysis identified 411 participants (50.2%) as having prediabetes or diabetes, without any previously known history of glucose homeostasis disturbance.

For the diet-gene interaction analysis, we included data from 819 subjects (Supplementary Figure S1). The general clinical characteristics of the studied population are presented in Table 1, and characteristics stratified by investigated genotypes are presented in Tables 2–4. No significant deviation from the Hardy-Weinberg equilibrium was observed for any of the investigated SNPs (*p* > 0.05). Among the investigated *FTO* SNPs, some of the loci were in very strong linkage disequilibrium (D' = 1.0 for rs8050136 and rs9939609) [25], so we present results for rs8050136.

Based on the demographic, anthropometric, behavioral (food intake and physical activity), and laboratory data, we observed that GG genotype carriers of rs3751812 (Table 2) and CC genotype carriers of rs8050136 (Table 3) presented significantly lower hip circumference. We also found that carriers of the TT rs8044769 genotype had the highest total cholesterol levels, and CT carriers presented the lowest percentage of daily energy intake from fat (Table 4).

We did not observe any other significant differences between studied genotypes; however, we noticed a tendency toward higher BMI, total body fat content, and waist circumference in TT genotype carriers of rs3751812 (Table 2) and AA genotype carriers of rs8050136 (Table 3). Between carriers of investigated genetic variants in rs8044769, we noted a tendency for differences in low-density lipoprotein cholesterol (LDL-cholesterol) concentration and daily physical activity level (Table 4).

3.1. Dietary Assessment

The 3-day food diaries were available from 662 subjects from the general cohort group and from 490 subjects who were genotyped for the investigated *FTO* SNPs. We did not find any differences between genotypes and dietary habits, except for the rs8044769 SNP. The heterozygous CT genotype carriers presented the lowest percentage of energy intake provided from dietary fat (Table 4).

We analyzed the interactions between dietary macronutrient intake and individual genotypes and their effect on continuous responses using multivariable linear regression models with the (dietary macronutrient quantile) × (genotype) interaction term. We observed that the association between selected genotypes and variables describing body composition (weight, BMI, and free fat mass) and the patient's glycemic status (fasting glucose levels) varied in different dietary groups, confirming the hypothesis that the effects of diet and genotypes interact. The differences in median values of the selected responses and the interquartile ranges (IQRs) in different genotypic and dietary strata are presented using boxplots in the figures. These results were significant after adjustment for age, sex, BMI (where applicable), and total energy intake.

Table 1. Study group characteristics.

Study Group Characteristics	
N	819
Age	42.1 (14.5)
Female/male (%)	52.5/47.5
BMI (kg/m^2)	28.5 (6.6)
<25.0	273 (33.9%)
25.0–29.9	278 (34.5%)
≥30.0	255 (31.6%)
Total body fat content (kg)	27.1 (13.8)
Total body fat content (%)	31.4 (9.6)
Waist circumference (cm)	96.2 (17.2)
Hip circumference (cm)	103.3 (12.7)
WHR	0.928 (0.088)
Visceral fat (cm^3)	108.4 (80.6)
Visceral fat (%)	37.1 (12.1)
Subcutaneous fat (cm^3)	167.9 (81.7)
Subcutaneous fat (%)	62.8 (12.3)
Visceral/subcutaneous fat ratio	0.669 (0.443)
Total cholesterol	195.4 (46.1)
HDL	59.7 (14.9)
LDL	112.0 (40.0)
TG	118.8 (95.1)
Fasting blood glucose level (mg/dL)	98.8 (23.9)
History of prediabetes or diabetes	
Yes	411 (50.2%)
No	408 (49.8%)
Dietary assessment (n)	490
Daily energy intake (kcal)	1792.5 (697.4)
Daily energy from protein (%)	18.9 (4.8)
Daily energy from fat (%)	31.2 (7.5)
Daily energy from carbohydrates (%)	47.6 (8.6)
Daily physical activity level	
Low	60 (7.3%)
Moderate	173 (21.1%)
High	586 (71.6%)

Data presented as mean and standard deviation (SD). BMI, body mass index; HDL, high-density lipoprotein; LDL, low-density lipoprotein; TG, triglycerides; WHR, waist–hip ratio.

3.2. Association of rs3751812 Genetic Variants with Obesity, Anthropometric Measures, Lipid Profile, and Dietary Intake

The comparison between genotypes showed that carriers of the GG genotype presented lower body weight (Figure 1A), BMI (Figure 1B), total body fat content (Figure 1C), and hip (Figure 1D) and waist (Figure 1E) circumference, but higher total cholesterol (Figure 1F) and LDL-cholesterol

(Figure 1G) levels, when compared to the GT genotype carriers, and lower body weight (Figure 1A), BMI (Figure 1B), and hip (Figure 1D) and waist (Figure 1E) circumference when compared to the TT genotype carriers.

Table 2. Characteristics of participants stratified by rs3751812 genotypes.

rs3751812	G/G	G/T	T/T	*p*-Value *
N	211	420	181	
Genotype frequency	0.26	0.52	0.22	>0.05
Age	40.5 (14.2)	41.2 (14.7)	39.5 (14.3)	0.33
Female (%)	53.8% (0.49)	53.8% (0.50)	47.8% (0.50)	0.36
BMI (kg/m^2)	27.6 (6.0)	28.7 (6.8)	28.9 (6.8)	0.060
<25.0	81 (38.8%)	136 (32.9%)	53 (29.9%)	
25.0–29.9	69 (33.0%)	141 (34.1%)	66 (37.3%)	0.410
≥30.0	59 (28.2%)	136 (32.9%)	58 (32.8%)	
Total body fat content (kg)	25.3 (12.4)	27.6 (13.8)	28.2 (15.2)	0.080
Total body fat content (%)	30.6 (9.1)	31.8 (9.6)	31.6 (10.3)	0.377
Waist circumference (cm)	94.3 (17.5)	96.7 (17.2)	97.5 (16.7)	0.054
Hip circumference (cm)	101.3 (12.4)	104.2 (13.0)	103.8 (12.5)	0.008
WHR	0.927 (0.091)	0.925 (0.088)	0.937 (0.085)	0.327
Visceral fat (cm^3)	103.0 (81.0)	110.0 (79.9)	112.3 (83.0)	0.379
Visceral fat (%)	36.4 (11.8)	37.5 (12.4)	37.2 (11.7)	0.587
Subcutaneous fat (cm^3)	163.5 (83.1)	167.2 (80.5)	175.0 (82.7)	0.401
Subcutaneous fat (%)	63.7 (11.7)	62.3 (12.9)	62.8 (11.7)	0.557
Visceral/subcutaneous fat ratio	0.642 (0.406)	0.687 (0.475)	0.665 (0.413)	0.554
Total cholesterol	202.7 (56.0)	191.7 (41.3)	194.0 (43.2)	0.070
HDL	60.7 (14.1)	59.8 (15.6)	59.5 (14.5)	0.662
LDL	117.3 (43.3)	109.4 (37.8)	111.2 (41.8)	0.095
TG	123.8 (143.9)	111.9 (69.7)	116.3 (61.9)	0.289
Blood glucose level during OGTT (mg/dL)				
0 min	96.8 (24.1)	95.6 (18.3)	97.1 (20.8)	0.914
30 min	147.0 (44.3)	145.0 (31.6)	150.1 (35.6)	0.312
60 min	132.3 (56.0)	129.5 (46.3)	134.2 (46.3)	0.380
120 min	100.7 (46.1)	99.1 (32.1)	98.8 (31.0)	0.621
History of prediabetes or diabetes				
Yes	103 (48.8%)	209 (49.8%)	95 (52.5%)	0.751
No	108 (51.2%)	211 (50.2%)	86 (47.5%)	
Dietary assessment (n)	126	259	101	
Daily energy intake (kcal)	1807.2 (732.3)	1766.9 (676.0)	1837.4 (713.4)	0.849
Daily energy from protein (%)	18.7 (4.4)	19.0 (4.9)	19.1 (4.9)	0.901
Daily energy from fat (%)	31.2 (7.2)	30.9 (7.5)	31.9 (7.8)	0.568
Daily energy from carbohydrates (%)	47.6 (7.7)	47.8 (9.1)	46.8 (8.5)	0.662
Daily physical activity level				
Low	16 (7.6%)	25 (6.0%)	18 (9.9%)	
Moderate	50 (23.7%)	83 (19.8%)	40 (22.1%)	0.302
High	145 (68.7%)	312 (74.3%)	123 (68.0%)	

Data presented as mean and standard deviation (SD), number of observations, and frequency. BMI, body mass index; HDL, high-density lipoprotein; LDL, low-density lipoprotein; OGTT, oral glucose tolerance test; TG, triglycerides; WHR, waist-hip ratio. * Holm-adjusted Kruskal-Wallis/ANOVA *p*-values.

Based on the analysis of the interactions between rs3751812 genotypes and carbohydrate intake, we observed that GG genotype carriers presented lower body weight (Figure 2A), BMI (Figure 2B), fat-free mass levels (Figure 2C), subcutaneous fat content (Figure 2D), and waist (Figure 2E) and hip (Figure 2F) circumference, as well as lower fasting blood glucose (Figure 2G) and higher HDL-cholesterol (Figure 2H) levels, when they were stratified to the group with higher than median carbohydrate intake. Moreover, we noted that TT carriers in the group with higher than median carbohydrate intake presented lower fasting insulin levels (Figure 2I) and HOMA-IR values (Figure 2J) compared to participants who were stratified to the group with lower than median carbohydrate intake. The interaction effect of (carbohydrate diet group) × (rs3751812 genotype) on body composition, anthropometric measures, and lipid profile was statistically significant with *p*-value < 0.05.

Table 3. Characteristics of participants stratified by rs8050136 genotypes.

rs8050136	C/C	A/C	A/A	*p*-Value *
N	209	424	182	
Genotype frequency	0.26	0.52	0.22	>0.05
Age	40.2 (14.1)	41.3 (14.8)	39.4 (14.3)	0.24
Females (%)	54.3% (0.49)	53.4% (0.49)	47.7% (0.50)	0.36
BMI (kg/m^2)	27.6 (6.1)	28.7 (6.8)	28.9 (6.8)	0.063
<25.0	80 (38.6%)	138 (33.2%)	54 (30.2%)	
25.0–29.9	69 (33.3%)	140 (33.7%)	67 (37.4%)	0.422
≥30.0	58 (28.0%)	138 (33.2%)	58 (32.4%)	
Total body fat content (kg)	25.3 (12.4)	27.5 (13.8)	28.2 (15.2)	0.095
Total body fat content (%)	30.6 (9.1)	31.8 (9.6)	31.6 (10.4)	0.465
Waist circumference (cm)	94.2 (17.6)	96.7 (17.2)	97.4 (16.6)	0.053
Hip circumference (cm)	101.2 (12.5)	104.1 (13.0)	103.8 (12.4)	0.008
WHR	0.927 (0.092)	0.925 (0.088)	0.936 (0.085)	0.382
Visceral fat (cm^3)	103.3 (81.5)	110.2 (80.0)	111.8 (82.5)	0.381
Visceral fat (%)	36.4 (11.7)	37.6 (12.5)	37.1 (11.6)	0.570
Subcutaneous fat (cm^3)	163.7 (83.5)	166.9 (80.8)	175.3 (83.1)	0.405
Subcutaneous fat (%)	63.7 (11.6)	62.2 (13.0)	62.9 (11.6)	0.540
Visceral/subcutaneous fat ratio	0.641 (0.404)	0.690 (0.477)	0.662 (0.410)	0.536
Total cholesterol	201.9 (56.1)	192.1 (41.4)	193.7 (43.1)	0.153
HDL	60.8 (14.0)	59.6 (15.7)	59.7 (14.4)	0.422
LDL	116.3 (43.3)	109.9 (37.9)	111.1 (41.8)	0.189
TG	124.1 (144.3)	113.2 (71.1)	115.1 (61.3)	0.491
Blood glucose level during OGTT (mg/dL)				
0 min	96.8 (24.2)	95.8 (18.5)	96.8 (20.4)	0.922
30 min	146.6 (44.5)	145.2 (31.7)	150.1 (35.4)	0.263
60 min	131.7 (56.4)	129.9 (46.1)	133.9 (46.1)	0.413
120 min	100.2 (46.3)	99.3 (32.2)	98.6 (30.9)	0.493
History of prediabetes or diabetes				
Yes	100 (47.8%)	213 (50.2%)	96 (52.7%)	0.628
No	109 (52.2%)	211 (49.8%)	86 (47.3%)	
Dietary assessment (n)	103	264	123	
Daily energy intake (kcal)	1820.5 (734.9)	1759.6 (673.3)	1853.9 (716.3)	0.645
Daily energy from protein (%)	18.7 (4.5)	19.0 (4.9)	19.0 (4.9)	0.855
Daily energy from fat (%)	31.2 (7.2)	30.9 (7.5)	31.9 (7.8)	0.506
Daily energy from carbohydrates (%)	47.6 (7.7)	47.9 (9.1)	46.8 (8.4)	0.572
Daily physical activity level				
Low	16 (7.7%)	25 (5.9%)	19 (10.4%)	
Moderate	50 (23.9%)	83 (19.6%)	40 (22.0%)	0.179
High	143 (68.4%)	316 (74.5%)	123 (67.6%)	

Data presented as mean and standard deviation (SD), number of observations, and frequency. BMI, body mass index; HDL, high-density lipoprotein; LDL, low-density lipoprotein; OGTT, oral glucose tolerance test; TG, triglycerides; WHR, waist-hip ratio. * Holm-adjusted Kruskal-Wallis/ANOVA *p*-values.

Table 4. Characteristics of participants stratified by rs8044769 genotypes.

rs8044769	C/C	C/T	T/T	*p*-Value *
N	270	406	138	
Genotype frequency	0.33	0.50	0.17	>0.05
Age	40.4 (14.8)	41.2 (14.6)	39.6 (13.7)	0.54
Females (%)	51.8% (0.50)	51.9% (0.50)	56.2% (0.49)	0.65
BMI (kg/m^2)	28.5 (6.8)	28.7 (6.7)	27.8 (6.1)	0.534
<25.0	88 (33.2%)	131 (32.8%)	51 (37.5%)	
25.0–29.9 (kg/m^2)	92 (34.7%)	143 (35.8%)	43 (31.6%)	0.862
≥30.0 (kg/m^2)	85 (32.1%)	126 (31.5%)	42 (30.9%)	
Total body fat content (kg)	27.7 (15.0)	27.2 (13.6)	25.8 (11.9)	0.676
Total body fat content (%)	31.7 (10.1)	31.4 (9.6)	31.1 (9.0)	0.893
Waist circumference (cm)	96.2 (17.2)	96.8 (17.3)	94.6 (16.8)	0.429
Hip circumference (cm)	103.3 (12.6)	104.0 (13.0)	101.5 (12.2)	0.189
WHR	0.928 (0.089)	0.927 (0.088)	0.929 (0.089)	0.980
Visceral fat (cm^3)	109.2 (78.3)	110.3 (83.7)	99.8 (72.5)	0.648
Visceral fat (%)	37.7 (11.8)	36.9 (12.3)	36.4 (12.0)	0.617
Subcutaneous fat (cm^3)	168.6 (83.3)	169.7 (82.4)	160.9 (74.2)	0.773
Subcutaneous fat (%)	62.3 (11.8)	63.0 (12.8)	63.7 (11.8)	0.598
Visceral/subcutaneous fat ratio	0.689 (0.492)	0.662 (0.421)	0.641 (0.408)	0.590
Total cholesterol	193.8 (40.1)	191.7 (42.7)	206.7 (62.7)	0.029
LDL	111.0 (38.6)	109.6 (38.7)	119.8 (46.5)	0.058
HDL	60.7 (14.7)	59.2 (15.5)	60.4 (13.8)	0.176
TG	110.5 (58.7)	114.6 (73.4)	132.6 (170.6)	0.689
Blood glucose level during OGTT (mg/dL)				
0 min	96.6 (20.0)	96.1 (21.0)	96.0 (20.1)	0.910
30 min	149.6 (35.1)	143.7 (32.0)	148.8 (47.8)	0.229
60 min	133.3 (45.4)	128.7 (47.3)	133.5 (59.3)	0.303
120 min	98.6 (31.8)	99.3 (30.7)	100.6 (53.0)	0.147
History of prediabetes or diabetes				
Yes	143 (53.0%)	201 (49.5%)	64 (46.4%)	0.420
No	127 (47.0%)	205 (50.5%)	74 (53.6%)	
Dietary assessment (n)	157	248	83	
Daily energy intake (kcal)	1775.7 (646.2)	1792.7 (735.8)	1816.0 (686.3)	0.791
% of daily energy from protein	18.9 (4.8)	19.2 (5.0)	18.4 (4.1)	0.608
% of daily energy from fat	32.5 (7.3)	30.1 (7.5)	31.9 (7.4)	0.005
% of daily energy from carbohydrates	46.5 (8.5)	48.3 (8.9)	47.1 (8.1)	0.164
Daily physical activity level				
Low	23 (8.5%)	26 (6.4%)	10 (7.2%)	
Moderate	55 (20.4%)	77 (19.0%)	41 (29.7%)	0.060
High	192 (71.1%)	303 (74.6%)	87 (63.0%)	

Data presented as mean and standard deviation (SD), number of observations, and frequency. BMI, body mass index; HDL, high-density lipoprotein; LDL, low-density lipoprotein; OGTT, oral glucose tolerance test; TG, triglycerides; WHR, waist-hip ratio. * Holm-adjusted Kruskal-Wallis/ANOVA *p*-values.

Our further analysis showed that GG carriers in the group with lower than median protein intake presented lower blood glucose levels at 60 (Figure 3A) and 120 min (Figure 3B) of OGTT, while TT genotype participants in the group with higher than median protein intake presented higher insulin levels at 60 min (Figure 3C). We also observed higher insulin levels at 120 min of OGTT in GG and TT genotype carriers stratified to the group with higher than median protein intake (Figure 3D). The heterozygous GT genotype carriers in the group with lower than median dietary protein intake presented lower body weight (Figure 3E), BMI (Figure 3F) and total body (Figure 3G) and subcutaneous (Figure 3H) fat content. Using linear modeling, we found a significant interaction effect of (protein diet group) × (rs3751812 genotype) on body composition (*p*-value < 0.05) and blood glucose and insulin levels (*p*-value < 0.01) at 60 and 120 min of OGTT.

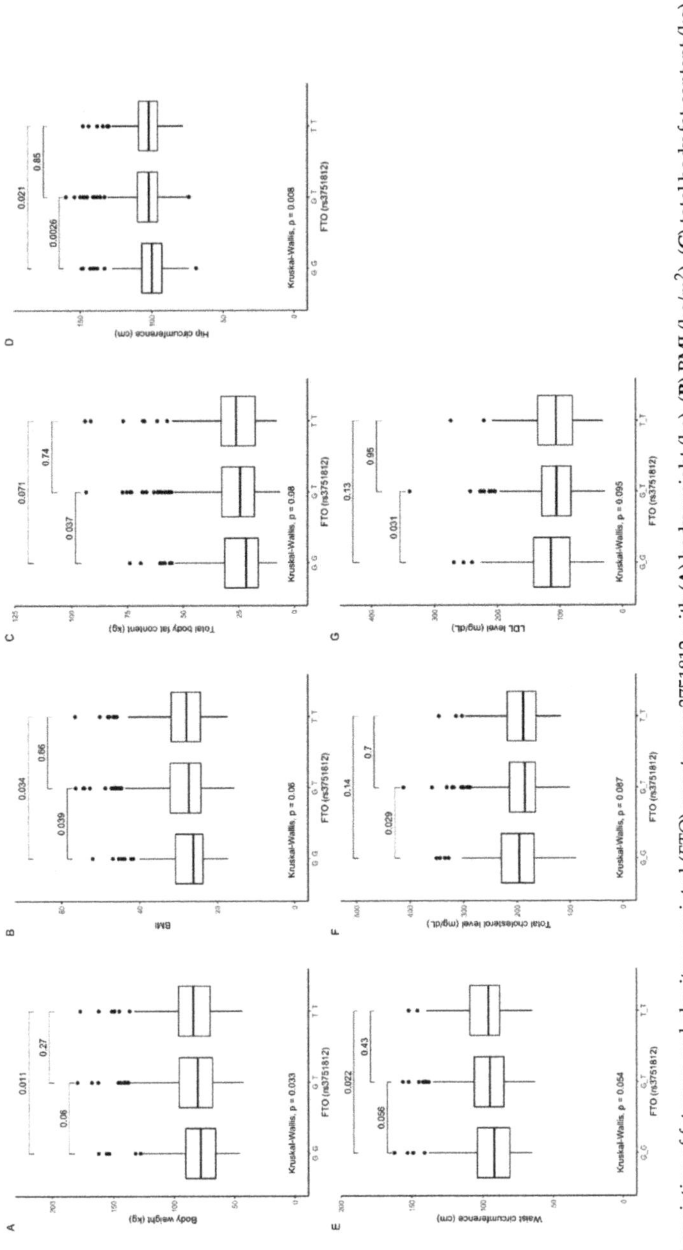

Figure 1. Association of fat mass and obesity-associated (*FTO*) genotype rs3751812 with (**A**) body weight (kg), (**B**) BMI (kg/m^2), (**C**) total body fat content (kg), (**D**) hip circumference (cm), (**E**) waist circumference (cm^3), (**F**) total cholesterol level (mg/dL), (**G**) LDL level (mg/dL).

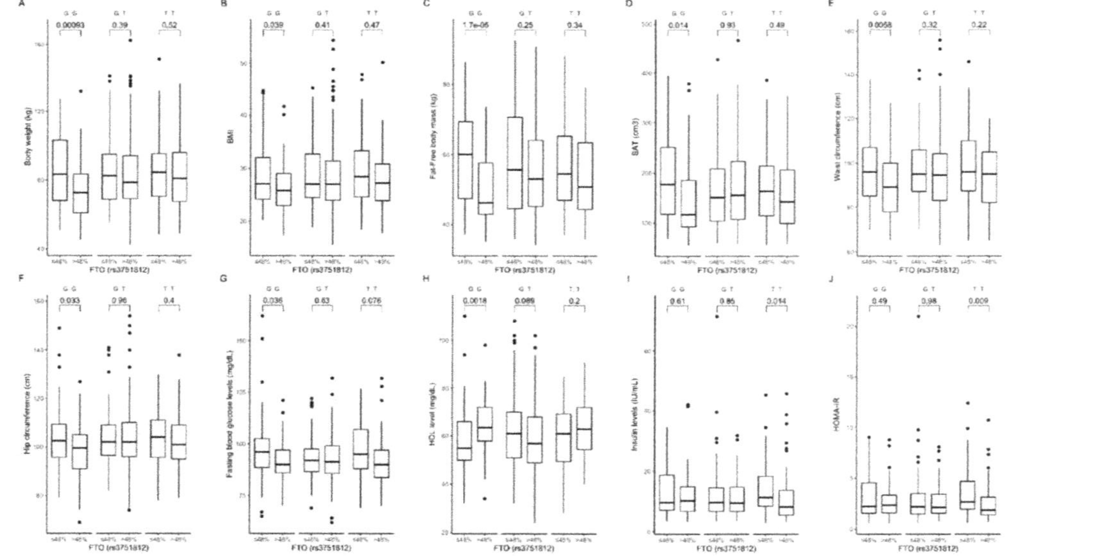

Figure 2. Association of *FTO* genotypes rs3751812 with (**A**) body weight (kg), (**B**) BMI (kg/m^2), (**C**) fat-free body mass (kg), (**D**) subcutaneous adipose tissue (SAT) (cm^3), (**E**) waist circumference (cm^3), (**F**) hip circumference (cm^3), (**G**) fasting blood glucose level (mg/dL), (**H**) HDL level (mg/dL), (**I**) fasting blood insulin level (IU/mL), and (**J**) HOMA-IR by dietary carbohydrate intake strata: ≤48% and >48% of total daily energy intake. HOMA-IR, Homeostatic Model Assessment for Insulin Resistance.

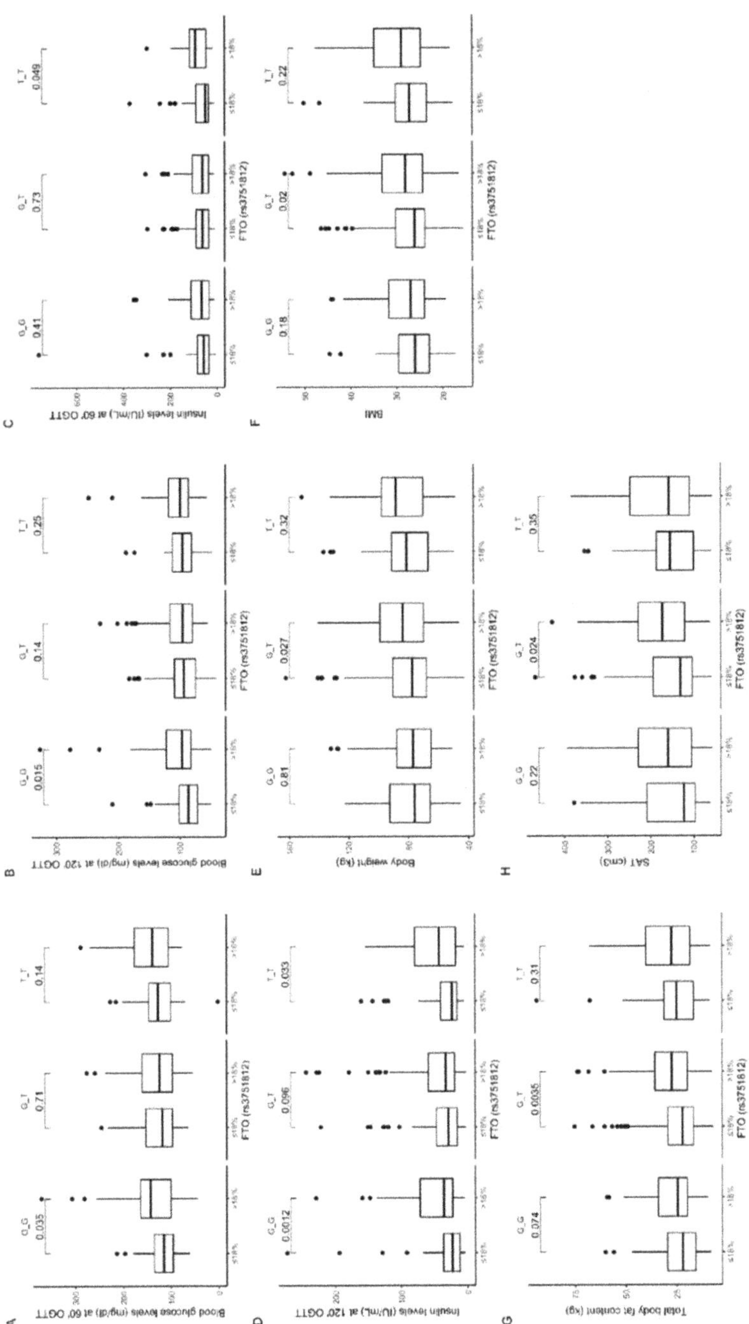

Figure 3. Association of dietary protein intake ≤18% and >18% of total daily energy intake with blood glucose level (mg/dL) at (**A**) 60 min and (**B**) 120 min of OGTT; insulin level (IU/mL) at (**C**) 60 min and (**D**) 120 min of OGTT; (**E**) body weight (kg); (**F**) BMI (kg/m^2); (**G**) total body fat content (kg); and (**H**) SAT (cm^3) in *FTO* rs3751812 genotype carriers.

Analyzing the dietary fat intake, we noted that carriers of the GG genotype stratified to the group with lower than median fat intake presented lower subcutaneous (Figure 4A) and visceral (Figure 4B) fat content. Surprisingly, we observed that GT genotype carriers showed higher HDL levels (Figure 4C) when they were stratified to the group with higher than median fat intake. We did not observe any other association with dietary fat intake. The interaction effect of (fat diet group) × (rs3751812 genotype) on subcutaneous and visceral fat content was statistically significant, with p-value < 0.01, as well as on HDL levels (p-value < 0.002).

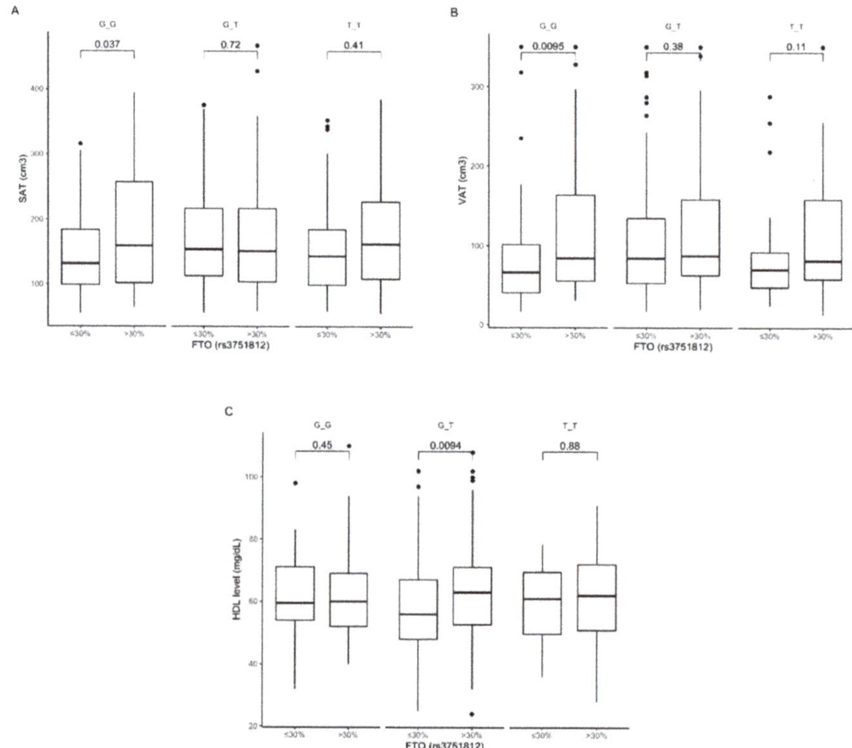

Figure 4. Association of dietary fat intake ≤30% and >30% of total daily energy intake with (**A**) SAT (cm^3), (**B**) VAT (cm^3), and (**C**) HDL level (mg/dL) in *FTO* rs3751812 genotype carriers.

3.3. Association of rs8050136 Genetic Variants with Obesity, Anthropometric Measures, Lipid Profile, and Dietary Intake

Our analysis showed that CC genotype carriers presented significantly lower body weight (Figure 5A), BMI (Figure 5B), and waist (Figure 5D) and hip (Figure 5E) circumference compared to TT, and significantly lower BMI (Figure 5B), total body fat content (Figure 5C), and hip (Figure 5E) circumference when compared to CT genotype carriers.

Based on the analysis of the interactions between rs8050136 genotypes and carbohydrate intake, we observed that CC genotype carriers in the group with higher than median carbohydrate intake presented lower body weight (Figure 6A), fat-free body mass level (Figure 6B), skeletal muscle mass content (Figure 6C), subcutaneous fat content (Figure 6D), and waist circumference (Figure 6E) and higher HDL-cholesterol level (Figure 6F). The interaction effect of (carbohydrate diet group) × (rs8050136 genotypes) on body composition (p-value < 0.05) and HDL (p-value < 0.01) was statistically significant.

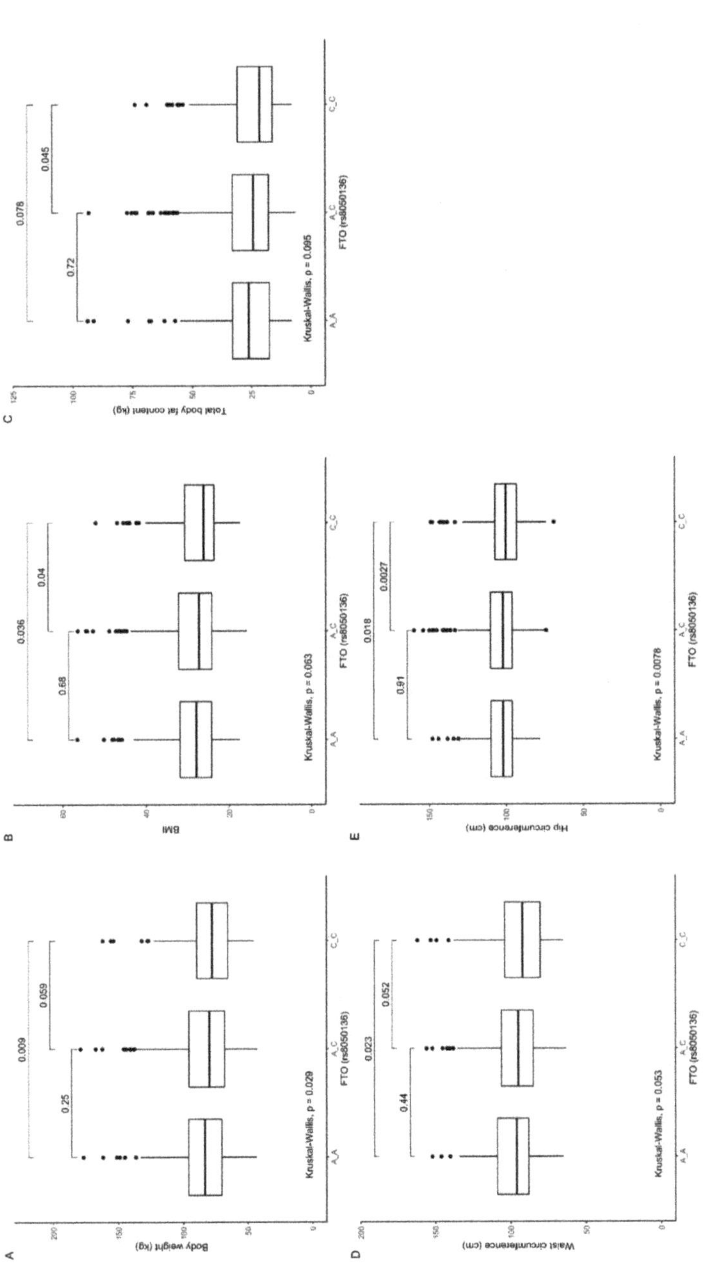

Figure 5. Association of *FTO* genotypes rs8050136 with (**A**) body weight (kg), (**B**) BMI (kg/m^2), (**C**) total body fat content (kg), (**D**) waist circumference (cm^3), (**E**) hip circumference (cm^3).

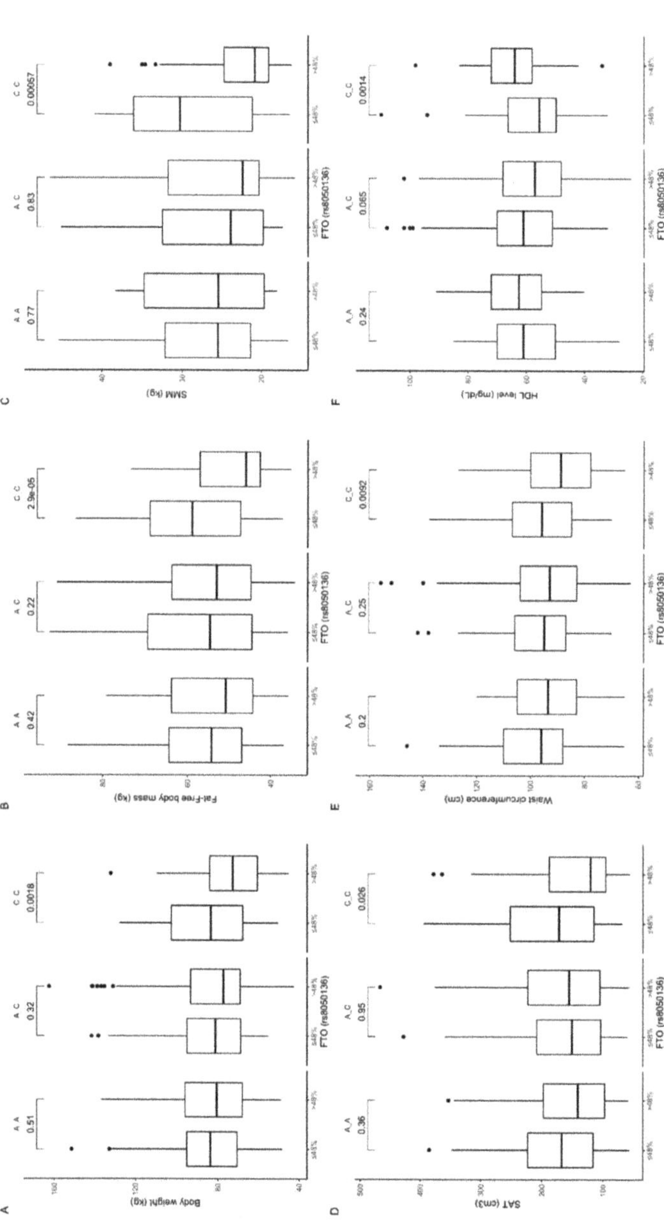

Figure 6. Association of *FTO* genotypes rs8050136 with (**A**) body weight (kg), (**B**) fat-free body mass (kg), (**C**) skeletal muscle mass (SMM) (kg), (**D**) SAT (cm^3), (**E**) waist circumference (cm^3), and (**F**) HDL level (mg/dL) by dietary carbohydrate intake strata: ≤48% and >48% of total daily energy intake.

We observed that AC genotype carriers stratified to the group with lower than median protein intake presented lower body weight (Figure 7A), BMI (Figure 7B), total body fat content (Figure 7C), and subcutaneous fat content (Figure 7D). Additionally, we noticed that CC genotype carriers who were stratified to the group with higher than median protein intake presented higher blood glucose levels at 60 min (Figure 7E) and 120 min (Figure 7F) of OGTT. Higher insulin levels were observed in AA genotype carriers at 60 min (Figure 7G) and 120 min (Figure 7H) of OGTT, and in CC genotype carriers at 120 min (Figure 7H) of OGTT. The interaction effect of (protein diet group) × (rs8050136 genotypes) on body composition, anthropometric measures, and lipid profile was statistically significant with p-value < 0.05.

The analysis of dietary fat intake showed that carriers of the CC genotype stratified to the group with lower than median fat intake presented surprisingly higher subcutaneous fat content (Figure 8A) and lower visceral fat content (Figure 8B), as well as lower VAT/SAT ratio (Figure 8C). In AA genotype carriers, we noticed similar tendencies (Figure 8A,C). In carriers of the AC genotype stratified to the group with higher than median fat intake, we observed higher HDL-cholesterol levels (Figure 8D) compared to those who were stratified to the group with lower than median fat intake. The interaction effect of (fat diet group) × (rs3751812 genotype) on subcutaneous and visceral fat content as well as it ratio was statistically significant with p-value < 0.01, the interaction effect on HDL was the largest, with the p-value of 0.001.

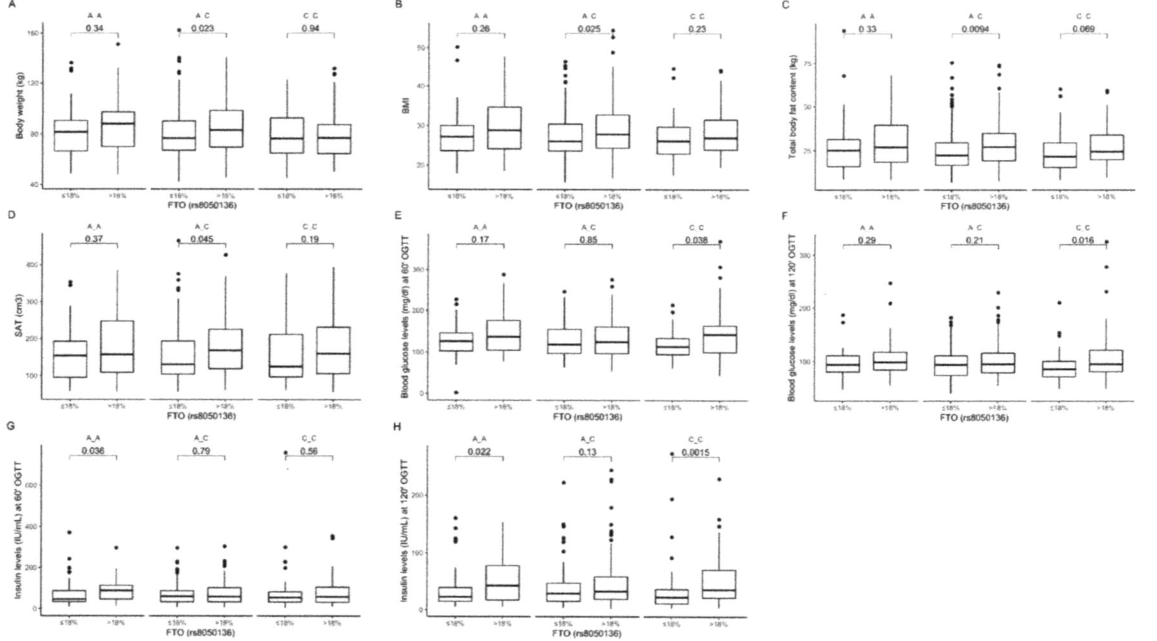

Figure 7. Association of dietary protein intake ≤18% and >18% of total daily energy intake with (**A**) body weight (kg); (**B**) BMI (kg/m^2); (**C**) total body fat content (kg); (**D**) SAT (cm^3); blood glucose level (mg/dL) at (**E**) 60 min and (**F**) 120 min of OGTT; and insulin level (IU/mL) at (**G**) 60 min and (**H**) 120 min of OGTT in *FTO* rs8050136 genotype carriers.

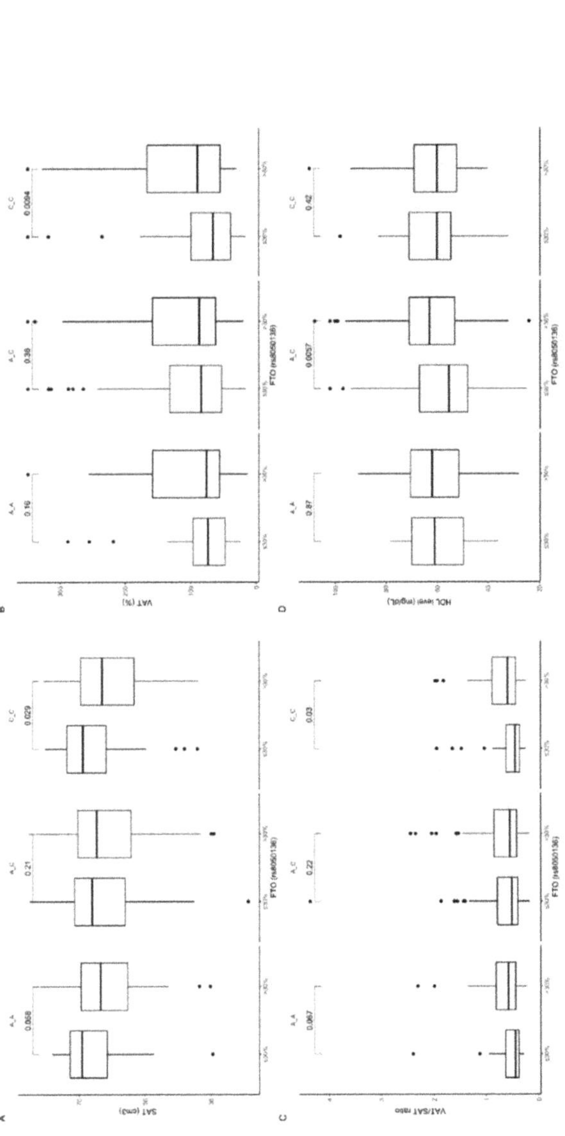

Figure 8. Association of dietary fat intake ≤30% and >30% of total daily energy intake with (**A**) SAT (cm^3), (**B**) VAT (%), (**C**) SAT/VAT ratio, and (**D**) HDL level (mg/dL) in *FTO* rs8050136 genotype carriers.

3.4. Association of rs8044769 Genetic Variants with Obesity, Anthropometric Measures, Lipid Profile, and Dietary Intake

The TT genotype carriers of rs8044769 presented significantly higher total cholesterol (Figure 9A) and LDL-cholesterol (Figure 9B) levels when compared to CT genotype carriers, and similar marginally significant results were noted between TT and CT genotype carriers (Figure 9A,B).

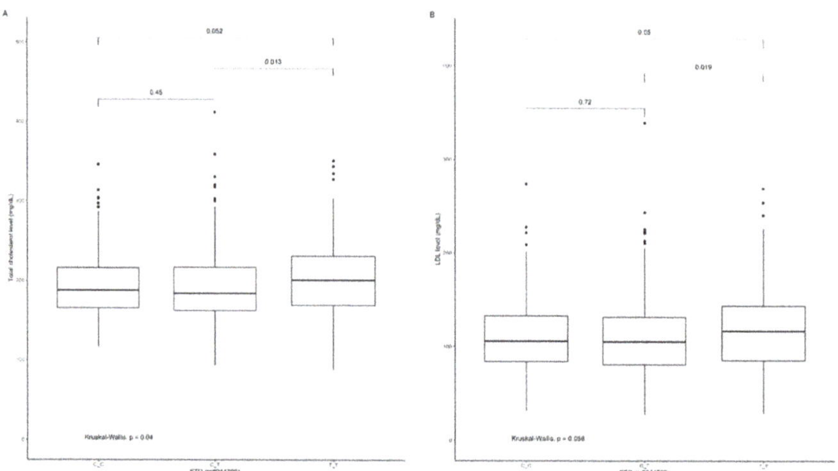

Figure 9. Association of *FTO* genotypes rs8044769 with (**A**) total cholesterol level (mg/dL) and (**B**) LDL level (mg/dL).

The gene-diet interaction analysis showed that TT and CT genotype carriers presented higher body weight (Figure 10A), BMI (Figure 10B), fat-free body mass (Figure 10C), and waist circumference (Figure 10D) when they were stratified to the group with lower than median carbohydrate intake. Homozygous TT carriers in the group with lower than median carbohydrate intake also presented lower HDL-cholesterol levels (Figure 10E) and surprisingly higher skeletal muscle mass content (Figure 10F). We did not notice any association between percentage of daily energy intake provided from carbohydrates and investigated metabolic parameters in CC genotype carriers. The interaction effect of (carbohydrate diet group) × (rs8044769 genotype) on body composition, anthropometric measures, and HDL was statistically significant with p-value < 0.05.

Among individuals in the group with higher than median protein intake, we observed that heterozygous CT carriers showed higher body weight (Figure 11A), BMI (Figure 11B), total body fat content (Figure 11C), subcutaneous fat content (Figure 11D), waist circumference (Figure 11E), and hip circumference (Figure 11F). In addition, we noted that both CC and TT homozygous carriers presented higher insulin concentration at 120 min (Figure 11G); however, higher fasting blood glucose levels (Figure 11H) and blood glucose levels at 30 min (Figure 11I), 60 min (Figure 11J), and 120 min (Figure 11K) of OGTT were noted only in TT genotype carriers stratified to the group with higher than median protein intake. We noted significantly higher values of HOMA-IR in TT genotype carriers, and the same tendency in CC genotype carriers, when dietary protein provided >18% of total energy compared to subjects who were stratified to the group with lower than median protein intake (Figure 11L). The interaction effect of (protein diet group) × (rs8044769 genotype) on body composition, and blood glucose and insulin levels was statistically significant with p-value < 0.05.

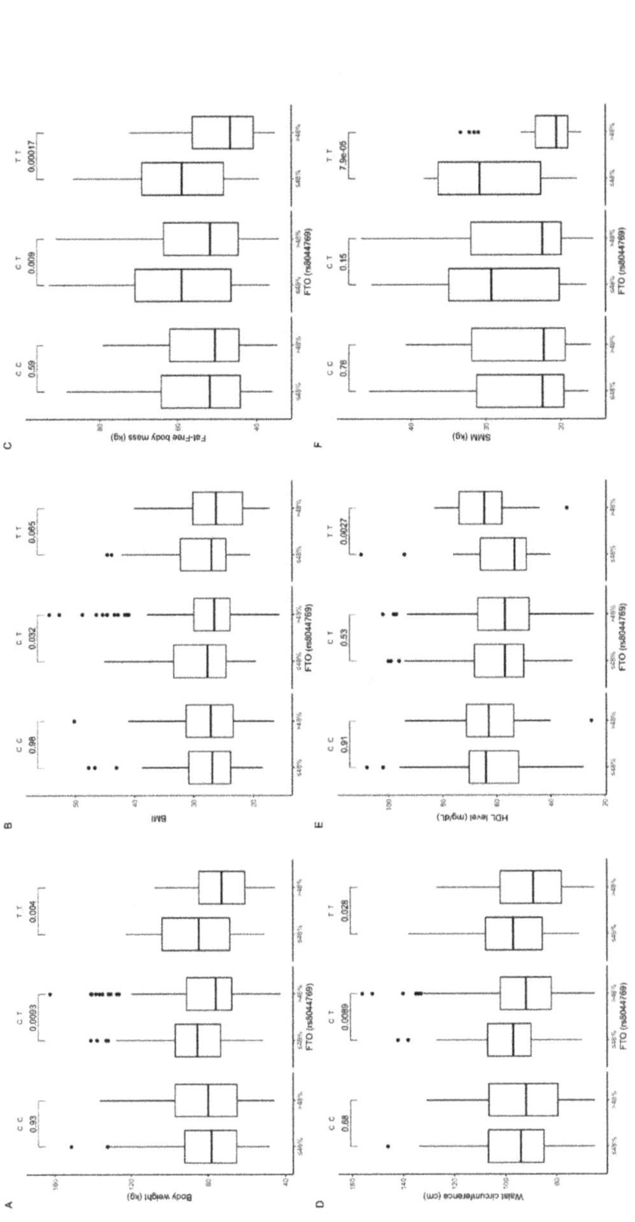

Figure 10. Association of *FTO* genotypes rs8044769 with (**A**) body weight (kg), (**B**) BMI (kg/m^2), (**C**) fat-free body mass (kg), (**D**) waist circumference (cm^3), (**E**) HDL level (mg/dL), and (**F**) SMM (kg) by dietary carbohydrate intake strata: ≤48% and >48% of total daily energy intake.

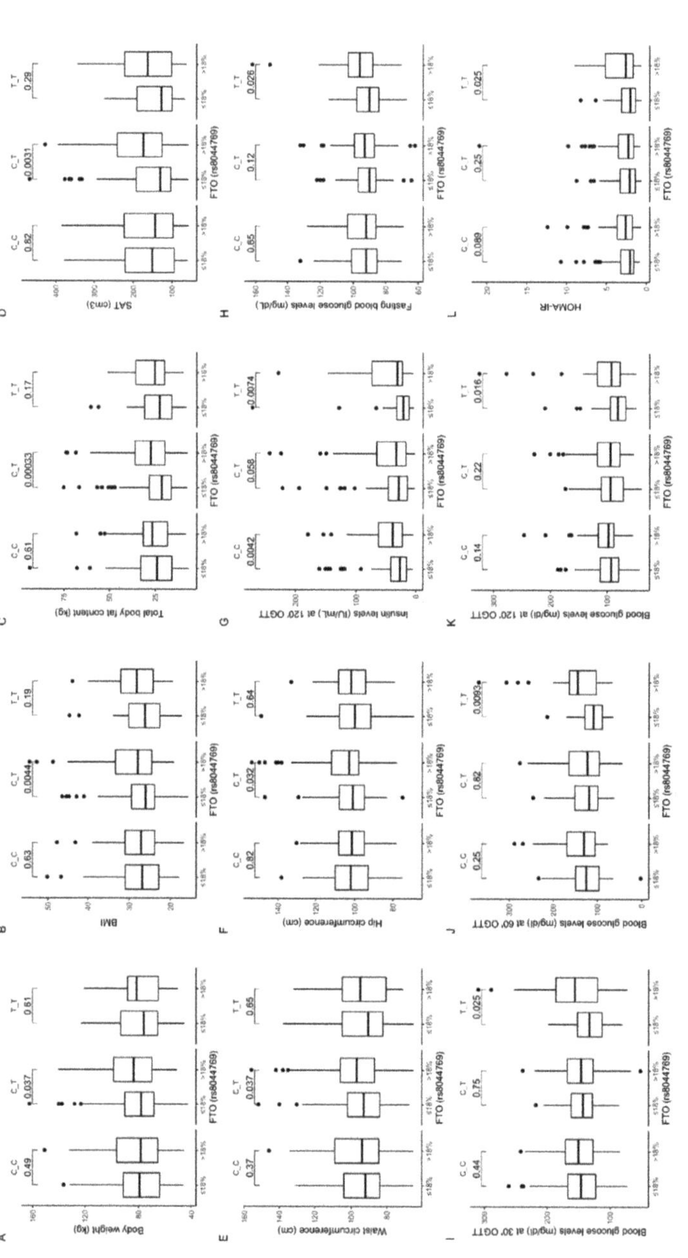

Figure 11. Association of dietary protein intake ≤18% and >18% of total daily energy intake with (**A**) body weight (kg); (**B**) BMI (kg/m^2); (**C**) total body fat content (kg); (**D**) SAT (cm^3); (**E**) waist circumference (cm); (**F**) hip circumference (cm^3); (**G**) insulin level (IU/mL) at 120 min of OGTT; (**H**) fasting blood glucose level (mg/dL); blood glucose level (mg/dL) at (**I**) 30 min, (**J**) 60 min, and (**K**) 120 min of OGTT; and (**L**) HOMA-IR in *FTO* rs8044769 genotype carriers.

An association with dietary fat intake was observed in carriers of the heterozygous CT genotype stratified to the group with higher than median fat intake, who presented higher fat-free body mass content (Figure 12A). TT genotype carriers had higher visceral fat content (Figure 12B) and lower subcutaneous fat content (Figure 12C) when they were stratified to the group with higher than median dietary fat intake compared to carriers of the same genotype in the group with lower than median dietary fat intake. The interaction effect of (fat diet group) × (rs8044769 genotype) on subcutaneous and visceral fat content was statistically significant with *p*-value < 0.05.

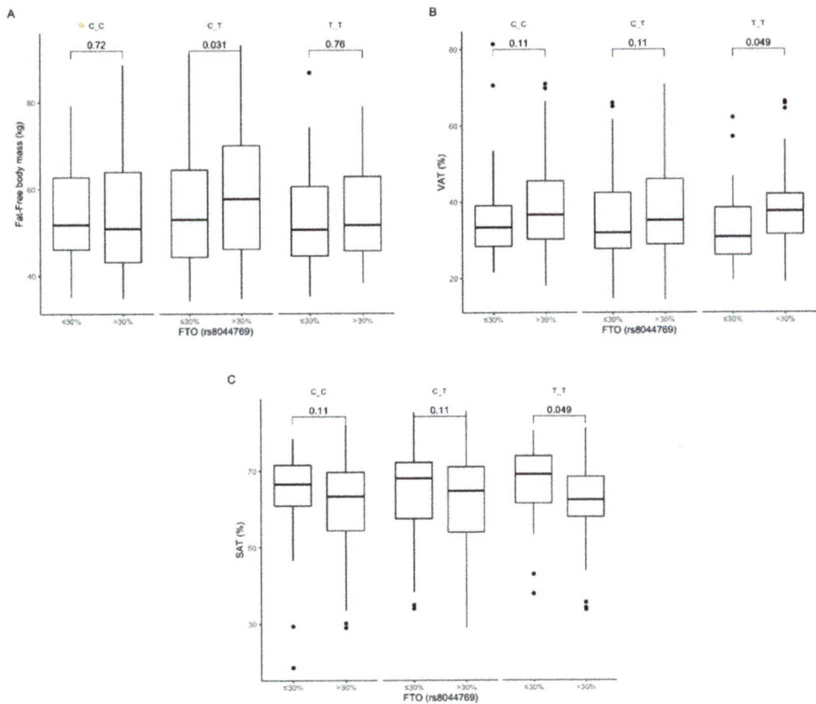

Figure 12. Association of dietary fat intake ≤30% and >30% of total daily energy intake with (**A**) fat-free body mass (kg), (**B**) SAT (cm^3), and (**C**) VAT (%) in *FTO* rs8044769 genotypes carriers.

4. Discussion

Over the past few decades, public awareness in the field of nutrition and physical activity has increased, but obesity and its comorbidities are still serious international health problems [26]. It is widely known that the *FTO* gene is an established genetic susceptibility locus for the risk of obesity development [27]. However, the association between the *FTO* gene and dietary factors is still unclear and there is a scientific need to investigate the associations between environmental and genetic risk factors and their interactions and roles in obesity development and treatment. In our study, we demonstrated an interplay between *FTO* genetic variants and dietary carbohydrate, protein, and fat intake, and the impact of these interactions on body weight, body fat content and distribution, and other anthropometric measures, as well as on glucose homeostasis and lipid profile, in a Polish population of adults. For our study, we chose some of the most common SNPs based on previously published results [25–27].

We observed a protective effect of the GG genotype of rs3751812 against obesity, but GG genotype carriers presented higher total cholesterol and LDL-cholesterol levels. It was shown previously that *FTO* rs3751812 risk allele T is related to increased BMI and body fat distribution compared to the protective allele G [28]. However, our results indicate that carrying the GG genotype leads to more beneficial effects if more than 48% of total diet energy comes from carbohydrates; then, we could observe significantly lower obesity-related parameters. We did not notice any difference in total body fat content, and lower body weight and BMI could be associated with noted lower fat-free mass. Nevertheless, we did not observe any adverse effects of lower fat-free mass content, since we also noted lower fasting glucose and higher HDL-cholesterol concentration. The impact of the TT genotype, which appears to be a risk genotype for obesity, seems to not be related to carbohydrate intake, except for the associations with fasting insulin concentration and HOMA-IR level.

We also noted associations between dietary protein intake, SNPs, and metabolic parameters for all investigated genetic variants of rs3751812, indicating that we can observe more beneficial results if dietary protein provides no more than 18% of total daily energy intake. These observations are worth underlining, especially in light of the current interest in high-protein diets. Based on the results that we noted for high-risk TT genotype carriers (rs3751812) in the group with higher than median protein intake, including higher post-absorptive insulin levels and higher fasting insulin concentrations and HOMA-IR levels in subjects in the group with lower than median carbohydrate intake, we can hypothesize that these individuals should avoid high-protein, low-carbohydrate diets. It is also worth noting that for TT genotype subjects, we did not observe any differences or adverse effects that would depend on dietary fat intake. Moreover, it has already been found that carriers of minor allele T rs3751812 present a lower risk of obesity when they adhere better to a Mediterranean diet, which consists of higher daily consumption of fats, especially from olive oil and nuts [29]. We noted an association with dietary fat intake only in GG and GT genotypes carriers for body fat distribution and HDL-cholesterol level. In these subjects, if daily energy intake derived from fat was more than 30%, we could observe higher visceral and subcutaneous fat content.

Our study also shows significant associations of *FTO* SNPs in rs8050136 with the investigated markers of obesity. The CC genotype has been shown to play a protective role, and CC genotype carriers presented significantly lower body weight, BMI, total body fat content, and waist and hip circumference. Our observations are in line with results from previous studies [30–33]. Although numerous studies have shown an association of SNPs in rs8050136 with higher values of BMI and waist and hip circumference, the impact of dietary intake on this association is still unclear. In our study, we did not notice any crucial differences in daily dietary macronutrient intake between genotypes. Nevertheless, we observed that CC genotype subjects were more susceptible to the beneficial effects when carbohydrate in their diets provided more than 48% of total daily energy intake and no more than 18% came from dietary protein. Moreover, dietary fat should be limited to less than 30% of total daily energy intake to avoid visceral fat accumulation in these subjects. We also observed that AC genotype carriers of rs8050136 stratified to the group with lower than median dietary fat intake presented lower HDL-cholesterol levels. Bego et al. [34] observed that the risk A allele of rs8050136 was significantly associated with decreased HDL-cholesterol levels in control subjects in type 2 diabetes studies. We did not notice differences between studied genotypes, but surprisingly, we observed lower HDL-cholesterol concentrations in AC genotype carriers only when they followed a diet with less than 30% of energy from dietary fat. In these subjects, when dietary fat provided more than 30% of total energy, then HDL concentrations were significantly higher, without any differences in total cholesterol or LDL-cholesterol levels. We did not evaluate the source of fat, if the diet was rich in saturated or unsaturated fatty acids, or what could explain our observations, because it is well known that various types of fatty acids have different impacts on plasma lipid concentrations [35].

We observed that dietary protein intake might have an impact on obesity-related parameters only in heterozygous CA carriers of rs8050136, while in AA and CC genotype carriers, on glucose homeostasis-related markers, and in all cases, beneficial effects were noted when dietary protein provided no more than 18% of total daily energy intake. It has already been reported by Park et al. that the association of the rs8050136 risk variant may be partially mediated by macronutrient intake [36], but only the association with percentage of energy derived from fat has been detected.

The analysis of differences between genetic variants of rs8044769 showed only that TT genotype carriers presented higher total and LDL-cholesterol levels. We did not notice any other differences between genotypes; however, TT and CT genotype carriers presented lower body weight, BMI, and waist circumference and higher HDL-cholesterol levels when more than 48% of total daily energy was derived from carbohydrates, even if fat-free mass and skeletal muscle mass were also significantly lower. The obesity-related parameters seemed to be associated with dietary protein intake only in heterozygous CT carriers and in TT genotype carriers with glucose homeostasis-related parameters. All of the noted associations were more beneficial if daily energy derived from dietary protein intake did not exceed 18%. Moreover, our study suggests that CT and TT genotype carriers should also not consume dietary fat exceeding 30% of total daily energy intake, to avoid visceral fat accumulation. We did not observe any significant associations for CC genotype carriers that could depend on dietary protein or fat intake, except insulin levels at 120 min of OGTT, which were significantly higher in subjects stratified to the group with higher than median protein intake. There is a very limited number of available studies on *FTO* rs8044769 genetic variants, and some of them present these variants as BMI-associated SNPs [37,38], but in others, the authors did not observe such a relationship [39]. Moreover, considering the fact that all associations can vary with ethnicity, gender, dietary intake, and some other factors, studies in larger and more diverse populations are needed.

The present study has several strengths. As far as we know, this is one of the first studies to present interactions between *FTO* SNPs rs3751812, rs8044769, rs8050136, and rs9939609 and macronutrient intake, and the effect of these relationships on obesity and obesity-related complications. Another strength of our study is that it is based on a relatively large population. It is also worth noting that it has been shown that *FTO* genetic variants may influence dietary factors [14,40–42] or dietary fat intake [36,43,44], while other studies did not confirm these associations [12,15]. In general, we did not observe any significant differences in macronutrient intake between studied genotypes, which can also be interpreted as a strength of our study, because we can exclude the possibility that our results might be affected by the impact of different macronutrient intake on gene expression and the activation of different metabolic pathways. Nevertheless, several limitations of our study also need to be addressed. Some parts of our results are based on self-reported data, such as three-day diaries of food intake, and it has been shown that obese people tend to underreport or misreport their total dietary intake, especially fatty foods and foods rich in carbohydrates [45]. However, dietary questionnaires are the only known implements available for large-scale population investigations so far. Moreover, only Caucasian individuals were recruited for our study; therefore, in order to verify our findings in other ethnic groups, the data should be replicated in other populations.

Our results, if confirmed in larger populations of different ethnic groups, may have also practical clinical implications. Based on our observations, we can recommend that carriers of GG genotype of rs3751812 and CC genotype of rs8050136 follow diets in which no less than 48% of daily energy intake is derived from carbohydrates and no more than 30% from dietary fat. Moreover, carriers of TT and CT genotypes of rs8044769 should avoid diets in which carbohydrates provide less than 48% of total energy, whereas carriers of TT genotype should avoid diets in which proteins provide more than 18% of total daily energy, to prevent glucose homeostasis disturbances. These recommendations seem to be highly important, since we noticed that the mean amounts of macronutrients in the diets of the investigated population were mostly less than 48% of total energy intake for carbohydrates, more than 30% for dietary fat, and more than 18% for proteins, which may have adverse effects for carriers of the above-mentioned genotypes.

5. Conclusions

In conclusion, our findings provide new insights into the role of the interactions between diet and *FTO* SNPs in the risk of obesity and its metabolic consequences. Advances in this field bring us closer to the development of genome-customized diet recommendations to prevent obesity. Detecting *FTO* risk genotype carriers and modifying dietary intake according to the genetic profile may be a novel, efficient strategy to prevent obesity development.

Supplementary Materials: The following are available online at http://www.mdpi.com/2072-6643/12/11/3255/s1, Figure S1: Study flowchart diagram.

Author Contributions: Conceptualization, E.A.-P., M.G., and A.K.; methodology, E.A.-P., M.G., and A.K.; formal analysis, W.B.; investigation, E.A.-P., W.B., J.F., U.K., and M.M.; writing-original draft preparation, P.C. and E.A.-P.; writing-review, E.A.-P., M.G., and A.K.; visualization, W.B.; supervision, M.G. and A.K.; project administration, E.A.-P.; funding acquisition, E.A.-P., M.G., and A.K. All authors have read and agreed to the published version of the manuscript.

Funding: This study was funded by the Polish Ministry of Science and Higher Education (4774/B/P01/2009/37) and by the Medical University of Bialystok. P.C. was granted № POWR.03.02.00-00-I051/16 from European Union funds, PO WER 2014-2020, grant № 03/IMSD/G/2019.

Acknowledgments: We thank the staff of the Medical University of Bialystok, Poland, for their contributions and help with data collection and laboratory analysis.

Conflicts of Interest: The authors declare no conflict of interest. The funders had no role in the design of the study; in the collection, analysis, or interpretation of data; in the writing of the manuscript, or in the decision to publish the results.

References

1. Jaacks, L.M.; Vandevijvere, S.; Pan, A.; McGowan, C.J.; Wallace, C.; Imamura, F.; Mozaffarian, D.; Swinburn, B.; Ezzati, M. The obesity transition: Stages of the global epidemic. *Lancet Diabetes Endocrinol.* **2019**, *7*, 231–240. [CrossRef]

2. Loh, M.; Zhou, L.; Ng, H.K.; Chambers, J.C. Epigenetic disturbances in obesity and diabetes: Epidemiological and functional insights. *Mol. Metab.* **2019**, *27S*, S33–S41. [CrossRef] [PubMed]

3. Fu, J.F. Big challenges: Obesity and type 2 diabetes in children and adolescents. *World J. Pediatr.* **2019**, *15*, 313–314. [CrossRef] [PubMed]

4. Kumar, S.; Kelly, A.S. Review of Childhood Obesity: From Epidemiology, Etiology, and Comorbidities to Clinical Assessment and Treatment. *Mayo Clin. Proc.* **2017**, *92*, 251–265. [CrossRef] [PubMed]

5. Saeedi, P.; Petersohn, I.; Salpea, P.; Malanda, B.; Karuranga, S.; Unwin, N.; Colagiuri, S.; Guariguata, L.; Motala, A.A.; Ogurtsova, K.; et al. Global and regional diabetes prevalence estimates for 2019 and projections for 2030 and 2045: Results from the International Diabetes Federation Diabetes Atlas, 9(th) edition. *Diabetes Res. Clin. Pract.* **2019**, *157*, 107843. [CrossRef] [PubMed]

6. Xu, Z.R.; Zhang, M.Y.; Ni, J.W.; Cheng, R.Q.; Zheng, Z.Q.; Xi, L.; Luo, F. Clinical characteristics and beta-cell function of Chinese children and adolescents with type 2 diabetes from 2009 to 2018. *World J. Pediatr.* **2019**, *15*, 405–411. [CrossRef] [PubMed]

7. Dal Canto, E.; Ceriello, A.; Rydén, L.; Ferrini, M.; Hansen, T.B.; Schnell, O.; Standl, E.; Beulens, J.W. Diabetes as a cardiovascular risk factor: An overview of global trends of macro and micro vascular complications. *Eur. J. Prev. Cardiol.* **2019**, *26* (Suppl. 2), 25–32. [CrossRef] [PubMed]

8. Thorleifsson, G.; Walters, G.B.; Gudbjartsson, D.F.; Steinthorsdottir, V.; Sulem, P.; Helgadottir, A.; Styrkarsdottir, U.; Gretarsdottir, S.; Thorlacius, S.; Jonsdottir, I.; et al. Genome-wide association yields new sequence variants at seven loci that associate with measures of obesity. *Nat. Genet.* **2009**, *41*, 18–24. [CrossRef]

9. Loos, R.J. Genetic determinants of common obesity and their value in prediction. *Best Pract. Res. Clin. Endocrinol. Metab.* **2012**, *26*, 211–226. [CrossRef]

10. Tan, L.J.; Zhu, H.; He, H.; Wu, K.H.; Li, J.; Chen, X.D.; Zhang, J.G.; Shen, H.; Tian, Q.; Krousel-Wood, M.; et al. Replication of 6 obesity genes in a meta-analysis of genome-wide association studies from diverse ancestries. *PLoS ONE* **2014**, *9*, e96149. [CrossRef]

11. Sonestedt, E.; Roos, C.; Gullberg, B.; Ericson, U.; Wirfält, E.; Orho-Melander, M. Fat and carbohydrate intake modify the association between genetic variation in the *FTO* genotype and obesity. *Am. J. Clin. Nutr.* **2009**, *90*, 1418–1425. [CrossRef]

12. Haupt, A.; Thamer, C.; Staiger, H.; Tschritter, O.; Kirchhoff, K.; Machicao, F.; Häring, H.-U.; Stefan, N.; Fritsche, A. Variation in the *FTO* gene influences food intake but not energy expenditure. *Exp. Clin. Endocrinol. Diabetes* **2009**, *117*, 194–197. [CrossRef]

13. Wardle, J.; Carnell, S.; Haworth, C.M.; Farooqi, I.S.; O'Rahilly, S.; Plomin, R. Obesity associated genetic variation in *FTO* is associated with diminished satiety. *J. Clin. Endocrinol. Metab.* **2008**, *93*, 3640–3643. [CrossRef] [PubMed]

14. Wardle, J.; Llewellyn, C.; Sanderson, S.; Plomin, R. The *FTO* gene and measured food intake in children. *Int. J. Obes.* **2009**, *33*, 42–45. [CrossRef] [PubMed]

15. Speakman, J.R.; Rance, K.A.; Johnstone, A.M. Polymorphisms of the *FTO* gene are associated with variation in energy intake, but not energy expenditure. *Obesity* **2008**, *16*, 1961–1965. [CrossRef] [PubMed]

16. Steemburgo, T.; Azevedo, M.J.; Gross, J.L.; Milagro, F.I.; Campion, J.; Martinez, J.A. The rs9939609 polymorphism in the *FTO* gene is associated with fat and fiber intakes in patients with type 2 diabetes. *J. Nutr. Nutr.* **2013**, *6*, 97–106.

17. Mu, M.; Xu, L.F.; Hu, D.; Wu, J.; Bai, M.J. Dietary Patterns and Overweight/Obesity: A Review Article. *Iran. J. Public Health* **2017**, *46*, 869–876. [PubMed]

18. Sales, N.M.; Pelegrini, P.B.; Goersch, M.C. Nutrigenomics: Definitions and advances of this new science. *J. Nutr. Metab.* **2014**, *2014*, 202759. [CrossRef]

19. Vimaleswaran, K.S.; Bodhini, D.; Lakshmipriya, N.; Ramya, K.; Anjana, R.M.; Sudha, V.; Lovegrove, J.A.; Kinra, S.; Mohan, V.; Radha, V. Interaction between *FTO* gene variants and lifestyle factors on metabolic traits in an Asian Indian population. *Nutr. Metab.* **2016**, *13*, 39. [CrossRef]

20. Kretowski, A.; Adamska, E.; Maliszewska, K.; Wawrusiewicz-Kurylonek, N.; Citko, A.; Goscik, J.; Bauer, W.; Wilk, J.; Golonko, A.; Waszczeniuk, M.; et al. The rs340874 PROX1 type 2 diabetes mellitus risk variant is associated with visceral fat accumulation and alterations in postprandial glucose and lipid metabolism. *Genes Nutr.* **2015**, *10*, 4. [CrossRef]

21. Maliszewska, K.; Adamska-Patruno, E.; Goscik, J.; Lipinska, D.; Citko, A.; Krahel, A.; Miniewska, K.; Fiedorczuk, J.; Moroz, M.; Gorska, M.; et al. The Role of Muscle Decline in Type 2 Diabetes Development: A 5-Year Prospective Observational Cohort Study. *Nutrients* **2019**, *11*, 834. [CrossRef]

22. Adamska-Patruno, E.; Goscik, J.; Czajkowski, P.; Maliszewska, K.; Golonko, A.; Wawrusiewicz-Kurylonek, N.; Citko, A.; Waszczeniuk, M.; Kretowski, A.; Gorska, M.; et al. The MC4R genetic variants are associated with lower visceral fat accumulation and higher postprandial relative increase in carbohydrate utilization in humans. *Eur. J. Nutr.* **2019**, *58*, 2929–2941. [CrossRef] [PubMed]

23. Hagstromer, M.; Oja, P.; Sjostrom, M. The International Physical Activity Questionnaire (IPAQ): A study of concurrent and construct validity. *Public Health Nutr.* **2006**, *9*, 755–762. [CrossRef]

24. R Core Team (2020). R: A language and Environment for Statistical Computing. R Foundation for Statistical Computing, Vienna, Austria. Available online: https://www.R-project.org/ (accessed on 22 July 2020).

25. Peng, S.; Zhu, Y.; Xu, F.; Ren, X.; Li, X.; Lai, M. *FTO* gene polymorphisms and obesity risk: A meta-analysis. *BMC Med.* **2011**, *9*, 71. [CrossRef] [PubMed]

26. Frayling, T.M.; Timpson, N.J.; Weedon, M.N.; Zeggini, E.; Freathy, R.M.; Lindgren, C.M.; Perry, J.R.B.; Elliott, K.S.; Lango, H.; Rayner, N.W.; et al. A common variant in the *FTO* gene is associated with body mass index and predisposes to childhood and adult obesity. *Science* **2007**, *316*, 889–894. [CrossRef]

27. Merritt, D.C.; Jamnik, J.; El-Sohemy, A. *FTO* genotype, dietary protein intake, and body weight in a multiethnic population of young adults: A cross-sectional study. *Genes Nutr.* **2018**, *13*, 4. [CrossRef]

28. Wing, M.R.; Ziegler, J.; Langefeld, C.D.; Ng, M.C.Y.; Haffner, S.M.; Norris, J.M.; Goodarzi, M.O.; Bowden, D.W. Analysis of *FTO* gene variants with measures of obesity and glucose homeostasis in the IRAS Family Study. *Hum. Genet.* **2009**, *125*, 615–626. [CrossRef]

29. Hosseini-Esfahani, F.; Koochakpoor, G.; Daneshpour, M.S.; Sedaghati-Khayat, B.; Mirmiran, P.; Azizi, F. Mediterranean Dietary Pattern Adherence Modify the Association between *FTO* Genetic Variations and Obesity Phenotypes. *Nutrients* **2017**, *9*, 1064. [CrossRef] [PubMed]

30. Chauhan, G.; Tabassum, R.; Mahajan, A.; Dwivedi, O.P.; Mahendran, Y.; Kaur, I.; Nigam, S.; Dubey, H.; Varma, B.; Madhu, S.V.; et al. Common variants of *FTO* and the risk of obesity and type 2 diabetes in Indians. *J. Hum. Genet.* **2011**, *56*, 720–726. [CrossRef] [PubMed]

31. Graff, M.; Gordon-Larsen, P.; Lim, U.; Fowke, J.H.; Love, S.A.; Fesinmeyer, M.; Wilkens, L.R.; Vertilus, S.; Ritchie, M.D.; Prentice, R.L.; et al. The influence of obesity-related single nucleotide polymorphisms on BMI across the life course: The PAGE study. *Diabetes* **2013**, *62*, 1763–1767. [CrossRef] [PubMed]

32. Wu, J.; Xu, J.; Zhang, Z.; Ren, J.; Li, Y.; Wang, J.; Cao, Y.; Rong, F.; Zhao, R.; Huang, X.; et al. Association of *FTO* polymorphisms with obesity and metabolic parameters in Han Chinese adolescents. *PLoS ONE* **2014**, *9*, e98984. [CrossRef] [PubMed]

33. Xiao, S.; Zeng, X.; Quan, L.; Zhu, J. Correlation between polymorphism of *FTO* gene and type 2 diabetes mellitus in Uygur people from northwest China. *Int. J. Clin. Exp. Med.* **2015**, *8*, 9744–9750.

34. Bego, T.; Causevic, A.; Dujic, T.; Malenica, M.; Velija-Asimi, Z.; Prnjavorac, B.; Marc, J.; Nekvindová, J.; Palička, V.; Semiz, S. Association of *FTO* Gene Variant (rs8050136) with Type 2 Diabetes and Markers of Obesity, Glycaemic Control and Inflammation. *J. Med. Biochem.* **2019**, *38*, 153–163. [CrossRef] [PubMed]

35. Fernandez, M.L.; West, K.L. Mechanisms by which dietary fatty acids modulate plasma lipids. *J. Nutr.* **2005**, *135*, 2075–2078. [CrossRef] [PubMed]

36. Park, S.L.; Cheng, I.; Pendergrass, S.A.; Kucharska-Newton, A.M.; Lim, U.; Ambite, J.L.; Caberto, C.P.; Monroe, K.R.; Schumacher, F.; Hindorff, L.A.; et al. Association of the *FTO* obesity risk variant rs8050136 with percentage of energy intake from fat in multiple racial/ethnic populations: The PAGE study. *Am. J. Epidemiol.* **2013**, *178*, 780–790. [CrossRef] [PubMed]

37. Wing, M.R.; Ziegler, J.M.; Langefeld, C.D.; Roh, B.H.; Palmer, N.D.; Mayer-Davis, E.J.; Rewers, M.J.; Haffner, S.M.; E Wagenknecht, L.; Bowden, D.W. Analysis of *FTO* gene variants with obesity and glucose homeostasis measures in the multiethnic Insulin Resistance Atherosclerosis Study cohort. *Int. J. Obes.* **2011**, *35*, 1173–1182. [CrossRef]

38. Zhao, J.; Bradfield, J.P.; Li, M.; Wang, K.; Zhang, H.; Kim, C.E.; Annaiah, K.; Glessner, J.T.; Thomas, K.; Garris, M.; et al. The role of obesity-associated loci identified in genome-wide association studies in the determination of pediatric BMI. *Obesity* **2009**, *17*, 2254–2257. [CrossRef]

39. Dai, J.; Ying, P.; Shi, D.; Hou, H.; Sun, Y.; Xu, Z.; Chen, D.; Zhang, G.; Ni, M.; Teng, H.; et al. *FTO* variant is not associated with osteoarthritis in the Chinese Han population: Replication study for a genome-wide association study identified risk loci. *J. Orthop. Surg. Res.* **2018**, *13*, 65. [CrossRef]

40. Cecil, J.E.; Tavendale, R.; Watt, P.; Hetherington, M.M.; Palmer, C.N. An obesity-associated *FTO* gene variant and increased energy intake in children. *N. Engl. J. Med.* **2008**, *359*, 2558–2566. [CrossRef]

41. Lappalainen, T.; Lindstrom, J.; Paananen, J.; Eriksson, J.G.; Karhunen, L.; Tuomilehto, J.; Uusitupa, M.; Jääskeläinen, T. Association of the fat mass and obesity-associated (*FTO*) gene variant (rs9939609) with dietary intake in the Finnish Diabetes Prevention Study. *Br. J. Nutr.* **2012**, *108*, 1859–1865. [CrossRef]

42. Ranzenhofer, L.; Mayer, L.E.; Davis, H.A.; Mielke-Maday, H.K.; McInerney, H.; Korn, R.; Gupta, N.; Brown, A.J.; Schebendach, J.; Tanofsky-Kraff, M.; et al. The *FTO* Gene and Measured Food Intake in 5- to 10-Year-Old Children Without Obesity. *Obesity* **2019**, *27*, 1023–1029. [CrossRef]

43. Bauer, F.; Elbers, C.C.; Adan, R.A.; Loos, R.J.; Onland-Moret, N.C.; E Grobbee, D.; Van Vliet-Ostaptchouk, J.V.; Wijmenga, C.; Van Der Schouw, Y.T. Obesity genes identified in genome-wide association studies are associated with adiposity measures and potentially with nutrient-specific food preference. *Am. J. Clin. Nutr.* **2009**, *90*, 951–959. [CrossRef] [PubMed]

44. McCaffery, J.M.; Papandonatos, G.D.; Peter, I.; Huggins, G.S.; A Raynor, H.; Delahanty, L.M.; Cheskin, L.J.; Balasubramanyam, A.; E Wagenknecht, L.; Wing, R.R.; et al. Obesity susceptibility loci and dietary intake in the Look AHEAD Trial. *Am. J. Clin. Nutr.* **2012**, *95*, 1477–1486. [CrossRef] [PubMed]

45. Heitmann, B.L.; Lissner, L. Dietary underreporting by obese individuals–is it specific or non-specific? *BMJ* **1995**, *311*, 986–989. [CrossRef] [PubMed]

Publisher's Note: MDPI stays neutral with regard to jurisdictional claims in published maps and institutional affiliations.

Article

Polygenetic-Risk Scores for A Glaucoma Risk Interact with Blood Pressure, Glucose Control, and Carbohydrate Intake

Donghyun Jee [1], ShaoKai Huang [2], Suna Kang [2] and Sunmin Park [2,*]

[1]   Division of Vitreous and Retina, Department of Ophthalmology, St. Vincent's Hospital, College of Medicine, The Catholic University of Korea, Suwon 16247, Korea; doj087@mail.harvard.edu

[2]   Food and Nutrition, Obesity/Diabetes Research Center, Hoseo University, Asan 31499, Korea; huangsk0606@gmail.com (S.H.); roypower003@naver.com (S.K.)

*   Correspondence: smpark@hoseo.edu; Tel.: +82-41-540-5345; Fax: +82-41-548-0670

Received: 18 September 2020; Accepted: 21 October 2020; Published: 26 October 2020

Abstract: Glaucoma, a leading cause of blindness, has multifactorial causes, including environmental and genetic factors. We evaluated genetic risk factors of glaucoma with gene-gene interaction and explored modifications of genetic risk with gene-lifestyles interaction in adults >40 years. The present study included 377 subjects with glaucoma and 47,820 subjects without glaucoma in a large-scale hospital-based cohort study from 2004 to 2013. The presence of glaucoma was evaluated by a diagnostic questionnaire evaluated by a doctor. The genome-wide association study was performed to identify genetic variants associated with glaucoma risk. Food intake was assessed using a semiquantitative food frequency questionnaire. We performed generalized multifactor dimensionality reduction analysis to construct polygenetic-risk score (PRS) and explored gene × nutrient interaction. PRS of the best model included LIM-domain binding protein-2 (*LDB2*) rs3763969, cyclin-dependent kinase inhibitor 2B (*CDKN2B*) rs523096, *ABO* rs2073823, phosphodiesterase-3A (*PDE3A*) rs12314390, and cadherin 13 (*CDH13*) rs12449180. Glaucoma risk in the high-PRS group was 3.02 times that in the low-PRS group after adjusting for confounding variables. For those with low serum glucose levels (<126 mg/dL), but not for those with high serum glucose levels, glaucoma risk in the high-PRS group was 3.16 times that in the low-PRS group. In those with high carbohydrate intakes (≥70%), but not in those with low carbohydrate intakes, glaucoma risk was 3.74 times higher in the high-PRS group than in the low-PRS group. The glaucoma risk was 3.87 times higher in the high-PRS group than in the low-PRS group only in a low balanced diet intake. In conclusion, glaucoma risk increased by three-fold in adults with a high PRS, and it can be reduced by good control of serum glucose concentrations and blood pressure (BP) with a balanced diet intake. These results can be applied to precision nutrition to reduce glaucoma risk.

Keywords: glaucoma; polygenetic-risk scores; gene-gene interaction; carbohydrate intake; gene-nutrient interaction; precision medicine

---

1. Introduction

Glaucoma is the second most common cause of irreversible vision loss worldwide [1]. Glaucoma is the result of optic neuropathies primarily caused by raised ocular pressure, and these neuropathies promote the progressive degeneration of retinal ganglion cells to induce visual loss [1]. Although the major risk factor of glaucoma is increased intraocular pressure, some glaucoma patients have recently been reported to have normal ocular pressure [2]. The risk factors of glaucoma include intraocular pressure, ocular perfusion pressure, ocular blood flow, myopia, central corneal thickness, and optic disc hemorrhages [3]. Increased intraocular pressure is considered to be associated with serum glucose concentrations and blood pressure, but these relations remain controversial. Accumulating evidence has shown that increased intraocular pressure is associated with T-cell-mediated autoimmunity [2]. Low-grade inflammation may initially induce an adaptive reaction of the retina to lead to excessive glial reactions that increase adaptive immune responses. Furthermore, it has been suggested that activated adaptive immunity contributes to progressive neural damage and the development of glaucoma [2].

Elevated intraocular pressure is associated with insulin resistance, which is a common feature of metabolic syndrome [4,5]. However, the components of metabolic syndrome exhibit different associations with intraocular pressure. Of these components, blood pressure is a well-known risk factor of increased intraocular pressure. High fasting glucose levels and obesity are also associated with the increment of intraocular pressure. On the other hand, the association between lipid profiles and intraocular pressure has not been studied. However, in East Asians, few studies have addressed the relationship between glaucoma and metabolic syndrome. Diabetes is an independent risk factor of glaucoma [6,7], which is linked to the neurodegeneration caused by central insulin resistance [8]. Therefore, associations between glaucoma risk and systemic insulin resistance and metabolic syndrome remain unclear due to inconsistencies between reported results.

Genetic and environmental factors influence the incidence of glaucoma. In particular, etiologic studies have demonstrated its incidence is dependent on ethnicity, which implies genetic variations play a critical role in glaucoma. More specifically, primary open-angle glaucoma is associated with inflammation-related genes such as toll-like receptor 4 (*TLR4*) rs4986791 and rs2149356 [9] and interleukin (IL)-10 rs1800871 and rs1800872 [10]. Glaucoma has been reported to be associated with cell-cell adhesion via the activation of transforming growth factor-β signaling [11]. Moreover, since open-angle glaucoma has been linked with insulin resistance, its incidence may have a positive association with insulin/insulin-like growth factor signaling. Some studies have been conducted to determine the effects of lifestyles, including nutrient intake, on glaucoma risk, and vitamin B1 and C and retinol intakes have been reported to be negatively associated with glaucoma risk, whereas magnesium intake is positively associated [12,13]. Interestingly, polyunsaturated fat intake with a higher ratio of *n*-3 and *n*-6 fatty acids has been shown to increase the risk of primary open-angle glaucoma and is suggested to be associated with the production of endogenous prostaglandin F2-α [14]. However, reports on the relation between polyunsaturated fatty acid intake and glaucoma are inconsistent, and smoking cessation, moderate aerobic exercise, a balanced diet, and coffee and tea intake have been shown to protect against the development and progression of primary open-angle glaucoma [15]. These findings demonstrate that environmental factors can reduce the risk of glaucoma. Furthermore, it is evident that environmental factors interact with genetic factors and that these gene-environmental interactions have potential utility in precision medicine.

Here, we aimed to explore the polygenetic variants of glaucoma risk related to inflammation and insulin resistance concerning gene-gene interactions and polygenetic variant and lifestyle interactions in the middle-aged and elderly individuals.

2. Materials and Methods

2.1. Participants Recruitment

During 2004–2013, a total of 20,274 men and 38,371 women (total 58,645) aged >40 years voluntarily participated in hospital-based city cohort studies (the Korean Genome and Epidemiology Study (KoGES)) organized by the Korean Center for Disease and Control. All procedures of the KoGES were performed according to the Declaration of Helsinki, and they were approved by the Institutional Review Board of the Korean National Institute of Health for the KoGES (KBP-2015-055) and Hoseo University (1041231-150811-HR-034-01). All subjects that participated in KoGES provided written informed consent.

2.2. Criteria of Glaucoma

Participants were asked whether they had received a diagnosis of glaucoma from a physician, and those that answered affirmatively were considered to have primary glaucoma. A total of 10,448 participants did not answer the question about the glaucoma diagnosis, and they were eliminated in the analysis.

2.3. Anthropometric and Biochemical Measurements

Information on age, education, income, smoking history and alcohol consumption, and physical activity was collected during a health interview. Education level was divided into three groups: less than high school, high school, and college or more. Household income (USD/month) was categorized as very low (<$1000), low ($1000–$2000), intermediate ($2000–$4000), and high (>$4000) [16]. Smoking status was divided into three categories: current smoker, past smoker, and never-smoker [16]. According to average daily alcohol consumption, the participants were categorized into nondrinker, light drinker (0–1 g), moderate drinker (1–20 g), and heavy drinker (>20 g) (Table 1) [16]. Dairy product consumptions (milk, yogurt, and cheese) were also obtained.

Table 1. Socioeconomic and lifestyle characteristics of the participants according to glaucoma presence.

Parameters Related to Glaucoma	Non-Glaucoma (n = 47,820)	Glaucoma (n = 377)	Adjusted OR for Glaucoma Risk (OR, 95% CI)
Age (years) [1]	53.7 ± 5.5	58.2 ± 5.4 ***	3.325 (2.623–4.213) ***
Gender (number, male %)	16,193 (33.9)	171 (45.4) ***	1.797 (1.430–2.257) ***
Education (number, %)			
<High school	6417 (18.5)	66 (25.7) **	1
High school	7689 (22.1)	60 (23.4)	1.060 (0.724–1.552)
College more	20,619 (59.4)	131 (51.0)	1.169 (0.821–1.665)
Income (number, %)			
<$1000/m	4508 (9.94)	66 (18.5) ***	1
$1000–$2000	9722 (21.4)	78 (21.9)	0.699 (0.496–0.986)
$2000–$4000	20,047 (44.2)	133 (37.4)	0.756 (0.542–1.053)
>$4000	11,084 (24.4)	79 (22.2)	0.895 (0.607–1.320)
Exercise (number, %)			
No	21,531 (45.2)	145 (38.7) *	1
Yes	26,144 (54.8)	230 (61.3)	1.216 (0.971–1.523)
Alcohol intake (number, %)			
No	27,131 (56.7)	238 (63.1) *	1
Mild drink (0–20 g)	1048 (2.19)	5 (1.33)	0.656 (0.269–1.600)
Moderate drink (≥20 g)	19,641 (41.1)	134 (35.5)	0.789 (0.612–1.016)
Coffee intake (number, %)			
Low (<3 cups/week)	18,037 (37.7)	156 (41.4)	1
Medium (3–16 cups/week)	29,329 (61.3)	218 (57.8)	1.030 (0.767–1.383)
High (≥16 cups/week)	454 (0.95)	3 (0.80)	0.903 (0.704–1.157)
Energy intake [2] (kcal)	1743 ± 531	1719 ± 516	0.848 (0.674–1.067)
CHO percent intake [3]	71.7 ± 20.8	71.6 ± 20.0	1.037 (0.816–1.317)
Fat percent intake [4]	13.9 ± 8.7	14.1 ± 8.0	1.137 (0.901–1.435)
Protein percent intake [5]	13.4 ± 5.8	13.3 ± 5.6	0.919 (0.699–1.209)

The values represent means ± standard deviations or number of the subjects (percentage of each group). The cutoff points of the parameters were as follows: [1] <55 years old, [2] <estimated energy intake, [3] <70 energy % of carbohydrate (CHO), [4] <15 energy % fat, and [5] <15 energy % protein. Adjusted odds ratio (ORs) after adjusting for covariates including gender, age, residence area, surveyed year, body mass index (BMI), smoking, alcohol, education, job, income, energy intake, and arthritis and dermatitis medicine intake in logistic regression models. * Significant differences by cataract at $p < 0.05$, ** at $p < 0.01$, *** $p < 0.001$.

Body weight, height, and waist circumference were measured using a standardized procedure [17]. Body mass index (BMI) was calculated by dividing weight in kilograms by the height (in meters) squared. Blood was collected after an overnight fast, and plasma and serum samples were used for biochemical measurements [17]. Fasting serum glucose and blood hemoglobin A1c (HbA1c; glycated hemoglobin) concentrations were determined using a Hitachi 7600 Automatic Analyzer (Hitachi, Tokyo, Japan). Blood pressure was measured on the right arms at the heart level in a sitting position.

2.4. Assessment of Food and Nutrient Intakes Using a Semiquantitative Food Frequency Questionnaire (SQFFQ)

Dietary intakes were estimated using an SQFFQ developed and validated for the KoGES [18]. This questionnaire requested information about the consumption of specified food items. The participants completed the SQFFQ. Based on the consumption of 103 food items, daily nutrient intakes were calculated using data from a food intake nutrient database maintained by the Korean Nutrition Society and the Computer-Aided Nutritional Analysis Program (CAN Pro) 3.0 (Seoul, Korea) [18].

2.5. Genotyping and Quality Control

Genotype data were provided by the Center for Genome Science, Korea National Institute of Health. Genomic DNA was extracted from whole blood, and genotypes were determined using the Affymetrix Korean Chip (Affymetrix, Santa Clara, CA, USA) made available for scientific studies. This chip has been previously used to study Korean genetic variants and included disease-related single-nucleotide polymorphism (SNP) [19]. Genotyping accuracy of the SNP results was examined by Bayesian robust linear modeling using the Mahalanobis distance (BRLMM) genotyping algorithm [20]. The genotype results met genotyping accuracy of ≥98%, a missing genotype call rate of <4%, heterozygosity of <30%, and show no gender bias. Genetic variants used for further analysis met the Hardy-Weinberg equilibrium (HWE) ($p > 0.05$).

2.6. Identification of the Best Model for Gene-Gene Interactions by Generalized Multifactor Dimensionality Reduction (GMDR) Method among the Genetic Variants Selected by Logistic Regression

A flow chart of the procedure used to calculate polygenetic-risk scores (PRSs) of glaucoma risk is presented in Figure 1.

Participants were categorized as having glaucoma ($n = 377$) or not ($n = 47,820$); 10,448 participants did not answer the question, and they were excluded. Logistic regression was performed to identify genetic variants associated with glaucoma risk ($p < 0.0001$). Corresponding genes were identified using scandb.org. Genes that interacted with insulin/insulin like-growth factor-1 (IGF-1) signaling related to insulin resistance and inflammation that influence glaucoma risk were selected using genemania.org. The 43 selected SNPs were subjected to generalized multifactor dimensionality reduction (GMDR) analysis to explore gene-gene interactions associated with glaucoma risk. Linkage disequilibrium (LD) analyses were performed on selected genetic variants in the same chromosome using Haploview 4.2 in PLINK (Boston, MA, USA). SNPs showing high LDs ($r^2 \geq 0.4$) were excluded as they provided similar information on glaucoma risk. The best model for predicting gene-gene interactions that influence glaucoma risk was selected by trained balanced accuracy (TRBA), testing balanced accuracy (TEBA), and cross-validation consistency (CVC) using GMDR. The final 10 potential genetic variants in the same chromosome included in the best model did not have a strong correlation in LD ($r^2 < 0.4$). PRSs for the best model were calculated by summing the number of risk alleles for each selected SNP in the best gene-gene interaction model. PRSs were divided into three categories by tertile; a high PRS indicated the individual concerned had a high number of risk alleles.

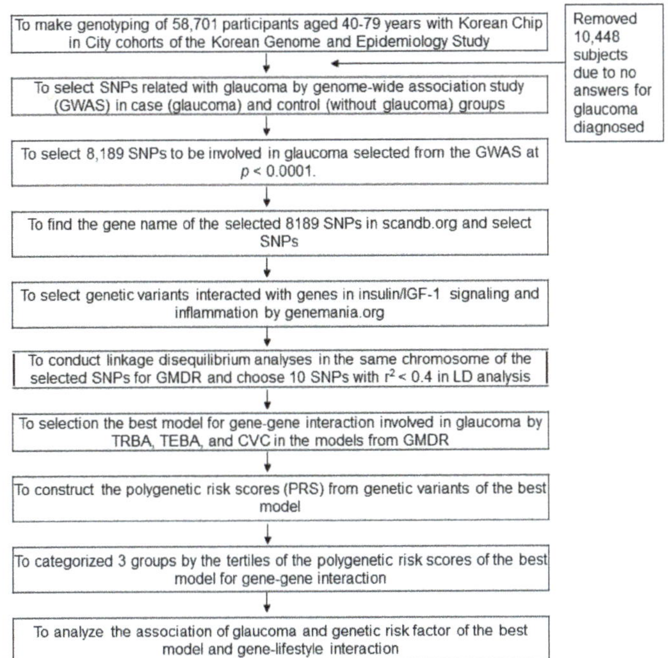

Figure 1. The flow chart to make polygenetic-risk scores to influence glaucoma risk. GMDR, generalized multifactor dimensionality reduction; TRBA, trained balanced accuracy; TEBA, testing balanced accuracy; CVC, cross-validation consistency.

2.7. Statistical Analyses

Statistical analysis was performed using PLINK v. 1.9. (http://zzz.bwh.harvard.edu/plink/; Boston, MA, USA) and SAS (v. 9.3.; SAS Institute, Cary, NC, USA). The best gene-gene interaction model was selected using a GMDR program and the signed-rank test using a *p*-value of <0.05 with TRBA and TEBA with or without adjusting for covariates of age, gender, education, income level, and body mass index [21]. Ten-fold cross-validation was used to check CVC since the sample size was greater than 1000 [21]. From the best model determined by GMDR analysis, the number of the risk allele in each SNP was counted in the selected best model [22]. For example, when the G allele had a positive association with an increased risk of glaucoma, TT, GT, and GG were assigned scores of 0, 1, and 2, respectively. PRSs were calculated by summing risk allele scores for each SNP included in the PRS for the best model.

The descriptive statistics of categorical variables (e.g., gender and lifestyle) were calculated by PRS tertile and designated as low, middle, or high. Frequency distributions of classification variables were analyzed using the chi-squared test. Means and standard deviations were calculated for continuous variables according to the PRS categories, and significant differences between groups were determined by one-way analysis of variance (ANOVA) with or without adjustment for covariates. Tukey's test was used to performed multiple group comparisons.

The association between PRSs and glaucoma risk was investigated using logistic regression analysis after adjusting for covariates. Odds ratios (ORs) and 95% confidence intervals (CI) were analyzed using logistic regression analysis with low-PRS as a reference, after either adjusting for age, gender, residence area, survey year, BMI, education, job, and income, or by adjusting age, gender, residence area, survey year, smoking status, alcohol intake, education, job, income, energy intake, physical activity, hypertension, milk intake, percent fat intake, percent carbohydrate intake, and arthritis and dermatitis medicine intake. To determine the interaction between PRSs and lifestyles and dietary intakes, participants were categorized into higher or lower intake groups using the criteria defined by 50th percentiles of each variable. A multivariate interaction model was used to evaluate interactions between PRSs and lifestyles and dietary intake after adjustment for covariates. *p*-values of ≤ 0.05 were considered statistically significant.

3. Results

3.1. General Characteristics of Participants with Glaucoma

Participants with glaucoma ($n = 377$) were older than those without glaucoma ($n = 47,820$). Participants aged ≥ 55 years old had a 3.3-fold greater risk of glaucoma than those aged <55 years old (Table 1), and men had a 1.8-fold higher risk than women (Table 1). Participants with a college degree and a family income of >\$2000/month had a lower prevalence of glaucoma than those with less education and a lower income. However, education and income status were not significantly associated with glaucoma risk after adjusting for covariates (gender, age, residence area, surveyed year, BMI, smoking, alcohol, education, job, income, energy intake, and arthritis and dermatitis medicine intake) (Table 1). Furthermore, daily regular exercise and alcohol intake had no significant association with glaucoma risk after adjusting covariates, and coffee, energy, and nutrient intakes were similar in those with or without glaucoma (Table 1).

3.2. Association between Glaucoma Risk and Metabolic Syndrome

Interestingly, participants with glaucoma had a 1.4-fold higher risk of metabolic syndrome than those without glaucoma (Table 2). Regarding components of metabolic syndrome, fasting serum glucose concentrations were significantly associated with glaucoma risk, but blood pressure, serum lipid, and waist circumference were not (Table 2). Fasting serum glucose concentration and glycated hemoglobin (HbA1C) were 1.5- and 1.7-fold higher, respectively, in those with glaucoma. Moreover, serum C-reactive protein-1 concentrations (>0.8 mg/dL) increased glaucoma risk by 2.1-fold in those with glaucoma (Table 2).

Table 2. The association of glaucoma risk and metabolic syndrome.

Components for Metabolic Syndrome	Non-Glaucoma ($n = 47,820$)	Glaucoma ($n = 377$)	Adjusted OR for Glaucoma Risk (OR, 95% CI)
Metabolic syndrome [1] (number, %)	6673 (14.0)	81 (21.5) ***	1.361 (1.032–1.793) [#]
BMI [2] (kg/m^2)	23.9 ± 2.8	23.8 ± 2.9	0.927 (0.736–1.167)
Waist circumferences [3] (cm)	80.6 ± 8.7	80.8 ± 8.4	0.971 (0.751–1.255)
Serum glucose [4] (mg/dL)	95.0 ± 20.2	100.1 ± 26.5 ***	1.539 (1.182–2.003) [##]
Blood HbA1c [5] (%)	5.7 ± 0.7	5.9 ± 0.9 ***	1.663 (1.170–2.364) [##]
Serum total cholesterol [6] (mg/dL)	197 ± 36	195 ± 38	1.108 (0.867–1.417)
Serum HDL [7] (mg/dL)	54.4 ± 13.3	53.7 ± 12.8	1.181 (0.929–1.501)
Serum TG [8] (mg/dL)	125 ± 86	119 ± 73	1.134 (0.901–1.428)
Serum BP [9] (number, %)	11,627 (24.3)	138 (36.6) ***	1.225 (0.968–1.551)
Serum CRP-1 [10] (mg/dL)	0.14 ± 0.38	0.18 ± 0.45	2.066 (1.221–3.496) [##]

The values represent adjusted means ± standard deviations. Adjusted and means odds ratio (ORs) after adjusting for covariates including gender, age, residence area, surveyed year, body mass index (BMI), smoking, alcohol, education, job, income, energy intake, and arthritis and dermatitis medicine intake in logistic regression models. The reference of cutoff points in the parameters were as follows: [1] no metabolic syndrome, [2] <25 kg/m^2 body mass index (BMI); [3] <90 and 85 cm waist circumferences for men and women, respectively; [4] <126 ml/dL fasting serum glucose plus diabetic drug intake; [5] <6.5% glycated hemoglobin (HbA1c) plus diabetic drug intake; [6] <230 mg/dL serum total cholesterol concentrations; [7] ≥40 and ≥50 mg/dL serum high density lipoprotein (HDL) cholesterol concentrations for men and women, respectively, and lipid-lowering drug; [8] <150 mg/dL serum triglyceride (TG) concentrations; [9] <130 mmHG systolic blood pressure (BP) and <90 mmHg diastolic BP and taking BP-lowering drug; [10] <0.8 mg/dL high-sensitive C-reactive protein (CRP-1). * Significantly different between the Non-glaucoma and Glucoma groups at *** $p < 0.001$. [#] Significantly different from the Non-glaucoma in multivariate logistic regression at $p < 0.05$, [##] at $p < 0.01$.

3.3. Selection of Genetic Variants Associated with Glaucoma Risk Using GMDR

Using the genetic variants selected for glaucoma risk by logistic regression, we used GMDR to investigate gene-gene interactions for genetic variants. Ten genetic variants were included in the GMDR analysis. The genetic characteristics of these 10 SNPs are presented in Table 3. Adjusted ORs of seven SNPs were significantly greater than 1 and those of three SNPs were between 0 and 1 (Table 3). All selected SNPs satisfied HWE criteria ($p > 0.05$).

Table 3. The characteristics of the ten genetic variants of genes related to glaucoma risk used for the generalized multifactor dimensionality reduction analysis.

CHR [1]	SNP [2]	Location	Mi [3]	OR [4]	p-Value for OR [5]	Genes	Feature	MAF [6]	HWE [7]
4	rs3763969	16648246	T	0.63 (0.48–0.83)	9.7.E–04	LDB2	intron	0.120	0.082
7	rs1852542	42096521	T	1.67 (1.24–2.25)	6.6.E–04	GLI3	intron	0.041	0.559
8	rs1020236	135543194	C	1.48 (1.19–1.83)	4.4.E–04	ZFAT	intron	0.093	0.732
9	rs523096	22019129	G	0.73 (0.57–0.93)	1.0.E–02	CDKN2B	intron	0.134	0.972
9	rs2073823	136132516	A	1.33 (1.13–1.56)	7.6.E–04	ABO	intron	0.215	0.941
12	rs12314390	20597977	T	1.70 (1.29–2.25)	1.8.E–04	PDE3A	intron	0.048	0.558
13	rs7335337	38221067	G	1.78 (1.34–2.37)	7.8.E–05	TRPC4	intron	0.041	0.230
15	rs1319859	99230263	G	1.32 (1.13–1.53)	3.5.E–04	IGF1R	intron	0.300	0.226
16	rs12449180	83547527	G	1.42 (1.18–1.69)	1.3.E–04	CDH13	intron	0.162	0.162
18	rs3902981	12658191	G	0.73 (0.62–0.87)	3.6.E–04	SPIRE1	near-gene-5	0.300	0.492

[1] Chromosome; [2] Single-nucleotide polymorphism; [3] Minor allele; [4] Odds ratio and lower and upper ends of 95% confidence interval; [5] p-value for OR after adjusting for age, gender, residence area, survey year, body mass index, daily energy intake, education, and income; [6] Minor allele frequency; [7] Hardy-Weinberg equilibrium.

Of the 10 SNPs, the best model for gene-gene interactions included five genetic variants, including LIM-domain binding protein 2 (*LDB2*) rs3763969, cyclin-dependent kinase inhibitor *2B* (*CDKN2B*) rs523096, *ABO* rs2073823, phosphodiesterase 3A (*PDE3A*) rs12314390, and cadherin 13 (CDH13) rs12449180 after adjusting for age and gender (adjustment 1) and age, gender, survey year, residence area, BMI (adjustment 2; Table 4). The model including these five SNPs had a TRBA of 0.6274 and a TEBA of 0.5927 after adjusting for age, gender, survey year, residence area, and BMI (adjustment 2; $p < 0.001$), and CVC was 10/10 (Table S1). The TRBA, TEBA, and CVC in model 1 with the adjusting for age and gender were similar to those in model 2 with adjustment of more covariates. In addition, the model including the 10 SNPs had 0.7603 and 0.5283 of TRBA and TEBA, respectively, and CVC was 10/10 ($p < 0.001$) after adjusting for covariates. The results indicated these 5 and 10 genetic variants exhibited gene-gene interactions that increased glaucoma risk. Since the model including five SNPs met the significant gene-gene interaction with fewer SNPs, the model with five SNPs was better for showing genetic impact for glaucoma risk.

Table 4. Adjusted odds ratios for glaucoma, age-related cataract, and metabolic syndrome according to the polygenetic risk scores (PRS) of the best model for gene-gene interaction after covariate adjustments.

Glaucoma-Related Diseases	Adjustment 1			Adjustment 2	
	Low-PRS ($n = 9245$)	Medium-PRS ($n = 31,227$)	High-PRS ($n = 6015$)	Medium-PRS ($n = 31,227$)	High-PRS ($n = 6015$)
Glaucoma	1	1.814 (1.280–2.573)	2.937 (1.965–4.389) ***	1.815 (1.213–2.715)	3.021 (1.898–4.809) ***
Cataract	1	0.871 (0.767–0.988)	0.935 (0.782–1.117)	0.898 (0.771–1.045)	0.983 (0.793–1.218)
Metabolic syndrome	1	0.992 (0.924–1.064)	1.004 (0.910–1.108)	0.984 (0.891–1.086)	1.037 (0.903–1.192)
Type 2 diabetes	1	0.879 (0.805–0.959) *	0.951 (0.841–1.075)	0.873 (0.799–0.954) *	0.930 (0.821–1.053)
Blood pressure	1	1.006 (0.949–1.067)	0.997 (0.919–1.082)	0.979 (0.908–1.054)	0.993 (0.894–1.103)

Values represent odd ratios and 95% confidence intervals. PRS was divided into three categories (0–3, 4–5, and >6) by tertiles as the low, medium, and high groups, respectively. Low-PRS was the reference for both model 1 and model 2. * Significantly different from low-PRS in logistic regression analysis at * $p < 0.05$, ** $p < 0.01$, *** $p < 0.001$. Adjustment 1: adjusted for age, gender, residence area, survey year, body mass index (BMI), education, job, and income. Adjustment 2: adjusted for age, gender, residence area, survey year, smoking, alcohol, education, job, income, energy, activity, hypertension, milk, fat percent intake, carbohydrate percent intake, and arthritis and atopic dermatitis medicine intake. * Significantly different from the Low-PRS in multivariate logistic regression at $p < 0.05$, *** at $p < 0.001$.

3.4. A Positive Association between PRSs and Glaucoma Risk

PRS for the best model was generated and used to quantify the genetic risk of glaucoma. A high-PRS increased glaucoma risk by 2.9- and 3.0-fold in models 1 and 2, respectively (Table 4). Adjustment 1 and 2 in the best model included different covariates: adjustment 1 included age, gender, residence area, survey year, BMI, education, job, and income as covariates, and adjustment 2 included age, gender, residence area, survey year, smoking, alcohol, education, job, income, energy, activity, hypertension, milk, fat percent intake, carbohydrate percent intake, and arthritis and atopic dermatitis medicine intake. However, a high-PRS had no association with cataract risk (Table 4). PRSs were not associated with the risk of metabolic syndrome or blood pressure in adjustment 1 or 2. Interestingly, a medium-PRS, but not a high-PRS, was negatively associated with type 2 diabetes since SNPs for the best model were selected from the insulin/IGF-signaling pathway. These results showed that a high-PRS had a positive association with glaucoma risk.

3.5. Interaction between PRSs and Lifestyle in Glaucoma Risk

Age showed an interaction with PRSs to glaucoma risk. In participants aged ≥55 years, a high-PRS had a much higher glaucoma risk than a low-PRS, but this was not observed in participants aged <55 years (Table 5). The prevalence of glaucoma was much greater for those with a high-PRS than for those with a low-PRS only in participants ≥55 years old (Figure 2A). Blood pressure ($p = 0.011$) and hyperglycemia ($p = 0.46$) interacted with PRS (Table 5). A high-PRS was associated with greater glaucoma risk than a low-PRS, particularly in participants with low blood pressure and normoglycemia (Table 5). Glaucoma incidence was greater in a high-PRS than in a low-PRS in all participants, but the incidence was much greater in the participants with hyperglycemia and hypertension (Figure 2B,C).

Table 5. Adjusted odds ratios for the glaucoma risk by polygenetic risk scores (PRS) of the best model and gene-environmental interactions after covariate adjustments.

Parameters for Glaucoma Risk	Low-PRS (n = 14,420)	Medium-PRS (n = 21,641)	High-PRS (n = 4201)	Gene-Nutrient Interaction p-Value
Less aged people	1	1.575 (0.737–3.364)	2.577 (1.063–6.246) [*]	0.0092
More aged people [1]		1.907 (1.185–3.068)	3.187 (1.844–5.506) [***]	
Low BP	1	1.695 (1.024–2.805)	3.659 (2.092–6.399) [***]	0.0106
High BP [2]		1.999 (1.021–3.915)	1.751 (0.723–4.242)	
Low serum glucose	1	1.791 (1.149–2.719)	3.165 (1.907–5.251) [***]	0.0460
High serum glucose [3]		2.020 (0.770–5.297)	2.195 (0.656–7.337)	
Low energy intake	1	2.456(1.404–4.293)	3.959 (2.113–7.417) [***]	0.1548
High energy intake [4]		1.210(0.669–2.191)	1.432 (0.526–3.894)[*]	
Low CHO intake	1	1.685 (0.855–3.321)	1.748 (0.722–4.236)	0.0083
High CHO intake [5]		1.892 (1.146–3.122)	3.741 (2.139–6.544) [***]	
Low protein intake	1	2.134(1.316–3.459)	3.370(1.909–5.950) [***]	0.2047
High protein intake [6]		1.171(0.554–2.472)	3.229(1.885–5.532)	
Low fat intake	1	1.743(1.053–2.887)	3.814(2.589–5.617) [***]	0.1850
High fat intake [7]		1.923(0.984–3.760)	2.440(1.076–5.536)	
Low Na intake	1	1.725 (1.081–2.752)	2.751 (1.594–4.749) [***]	0.7924
High Na intake [8]		2.030 (0.914–4.509)	3.780 (1.542–9.266) [**]	
Low BD intake	1	2.244 (1.410–3.896)	3.872(2.184–6.863) [***]	0.0464
High BD intake [9]		0.980 (0.495–1.940)	1.700 (0.730–3.956)	
Low NBR intake	1	1.534 (0.946–2.488)	3.325 (1.934–5.717) [***]	0.1151
High NBR intake [9]		2.524 (1.207–5.281) [*]	2.263 (0.903–5.672)	
Low RD intake	1	1.794 (1.085–2.965)	3.477 (1.970–6.139) [***]	0.4685
High RD intake [9]		1.855 (0.945–3.641)	2.262 (0.997–5.132)	

Values represent odds ratios and 95% confidence intervals. PRS was divided into three categories (0–3, 4–5, and >6) by tertiles as the low, medium, and high groups of the best model of generalized multifactor dimensionality reduction (GMDR). The reference of cutoff points in each parameter was as following: [1] <55 years old, [2] <130 mmHg SBP and ≥90 mmHg DBP, [3] <126 mg/dL serum glucose concentrations plus hypoglycemic medicine, [4] < estimated energy intake, [5] <70% carbohydrate (CHO), [6] <15% protein, [7] <15% fat, [8] <1600 mg/1000 kcal Na, and [9] <67 percentile of dietary patterns. BD, Balanced diet; NBR, noodle, bread, and red meat diet; RD, rice-rich diet. Multivariate regression models include the corresponding main effects, interaction terms of gene and main effects (energy and nutrient intake), and potential confounders such as BMI, gender, age, smoking, alcohol, education, job, income, energy, physical activity, hypertension, milk, fat percent intake, carbohydrate percent intake, and arthritis and dermatitis medicine intake. Reference was the low-PRS. [*] Significantly different from low-PRS in logistic regression analysis at [*] $p < 0.05$, [**] $p < 0.01$, [***] $p < 0.001$.

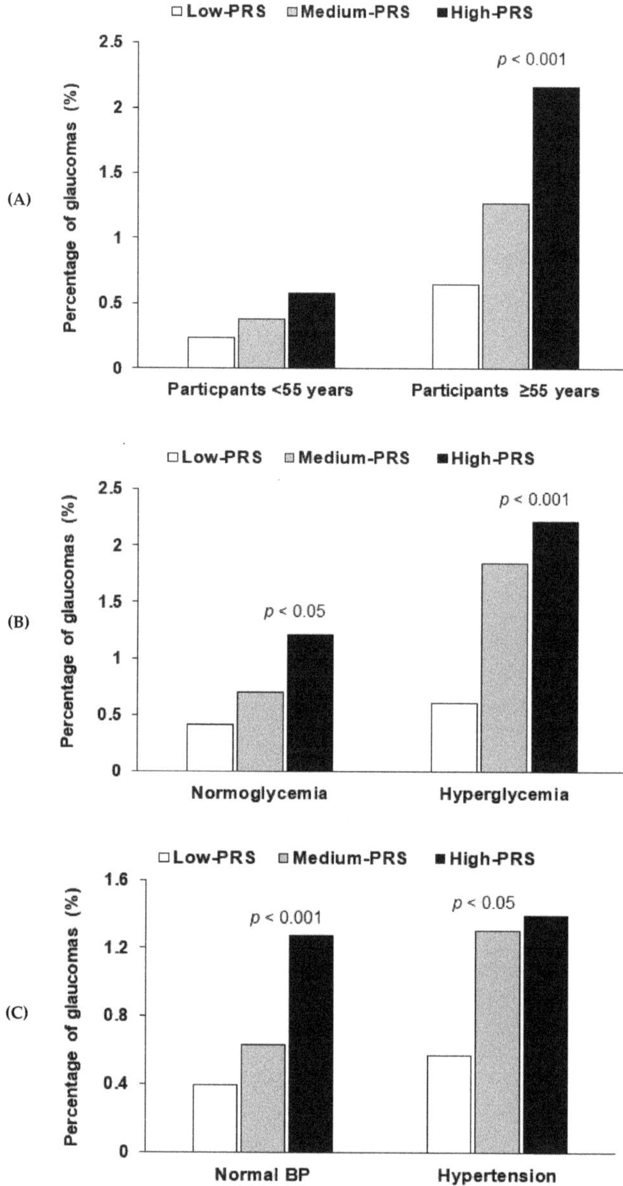

Figure 2. The frequency distribution of glaucoma in the three groups of polygenetic-risk scores (PRS) of the best model including *LDB2* rs3763969, *CDKN2B* rs523096, *ABO* rs2073823, *PDE3A* rs12314390, and *CDH13* rs12449180 according to the metabolic status. (**A**) According to age (cutoff point: 55 years old). (**B**) According to serum glucose concentrations (cutoff point: 126 mg/dL serum glucose concentrations). (**C**) According to the blood pressure (cutoff point: 130 mmHg for systolic blood pressure (SBP) and 90 mmHg for diastolic blood pressure (DBP)). PRS was calculated by the summation of each genetic-risk score of the best model, and PRS was categorized into three groups by the tertiles (Low-PRS, Medium-PRS, and High-PRS). BP, blood pressure.

For energy and nutrient intakes, only carbohydrate intake showed an interaction with PRS in terms of glaucoma risk ($p = 0.008$; Table 5). No interact was found for energy intake ($p = 0.155$), protein ($p = 0.205$), fat ($p = 0.185$), or Na intake ($p = 0.792$). For participants with high carbohydrate intake, glaucoma risk was 3.7-fold higher for those with a high-PRS than for those with a low-PRS (Table 5), and in those with high carbohydrate intakes, the prevalence of glaucoma was much higher among those with a high-PRS (Figure 3A). There was no interaction of PRS with coffee ($p = 0.687$) and alcohol intake ($p = 0.457$). No interaction was evident between daily regular exercise and PRSs for glaucoma risk ($p = 0.095$; Table 5).

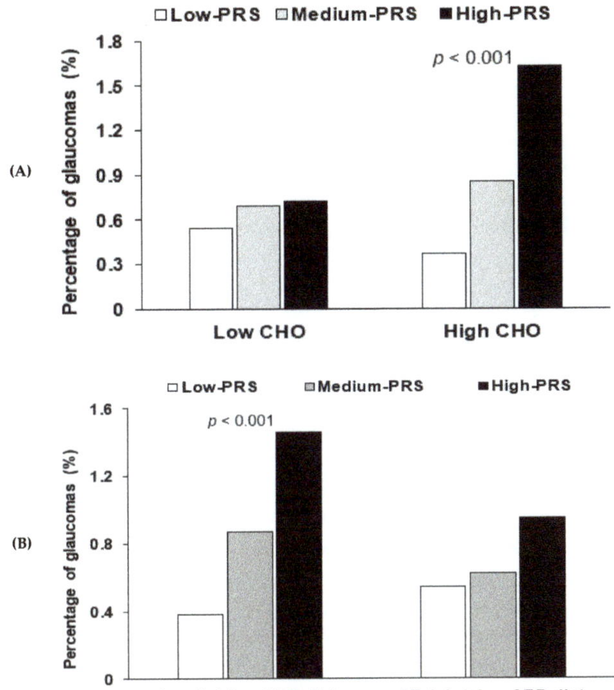

Figure 3. The frequency distribution of glaucoma in the three groups of polygenetic-risk scores (PRS) of the best model including *LDB2* rs3763969, *CDKN2B* rs523096, *ABO* rs2073823, *PDE3A* rs12314390, and *CDH13* rs12449180 according to the nutrient and food intake. (**A**) According to the carbohydrate intake (cutoff point: 70 energy %). CHO, carbohydrate. (**B**) According to the intake of a balanced diet pattern (cutoff point: 70th percentile). PRS was calculated by the summation of polygenetic-risk scores of the best model, and PRS was categorized into three groups by the tertiles (Low-PRS, Medium-PRS, and High-PRS). BD, balanced diet.

A balanced dietary (BD) pattern explained 40.1% of Korean diet patterns in three different dietary patterns. A BD pattern includes the consumption of beans, potatoes, kimchi, green and white vegetables, mushrooms, fatty and white fish, seaweeds, fruits, and pickles (loading ≥0.4). A BD pattern had an interaction with glaucoma risk ($p = 0.046$). A low BD pattern intake had a positive association by 3.87-fold with PRS ($p < 0.0001$), but a high BD pattern intake did not have a significant association with PRS (Table 5). Glaucoma prevalence was much higher in participants with a low BD intake and a high-PRS than in those with a low-PRS (Figure 3B). It suggested that adults with high-PRS needed to consume a high BD pattern to reduce the glaucoma risk.

4. Discussion

In the present study, we constructed a glaucoma PRS model by generalized multifactor dimensionality reduction and evaluated whether genetic impact represented by PRS was associated with the presence of glaucoma. Moreover, we assessed gene-nutrient interactions by evaluating the effect of nutrition on the risk of glaucoma using PRSs. The risk of glaucoma was found to be 3.01 times higher in participants with a high-PRS than in those with a low-PRS after adjusting for covariates. The genetic risk of glaucoma was modified by serum glucose levels, carbohydrate intakes, and consumption of a balanced diet. More specifically, the genetic risk of glaucoma was significantly higher in participants with low blood pressure, low serum glucose, high carbohydrate intake, and low balanced diet intake.

High levels of fasting serum glucose and HbA1c had a higher association with the risk of glaucoma by 1.53 and 1.66 times, respectively. High serum glucose and HbA1c are well-known risk factors of glaucoma. A meta-analysis involving 47 studies and 2,981,342 individuals from 16 countries demonstrated that the pooled relative risk of glaucoma was 1.48 in patients with diabetes as compared with normal controls [7]. Another cross-sectional study that included 3229 individuals involved in a National Health and Nutritional Examination Survey reported that diabetes was strongly associated with the prevalence of glaucoma (OR = 2.12) [23]. However, in the present study, low serum glucose levels were found to increase the genetic risk of glaucoma, and in participants with a low serum glucose level, those with a high-PRS had a 3.165 times higher risk of glaucoma than those with a low-PRS, whereas no such difference was observed for participants with a high serum glucose level. This finding suggests that the genetic risk of glaucoma is elevated in those with a low serum glucose level.

The genetic risk of glaucoma was significant only in participants with high carbohydrate intake, indicating that there was no significant genetic risk for glaucoma in subjects with low carbohydrate intake. A recent case-control study involving 37 glaucoma patients and 36 controls showed that carbohydrate ingestion induced autonomic dysregulation, which may lead to the development of glaucoma [24]. One possible explanation for this finding is that high carbohydrate intake might enhance the genetic risk of glaucoma development through autonomic dysregulation. A further longitudinal epidemiologic study is warranted to confirm that individuals with a high carbohydrate intake are more susceptible to glaucoma development.

We found that the genetic risk of glaucoma was significant only in subjects with a low balanced diet intake. The risk of glaucoma for those with a low balanced diet intake was 3.87 times higher in the high-PRS group than in the low-PRS group, whereas the risk of glaucoma for those with a high intake of a balanced diet was not significantly different in these two groups. Several studies have demonstrated that a balanced diet, including green leafy vegetables, omega fatty acids, and moderate intake of hot tea and coffee, possibly protects against the development or progression of glaucoma [15]. We suggest this finding is due to the protective effects of a high balanced intake attenuating the effect of genetic risk on glaucoma development.

We compared genetic impacts for glaucoma and cataract using PRSs. The risk of glaucoma in participants with a high-PRS was 3.02 times higher than in those with a low-PRS. However, the risk of cataract was not significantly different in these groups. One possible explanation for this finding is that glaucoma development is more dependent on genetic risk than cataract development. Age-related cataract development is mainly caused by environmental factors such as sunlight exposure, lens aging, inflammation, and oxidation. These factors are all associated with the free-radical generation, which leads to the aggregation of lens proteins and subsequent cataract formation [15,25,26].

The present study has several strengths and limitations. The strength of this study is that it involved a large number of participants, that is, 377 patients with glaucoma and 47,820 subjects without, which undoubtedly increased study reliability. This study could demonstrate a new scenario of therapeutic approaches to counteract the potential genetic risk of glaucoma development by the control of lifestyle. However, it also has several limitations. First, the presence of glaucoma was diagnosed in an individual by a doctor, and a questionnaire was used to answer if glaucoma was present. However, it was not reevaluated for the type of glaucoma by an ophthalmologist for the present study. Second, the study is inherently limited by its cross-sectional design, which prevented our inferring causality. However, it is unlikely that the presence of glaucoma changed eating patterns. Finally, the SQFFQ method used to evaluate usual food intake also has limitations, and at the individual level, nutrition intakes might be overestimated, because data was dependent on participant recall. Nevertheless, the SQFFQ was developed and validated for KoGES and has been widely utilized [18,27,28].

In conclusion, we constructed PRS for glaucoma by generalized multifactor dimensionality reduction and found that the risk of glaucoma was 3.01 times higher for the participants with a high-PRS. The present study demonstrates the existence of a nutrition-gene interaction in the risk of glaucoma development. Such genetic risk with high-PRS can be reduced by good control of serum glucose concentrations and blood pressure with a balanced diet intake. The results can be applied for precision medicine to protect against glaucoma in the person with genetic risk. Given that glaucoma is the leading cause of blindness worldwide, further study is required to evaluate nutrition-gene interactions to reduce the risk of glaucoma development.

Supplementary Materials: The following are available online at http://www.mdpi.com/2072-6643/12/11/3282/s1, Table S1: Generalized multifactor dimensionality reduction (GMDR) results of multi-locus interaction with genes related to glaucoma.

Author Contributions: S.P. and D.J. formulated the research question, interpreted the data, and wrote the first draft of the manuscript. S.H. and S.K. designed this study and analyzed the data. All authors have read and agreed to the published version of the manuscript.

Funding: This work was supported by the Technology Development Program (2019-S2828019) funded by the Ministry of SMEs and Startups (MSS, Korea).

Conflicts of Interest: The authors declare no conflict of interest.

References

1. Weinreb, R.N.; Aung, T.; Medeiros, F.A. The Pathophysiology and Treatment of Glaucoma. *JAMA* **2014**, *311*, 1901–1911. [CrossRef]
2. Jiang, S.; Kametani, M.; Chen, D.F. Adaptive Immunity: New Aspects of Pathogenesis Underlying Neurodegeneration in Glaucoma and Optic Neuropathy. *Front. Immunol.* **2020**, *11*, 65. [CrossRef] [PubMed]
3. Grzybowski, A.; Och, M.; Kanclerz, P.; Leffler, C.; De Moraes, C.G. Primary Open Angle Glaucoma and Vascular Risk Factors: A Review of Population Based Studies from 1990 to 2019. *J. Clin. Med.* **2020**, *9*, 761. [CrossRef] [PubMed]
4. Chun, Y.H.; Han, K.; Park, S.H.; Park, K.-M.; Yim, H.W.; Lee, W.-C.; Park, Y.G.; Park, Y.-M. Insulin Resistance Is Associated with Intraocular Pressure Elevation in a Non-Obese Korean Population. *PLoS ONE* **2015**, *10*, e112929. [CrossRef] [PubMed]
5. Fujiwara, K.; Yasuda, M.; Ninomiya, T.; Hata, J.; Hashimoto, S.; Yoshitomi, T.; Kiyohara, Y.; Ishibashi, T. Insulin Resistance Is a Risk Factor for Increased Intraocular Pressure: The Hisayama Study. *Investig. Opthalmology Vis. Sci.* **2015**, *56*, 7983–7987. [CrossRef] [PubMed]
6. Song, B.J.; Aiello, L.P.; Pasquale, L.R. Presence and Risk Factors for Glaucoma in Patients with Diabetes. *Curr. Diabetes Rep.* **2016**, *16*, 124. [CrossRef]
7. Zhao, D.; Cho, J.; Kim, M.H.; Friedman, D.S.; Guallar, E. Diabetes, Fasting Glucose, and the Risk of Glaucoma. *Ophthalmol.* **2015**, *122*, 72–78. [CrossRef]
8. Dada, T. Is Glaucoma a Neurodegeneration caused by Central Insulin Resistance: Diabetes Type 4? *J. Curr. Glaucoma Pr. Dvd.* **2017**, *11*, 77–79. [CrossRef]
9. Liu, H.; Qi, S.; He, W.; Chang, C.; Chen, Y.; Yu, J. Association of single-nucleotide polymorphisms in TLR4 gene and gene-environment interaction with primary open angle glaucoma in a Chinese northern population. *J. Gene Med.* **2019**, *22*, e3139. [CrossRef]
10. Zhang, Y.-H.; Xing, Y.-Q.; Chen, Z.; Ma, X.-C.; Lu, Q. Association between interleukin-10 genetic polymorphisms and risk of primary open angle glaucoma in a Chinese Han population: A case-control study. *Int. J. Ophthalmol.* **2019**, *12*, 1605–1611. [CrossRef]

11. Karmiris, E.; Kourtis, N.; Pantou, M.P.; Degiannis, D.; Georgalas, I.; Papaconstantinou, D. The Association between TGF-beta1 G915C (Arg25Pro) Polymorphism and the Development of Primary Open Angle Glaucoma: A Case-Control Study. *Med. Hypothesis Discov. Innov. Ophthalmol.* **2018**, *7*, 25–31. [PubMed]

12. Ramdas, W.D.; Wolfs, R.C.W.; Jong, J.C.K.-D.; Hofman, A.; De Jong, P.T.V.M.; Vingerling, J.R.; Jansonius, N.M. Nutrient intake and risk of open-angle glaucoma: The Rotterdam Study. *Eur. J. Epidemiol.* **2012**, *27*, 385–393. [CrossRef] [PubMed]

13. Ramdas, W.D.; Schouten, J.; Webers, C.A.B. The Effect of Vitamins on Glaucoma: A Systematic Review and Meta-Analysis. *Nutrients* **2018**, *10*, 359. [CrossRef] [PubMed]

14. Kang, J.H.; Pasquale, L.R.; Willett, W.C.; Rosner, B.; Egan, K.M.; Faberowski, N.; Hankinson, S. Dietary fat consumption and primary open-angle glaucoma. *Am. J. Clin. Nutr.* **2004**, *79*, 755–764. [CrossRef] [PubMed]

15. Perez, C.I.; Singh, K.; Lin, S. Relationship of lifestyle, exercise, and nutrition with glaucoma. *Curr. Opin. Ophthalmol.* **2019**, *30*, 82–88. [CrossRef] [PubMed]

16. Park, S.; Ahn, J.; Lee, B.-K. Self-rated Subjective Health Status Is Strongly Associated with Sociodemographic Factors, Lifestyle, Nutrient Intakes, and Biochemical Indices, but Not Smoking Status: KNHANES 2007-2012. *J. Korean Med. Sci.* **2015**, *30*, 1279–1287. [CrossRef]

17. Kim, Y.; Han, B.-G. The KoGES group Cohort Profile: The Korean Genome and Epidemiology Study (KoGES) Consortium. *Int. J. Epidemiol.* **2017**, *46*, 1350. [CrossRef]

18. Ahn, Y.; Kwon, E.; E Shim, J.; Park, M.K.; Joo, Y.; Kimm, K.; Park, C.; Kim, D.H. Validation and reproducibility of food frequency questionnaire for Korean genome epidemiologic study. *Eur. J. Clin. Nutr.* **2007**, *61*, 1435–1441. [CrossRef]

19. Liu, M.; Jin, H.S.; Park, S. Protein and fat intake interacts with the haplotype of PTPN11_rs11066325, RPH3A_rs886477, and OAS3_rs2072134 to modulate serum HDL concentrations in middle-aged people. *Clin. Nutr.* **2020**, *39*, 942–949. [CrossRef]

20. Rabbee, N.; Speed, T.P. A genotype calling algorithm for affymetrix SNP arrays. *Bioinform.* **2005**, *22*, 7–12. [CrossRef]

21. Jyothi, K.U.; Reddy, B.M. Gene-gene and gene-environment interactions in the etiology of type 2 diabetes mellitus in the population of Hyderabad, India. *Meta Gene* **2015**, *5*, 9–20. [CrossRef] [PubMed]

22. Hong, K.-W.; Kim, S.H.; Zhang, X.; Park, S. Interactions among the variants of insulin-related genes and nutrients increase the risk of type 2 diabetes. *Nutr. Res.* **2018**, *51*, 82–92. [CrossRef] [PubMed]

23. Zhao, D.; Cho, J.; Kim, M.H.; Friedman, D.; Guallar, E. Diabetes, Glucose Metabolism, and Glaucoma: The 2005–2008 National Health and Nutrition Examination Survey. *PLoS ONE* **2014**, *9*, e112460. [CrossRef] [PubMed]

24. Cao, L.; Graham, S.L.; Pilowsky, P.M. Carbohydrate ingestion induces differential autonomic dysregulation in normal-tension glaucoma and primary open angle glaucoma. *PLoS ONE* **2018**, *13*, e0198432. [CrossRef]

25. Balog, Z.; Sikić, J.; Vojniković, B.; Balog, S. Senile cataract and the absorption activity of cytochrome C oxidase. *Coll. Antropol.* **2001**, *25*.

26. Liu, Y.C.; Wilkins, M.; Kim, T.; Malyugin, B.; Mehta, J.S. Cataracts. *Lancet* **2017**, *390*, 600–612. [CrossRef]

27. Kim, S.H.; Choi, H.N.; Hwang, J.-Y.; Chang, N.; Kim, W.Y.; Chung, H.W.; Yang, Y.J. Development and evaluation of a food frequency questionnaire for Vietnamese female immigrants in Korea: The Korean Genome and Epidemiology Study (KoGES). *Nutr. Res. Pr.* **2011**, *5*, 260–265. [CrossRef]

28. Kim, D.; Kim, J. Dairy consumption is associated with a lower incidence of the metabolic syndrome in middle-aged and older Korean adults: The Korean Genome and Epidemiology Study (KoGES). *Br. J. Nutr.* **2017**, *117*, 148–160. [CrossRef]

Publisher's Note: MDPI stays neutral with regard to jurisdictional claims in published maps and institutional affiliations.

 *nutrients*

Review

Insulin Resistance in Obese Children: What Can Metabolomics and Adipokine Modelling Contribute?

Francisco J. Rupérez [1], Gabriel Á. Martos-Moreno [2,3,4,5], David Chamoso-Sánchez [1], Coral Barbas [1,*] and Jesús Argente [2,3,4,5,6,*]

1   Centro de Metabolómica y Bioanálisis (CEMBIO), Facultad de Farmacia, Universidad San Pablo-CEU, CEU Universities, Urbanización Montepríncipe, Boadilla del Monte, 28660 Madrid, Spain; ruperez@ceu.es (F.J.R.); davidchamoso@gmail.com (D.C.-S.)
2   Departments of Pediatrics & Pediatric Endocrinology, Hospital Infantil Universitario Niño Jesús, 28009 Madrid, Spain; gabrielangelmartos@yahoo.es
3   La Princesa Research Institute, 28006 Madrid, Spain
4   Department of Pediatrics, Universidad Autónoma de Madrid, 28006 Madrid, Spain
5   Centro de Investigación Biomédica en Red de Fisiopatología de la Obesidad y Nutrición (CIBEROBN), Instituto de Salud Carlos III, 28029 Madrid, Spain
6   IMDEA Food Institute, CEI UAM & CSIC, 28049 Madrid, Spain
*   Correspondence: cbarbas@ceu.es (C.B.); jesus.argente@uam.es (J.A.)

Received: 18 September 2020; Accepted: 27 October 2020; Published: 29 October 2020

Abstract: The evolution of obesity and its resulting comorbidities differs depending upon the age of the subject. The dramatic rise in childhood obesity has resulted in specific needs in defining obesity-associated entities with this disease. Indeed, even the definition of obesity differs for pediatric patients from that employed in adults. Regardless of age, one of the earliest metabolic complications observed in obesity involves perturbations in glucose metabolism that can eventually lead to type 2 diabetes. In children, the incidence of type 2 diabetes is infrequent compared to that observed in adults, even with the same degree of obesity. In contrast, insulin resistance is reported to be frequently observed in children and adolescents with obesity. As this condition can be prerequisite to further metabolic complications, identification of biological markers as predictive risk factors would be of tremendous clinical utility. Analysis of obesity-induced modifications of the adipokine profile has been one classic approach in the identification of biomarkers. Recent studies emphasize the utility of metabolomics in the analysis of metabolic characteristics in children with obesity with or without insulin resistance. These studies have been performed with targeted or untargeted approaches, employing different methodologies. This review summarizes some of the advances in this field while emphasizing the importance of the different techniques employed.

Keywords: obesity; childhood; insulin resistance; adipokine; metabolomics

---

1. Introduction

In the 21st century new holistic approaches have been developed to tackle the systems biology challenge of discerning all the processes that characterize a living system at the molecular level. Genomics, proteomics, transcriptomics and more recently metabolomics have improved our understanding of what occurs in a biological system, as well as how it occurs. In the context of the study of clinical alterations, such as those that occur in obesity or insulin resistance, metabolomics can provide information about the actual metabolic phenotype in a given condition (alteration or illness), and how this metabolite set differs from a control (i.e., healthy) state.

Obesity, insulin resistance, metabolic syndrome, type 2 diabetes mellitus and many other metabolic alterations have been studied with different metabolomic approaches and technologies.

However, most of these studies have been focused on adults. There is therefore a need to further our knowledge about how these alterations, which are at times hard to characterize in growing children, debut and how they develop in the pediatric population. Moreover, more biomarkers are needed to help in the adequate diagnosis of these conditions and for monitoring treatment efficiency and disease progression.

In light of our previous experience in the field, we have reviewed the current scientific literature on metabolomic studies in children and adolescents with obesity and/or insulin resistance. In order to provide insight into the type of results that can be obtained by using different metabolomics procedures, we have organized the available literature according to the methodology (i.e., untargeted, semi-targeted, targeted) used to perform the metabolomics study, highlighting the main results obtained with each approach.

2. Childhood Obesity and the Development of Insulin Resistance

2.1. Childhood Obesity

The secondary complications of obesity are of major concern as quality of life is diminished and the mortality rate is increased. These concerns are amplified in the study of obesity in children. Not only has childhood obesity increased worldwide in recent years, but these cases are more severe and more precocious, resulting in the onset of health threatening complications at earlier ages [1,2]. Not only is there a dramatic need for programs/therapies to reduce the incidence of childhood obesity, but better diagnostic tools to identify those children at greater risk of developing severe complications are of utmost importance. One recurring problem in the approach to this problem is that observations in adult patients cannot always be directly applied to pediatric patients, especially prepubertal children.

Obesity is the excessive accumulation of adipose tissue that results in impairment of the patient's physical and/or psychological function. Direct quantification of a patient's body fat can be precisely performed by using methodologies such as bioimpedanciometry, dual X-ray densitometry, plethismography or hydrodensitometry. However, these advanced tools are not widely accessible in the usual clinical setting, setting aside investigation facilities or specialized clinical units. Consequently, the diagnosis of overweight or obesity is commonly established on the basis of an indirect estimation of patient's body fat content by using body mass index (BMI = weight (kg)/(height (m)^2), which has been shown to exhibit a good correlation with body fat content [3], although with some limitations (i.e., with extreme muscular mass development). In adult patients, 25 and 30 kg/m^2 are widely accepted as the thresholds to diagnose overweight and obesity, respectively [4,5].

In children, agreement on a precise definition of obesity is more difficult than in adults. In children and adolescents, the BMI standardized for age and sex must be used, not its raw calculation. This has raised intense controversy regarding the establishment of "cut-off points" to define overweight and obesity and the population references that should be used [6]. In general, a child is considered to have excess body fat when their BMI is greater than the 95th percentile for his/her age and sex [3]. However, an optimum definition can be obtained by applying a cut-off point of BMI z-score above 2 compared with references from the same population, age and sex, thus meeting the proposal by the World Health Organization (WHO) [7]. Likewise, there is no consensus on the definition of morbid obesity in children and adolescents, with some authors suggesting a BMI z-score above 3 or 200% of the ideal body weight for height as possible cut-off points [8,9]. There is also no agreement on the definition of "early-onset" obesity, with ages below 5 or 2 years at the onset of the disease having been suggested by different authors [10]. This subgroup of patients with early-onset obesity is of particular interest because the excess of weight can be part of a syndrome or a monogenic disease. Another important difference between childhood and adult obesities is that in children, the development of the deleterious comorbidities associated with excess body fat may be a later event when considering the degree of obesity. This difference is in part due to the greater capacity of tissue turnover/regeneration in children.

2.2. Adipose Tissue in Obesity: The Importance of Age at Onset

The sequence of events in adipogenesis leading to the development of mature adipocytes from their embryonic pluripotential undifferentiated precursors requires their commitment towards the adipogenic lineage for the sequential formation of type I adipoblasts and after clonal expansion, type II preadipocytes. Growth arrest of type II preadipocytes will result in them becoming mature adipocytes and their accumulation of lipid droplets. In this cell linage the ability to synthesize and secrete adipokines is almost exclusively restricted to mature adipocytes [11–13].

As children continue to be in a period of growth and development, they possess a greater ability to adapt tissue morphology and function to their environment compared to adults. This is particularly relevant in the case of white adipose tissue, as its expandability is important in the metabolic complications in response to obesity [14]. Obesity results in histological, metabolic and endocrine changes in white adipose tissue [15]. These changes are determined by several factors: (1) The metabolic capacity of adipocytes to take-up free fatty acids (FFA) from the bloodstream, thus avoiding their ectopic deposition (lipotoxicity) [16]; (2) the production of chemoattracting proteins (chemokines) that results in an increase in specific proinflammatory populations of monocytes and macrophages [17–19] that substantially contribute to modifications in the adipokine secretion pattern of the tissue [20]; (3) the ability to recruit new adipocytes from preadipocytes, which has been postulated to occur once the former have reached a critical size [21] and (4) the change in the pattern of paracrine and endocrine adipokine secretion by hypertrophic adipocytes, as compared to normal-sized ones [22,23]. Each of these factors varies throughout the different stages of human development.

In adults, either lean or obese, the adipocyte population in white adipose tissue remains relatively stable due to a balance between adipogenesis and apoptosis. In contrast, children and adolescents progressively increase the number of adipocytes in their adipose tissue, with this rise being even greater in obese compared to lean subjects due to a higher proliferation rate [24]. Consequently, early onset obesity is associated with an increase in adipocytes that could allow, at least transiently, for a limited degree of adipocyte hypertrophy and thus reducing the impact of obesity on the adipokine secretion profile and metabolic impairment at early ages, but conversely increasing the risk to develop severe obesity and metabolic comorbidities at later stages of life [3,24]. This is reminiscent of the "hyperplasic" model of obesity in children where there is an increased number of non-hypertrophic adipocytes versus a "hypertrophic" model of obesity in adults, with an increase in the volume of pre-existing adipocytes, resulting in a postulated difference in the impairment of the adipokine secretion profile between these two types of obesity [23,25].

2.3. Adipokines in Childhood Obesity

Leptin and adiponectin are the main adipokines involved in energy homeostasis and insulin sensitivity, respectively, among the extensive and continuously growing list of peptides secreted by adipose tissue. Additionally, a number of adipokines with proinflammatory actions (e.g., resistin, IL-6 and TNF-α, among many others) are produced in white adipose tissue, mainly by mononuclear stromal cells with the cellular composition of this tissue changing during the progression of obesity.

Leptin, found in the bloodstream both free and bound to the soluble isoform of its specific receptor, is mainly produced by mature adipocytes and acts as an adiposity signal. Leptin modulates the activity of several neuronal populations involved in the regulation of food intake and energy homeostasis in the central nervous system, including the proopiomelanocortin (POMC) producing neurons in the hypothalamic arcuate nucleus, exerting its main activity as a signal of energy sufficiency [26] with reported cases of severe human obesity due to leptin deficiency, reversible after recombinant leptin administration [27].

The circulating levels of leptin are directly correlated with body fat mass and adipocyte triglyceride content [28]. Gestational age and birth weight are the main determinant for its levels and bioavailability in the newborn [29]. Leptin levels significantly increase throughout pubertal development in females and decrease at the final stage of puberty in males. In contrast, circulating levels of leptin's soluble receptor decrease in both sexes after pubertal onset, resulting in a puberty-related increase in free leptin that is more pronounced in adolescent females [30]. Obesity determines an increase in free leptin levels as a result of the increase in leptin and decrease of its soluble receptor, that is not reproduced in the spinal fluid, leading to "leptin resistance" [31].

Adiponectin is produced exclusively in mature adipocytes and circulates as polymers, with high molecular weight (HMW, 400–600 kDa) adiponectin postulated to be more metabolically relevant, particularly regarding its insulin sensitizing action [32]. This peptide acts through two specific receptors, adipoR1 and adipoR2, widely distributed but mainly located in muscle and liver, respectively. In muscle, liver and white adipose tissue adiponectin enhances insulin sensitivity and the promotion of fatty acid oxidation, with an increase indicating a beneficial apolipoprotein profile [33].

In newborns, adiponectin levels are higher than at later periods of life and positively correlate with gestational age and birth weight, with females having higher serum levels [29]. Postnatally serum adiponectin levels fall and its positive correlation with fat mass disappears around age 2 years, coincident with the increase in body fat [34]. In prepubertal children, most studies report a lack of differences between sexes, but males show lower adiponectin levels from mid-puberty onwards [30,32–34]. As opposed to leptin, serum adiponectin levels in adults are inversely related to the amount of body fat, with adult patients with obesity having decreased circulating adiponectin levels [35]. However, this inverse correlation between body fat and serum adiponectin levels is not present in all patients and is influenced by adipocyte size [36]. In adolescents with obesity, an inverse correlation between adiponectin levels, body fat and insulin resistance, similar to that reported for adults, has been demonstrated [34,37].

2.4. Carbohydrate Metabolism Impairment in Childhood Obesity: Insulin Resistance

As stated above, overt metabolic impairment in children with obesity can be delayed due, at least in part, to the singularities of young adipose tissue. This is particularly evident regarding carbohydrate metabolism. Although there is wide geographic and ethnic variability, we recently demonstrated a minimal incidence of type 2 diabetes mellitus (T2DM) in children despite the high number of patients affected with severe obesity in our country [38]. In contrast, the prevalence of initial glycemic alterations (defined as "prediabetic conditions") and more importantly, the number of patients showing peripheral resistance to insulin induced glucose caption ("insulin resistance" [IR]) is much higher [38].

However, the definition of IR in the clinical setting is extremely controversial, particularly in pediatrics, and continues to be a matter of intense debate in the international community. Although the clamp tests used in investigational facilities are considered the "gold standard" for IR validation, common clinical determinations can be used for the calculation of IR indexes on the basis of fasting (HOMA-IR index) and/or postprandial (insulinogenic and insulin sensitivity indexes) measurement of glycaemia and insulinemia [39]. However, these indexes have been shown to correlate well with those derived from studies using euglycemic-hyperinsulinemic clamps [40].

In the first definition of "X" or metabolic syndrome, IR was suggested as the pathophysiological basis of the remaining obesity-associated metabolic derangements [41]. The definition of metabolic syndrome has been subsequently modified, conferring a primary role to the presence of abdominal obesity and in particular visceral adipose tissue, with waist circumference showing a better association to cardiovascular risk than BMI itself [42]. This observation has been extended also to the pediatric and adolescent population and, consequently, abdominal circumference and not BMI has been considered as the anthropometric criterion for the definition of metabolic syndrome in children above 10 years of age [43] Surprisingly, in children the criteria to define metabolic syndrome regarding carbohydrate metabolism only take into consideration the presence of impaired fasting glucose (IFG) and/or T2DM, whereas IR is not considered [43]. Similarly, hyperinsulinemia/IR is not usually considered in the definition of "metabolically healthy" obesity in childhood [44], although some authors point out its relevance [45,46]. These consensus statements and criteria, originating mainly from studies in the adult population, should be revised for children and adolescents. Indeed, glycemic alterations are late, or frequently absent, findings in children with obesity in our environment, whereas a rise in both fasting and postprandial insulinemia can be identified as the initial steps of carbohydrate metabolism impairment, particularly in young children with obesity [38].

Additionally, a bidirectional influence between leptin and insulin exists, with hyperinsulinemia enhancing leptin production and increased free leptin levels increasing insulin resistance, with "leptin resistance" and IR usually coexisting in obesity, even at young ages [47]. The development of these hormonal derangements is gradual and identification of biomarkers to precociously predict the risk of developing serious metabolic complications is of great importance. Consequently, the application of new techniques, such as metabolomics, in combination with adipokine measurements can afford additional information to address the pathophysiological relevance of this controversial condition of IR in childhood, even when its definition is based upon analytical criteria derived from usual clinical practice.

3. Adipokine Modelling

As stated above, obesity-induced modifications in adipose tissue differ between children and adults and there is insufficient information regarding how these changes affect the development of further complications and metabolic syndrome in pediatric patients with obesity.

We have recently reported a novel approach to identifying biomarkers of insulin resistance in children, both prepubertal [48] and pubertal [49], with obesity. We found that combined sets of cytokines, adipokines and chemokines can be used as models to predict insulin resistance. In both pediatric age groups specific factors, including tumor necrosis factor (TNF)α, eotaxin, insulin-like growth factor (IGF)-1, leptin, triglycerides (TGL), monocyte chemeoattractant protein (MCP)1 and brain derived neurotrophic factor (BDNF), were identified as biomarkers involved in insulin resistance. Interestingly, in adolescents with obesity the presence of insulin resistance is influenced by the chemokines MCP-1 and eotaxin, as well as the growth factor platelet derived growth factor (PDGF)-BB, in a sex independent manner. These three biomarkers are part of the main component that together with stromal cell derived factor (SDF)1α and BDNF, determine 27.7% of the variance associated with insulin resistance. In prepubertal obese children, we defined two predictive models that include the combination of leptin, TG/HDL, IGF-1, TNFα, MCP1 and PDGF-BB with an optimal sensitivity and specificity of 93.2%. Hence, we suggest that the combination of these circulating parameters from a single fasting sample could be useful to predict insulin resistance in prepubertal children with obesity. These adipokines in combination with other biomarkers, such as specific metabolites, could possibly serve as an even more powerful predictive model.

4. Metabolomics

Metabolomics, in which potentially all small-molecule metabolites (the metabolome) are identified and at some level quantified, is generally acknowledged to be the omics discipline that supplies the most rapid and clearest information about the phenotype. For this reason, it is greatly appreciated for its role in biomarker discovery. The rationale underlying "the study of the metabolome" (i.e., metabolomics) is based on the assumption that the metabolome is the reflection of all the processes that might be occurring at one moment (time-course changes) or be altered under one condition (changes due to disease, treatment, etc.). The information for such study can be gathered through different experimental approaches that receive different names.

Unfortunately, in the field of metabolomics the terminology is not yet fully standardized, and clear unequivocal terms have not been assigned to each methodology. This lack of standardization might lead to confusion, as the same term may be used for different types of studies, whereas the same type of information might be obtained by using similar methodologies that have received different names. For clarity we have classified the studies as untargeted, targeted and semi-targeted, although the limits are sometimes diffuse, and methodologies are very often not accurately described as to establish a clear classification. Metabolomics studies related to obesity and/or insulin resistance in non-adult subjects are shown in Table 1. Analysis of the studies performed so far indicate that they include a limited number of individuals, compare different experimental groups (obese, normal weight, different ages and racial origin) with the only constant of IR and therefore, it is not surprising that results are also heterogeneous.

4.1. Untargeted Metabolomics

Untargeted metabolomics is an approach that highly contrasts with targeted analysis (the classical way to study metabolism), where a limited number of specific known compounds are analyzed. The rationale to employ an untargeted approach is that modern spectrometric/spectroscopic techniques can generate a huge amount of information in the form of signals coming from all the metabolites in the samples. Such signals can be statistically compared between case/control groups to identify the characteristics that differentiate these conditions without a priori hypothesis. Once isolated the specific spectral characteristics are associated with metabolites (as "identified" or more precisely, "annotated"). The analytical workflow in untargeted metabolomics is still far from being exhaustively protocolized, but it can be described as a sequential process that starts with the biological question, and then follows a series of steps of experimental design, sample treatment, instrumental analysis, signal processing, data pretreatment, statistical analysis, metabolite annotation, validation and biological interpretation [50]. This untargeted approach is of great interest in what can be called the "discovery phase" resulting in the possibility to unveil new, not previously described compounds that can be related to the characteristic of interest, for example metabolic alterations of insulin resistance in the context of childhood obesity.

To gather information about the metabolome, the instrumental analysis is generally conducted through either nuclear magnetic resonance (NMR) or mass spectrometry (MS). In recent years MS has become the most employed technique in metabolomics [50]. It can be hyphenated to a separation technique (gas chromatography, GC; liquid chromatography, LC; capillary electrophoresis, CE; or supercritical fluid chromatography, SFE) or not (Direct MS, DMS). MS benefits from detection permitted at high sensitivity and structural elucidation based on spectral libraries and tandem MS, even in complex biological samples. Nevertheless, as the diversity of chemical characteristics of the metabolites is so broad, there is no single technique that can cover the full range of metabolites in a sample. For this reason, the results obtained are strongly dependent on the technique and the methodology that have been used in the analysis and obtaining really non-biased results requires a "multiplatform" approach [51,52].

GC-MS is usually combined with LC-MS [51,53–56] as the biochemical information is complementary [57]: GC-MS is very well suited for the analysis of metabolites related to central carbon metabolism such as short chain organic acids, amino acids, monosaccharides, fatty acids, disaccharides and cholesterol, and LC-MS is adequate for less polar molecules, i.e., lipids [58]. GC-MS can even stand alone, and provide information of a large set of compounds, even of a single family such as 75 steroid-related metabolites in the context of childhood obesity [59]. However, the most widely employed technique for metabolic fingerprinting is reversed-phase LC-MS, which involves the minimum requirement for sample treatment and alteration or hydrolysis of the metabolites during the analysis among the hyphenated techniques [50]. CE-MS shows clear benefits for multiplatform untargeted metabolomics [60], and the analysis of amino acid-related compounds. Mastrangelo et al. [51] found differences in amino acids, acylcarnitines, polyamines and xanthines, among the metabolites that could be measured with this technique, as part of a multiplatform approach. The NMR profile can contain qualitative and quantitative information on hundreds of different small molecules present in the sample. Although sensitivity is poorer in NMR than in MS, the robustness and elucidation capabilities are usually claimed as its main advantages.

Currently, the main bottleneck in metabolomics is the identification of the metabolites of interest when they are found in LC-MS, CE-MS or DMS measurements. The identification process starts with the quest for information in publicly available databases, with the input being the exact mass and the output a chemical identity. However, databases for metabolomics are not fully standardized and the different databases vary in the number of records (compounds), in the fields for each record (compound properties), as well as the searchable fields (mass, m/z, MS/MS spectrum, name, etc.). To facilitate the simultaneous query in different databases, tools such as CEU Mass Mediator (http://ceumass.eps.uspceu.es/) allow one to obtain the information from the available databases, together with other utilities [61]. Given the fact that the discovery phase in untargeted metabolomics is usually performed measuring thousands of signals in a limited set of samples, validation of results with a different analytical technique and in a different cohort strongly increases their reliability.

4.2. Untargeted Metabolomics Applied to Obesity and Insulin Resistance in Children and Adolescents

Researchers have applied different untargeted approaches in the study of obesity and IR. The broadest metabolite coverage when studying the effect of obesity and IR was obtained by our group by using three analytical techniques (LC-MS, GC-MS and CE-MS) [51]. Amino acids, gut microbiota by-products and lipids were found altered, and the study also highlighted that these modifications were sex specific, with differences in boys and girls even though the children in these studies were prepubertal. The most relevant findings were later validated with a target method in a bigger cohort [62], which showed an increase in branched chain amino acids (BCAA) and aromatic amino acids (Phe, Tyr and Trp), with the most altered pathways being the urea cycle, alanine metabolism and the glucose-alanine cycle.

The robustness and capability to perform studies with large sample number should be one of the potential advantages of NMR, although it has been applied only to small studies: Tricó et al. [63] showed the relevance of specific patterns of amino acids and carbohydrates to predict future (2.3 years) worsening in glycemia, whereas Hosking et al. [64] demonstrated that these types of metabolites were different between boys and girls, and insulin resistance was worse in girls than in boys. Both studies were performed in normal weight adolescents.

4.3. Semi-Targeted Metabolomics

Advances in instrumentation and software processing tools, together with massive data storage capabilities have permitted the development of what we could call semi-targeted analysis: A non-biased sample treatment and generic chromatographic conditions, coupled to an MS device that obtains fragmentation spectra of each and every compound that can be detected. Such spectra are compared with those included in a database built with the analysis of real standards [65]. This methodology increases the throughput of the process, because only those compounds previously included in the database are sought in the samples. This reduces the time devoted to the elucidation of the unknown compounds.

4.4. Semi-Targeted Metabolomics Applied to Obesity and Insulin Resistance in Children and Adolescents

One of the drawbacks of untargeted MS metabolomics approaches is that, due to inherent variability of the signals from sample to sample, the discovery phase studies are usually performed in small cohorts, usually less than 125 individuals per group, thus less than 250 for the entire study (See Table 1). Nevertheless, with the most commonly cited semi-targeted approach with LC-MS/MS combined with GC-MS [66], samples of over 700 Hispanic children were successfully compared (obese/non obese, boys/girls) [56]. Moreover, the use of such common methodology applied to four different untargeted studies permitted the performance of an individual participant meta-analysis about obesity and insulin resistance in children: one sphingomyelin (SM(d18:2/14:0)) was positively associated with obesity, whereas association with HOMA was found for alanine (positive) and acylcarnitines and non-esterified fatty acids (negative) [67]. Perng et al. applied such data-driven LC-MS/MS approach combined with GC-MS [66] to find differences in a set of more than 3000 compounds in studies related to obesity [53] and metabolic risk [55], and they also found that the most prominent changes were related to branched chain amino acids (BCAA) and acylcarnitines, although not always with a consistent trend, because BCAA was not associated with worsening metabolic health during early adolescence and the relationship of BCAA with fasting glucose or serum triglycerides was different in boys and girls [55].

4.5. Targeted Metabolomics

While untargeted metabolomics is the choice for the discovery of new previously unknown compounds, the capabilities of available analytical instrumentation also allows several hundreds of well-characterized compounds to be measured simultaneously. However, when the systems are programmed to measure one set of metabolites, there will be no signal from other compounds. This approach, very popular in clinical studies, is considered as targeted analysis. By measuring a large set of metabolites with this targeted approach researchers can perform "targeted metabolomics." But metabolomics does not mean just measuring a large number of metabolites, and this denomination alone is far from being a clear indication of the methodology employed in its analytical determination.

In this type of studies, the keyword "metabolomics" seems to indicate only that a large set of metabolites has been simultaneously determined, but this is far from being a clear indication of the methodology employed in its analytical determination.

These studies are called "metabolomics" because they generate a multivariate space, with all the metabolites that can be measured. This has become useful to find mathematical associations between metabolites, which can help to define possible single theragnostic biomarkers such as asymmetric dimethyl arginine (ADMA), which was found to be associated to insulin resistance in adolescents [68]. Moreover, the possibility of identifying a set of metabolites that together show predictive power is one of the biggest achievements of the multivariate metabolomic approach. Such strategy has resulted in the proposal of the so-called metabolic signature [69] (later called BCAA-related signature [70]), metabolomic signature [71], metabolite profiling [72], metabolomic profile [53,73–79], or metabolic phenotype [80,81] and although the names differ, they share the same underlying concept.

Something common to all metabolomics studies, whether untargeted or targeted, is the application of multivariate statistical analysis, both unsupervised (e.g., principal component analysis, PCA) and supervised (e.g., projection on latent structures/partial least squares-discriminate analysis, PLS-DA). Ideally, this should lead to the proposal of strong biomarkers of insulin resistance such as different adipokines [82]. Nevertheless, and despite the accumulated evidence [83], biomarkers from metabolomics studies such as BCAA, aromatic amino acids, acylcarnitines, or some lipids (Table 1) are not measured in the routine of the clinics of obesity or insulin resistance.

4.6. Targeted Metabolomics Applied to Obesity and Insulin Resistance in Children and Adolescents

In the field of obesity and insulin resistance, the concentrations of BCAA and acylcarnitines have been extensively studied by using targeted metabolomics since Newgard et al. described "a BCAA metabolic signature" associated to obesity and insulin resistance [69]. In recent years, several studies have been performed to gain more evidence concerning the relationship between obesity and BCAA. In adults, most of these studies have consistently shown the association of obesity, insulin resistance and type 2 diabetes with elevated BCAA, aromatic amino acids, C3 and C5 acylcarnitines and glutamate and alanine [70], and BCAA have been proposed as good biomarkers of obesity and insulin resistance in adult individuals [83]. Along with BCAA, an alteration in the levels of acylcarnitines could also be used to discriminate between children with or without insulin resistance [51,56,62,67,76,84,85].

However, its usefulness as a biomarker in childhood obesity and insulin resistance remains to be elucidated. As previously mentioned, the studies shown in Table 1 indicate some of the difficulties for performing meta-analysis with them, as not only do they present differences in terms of the methodology used to carry out the instrumental analyses, but the size, origin and characteristics of the cohorts are also different. These differences in the design of the studies might justify some apparent discrepancies: In most of the studies, BCAA is increased in obese children and adolescents [53,56,72,76,85]. In addition, prepubertal children with obesity and insulin resistance present an increase in BCAA compared to obese prepubertal children without insulin resistance [51,62]. The same trend was shown in adolescents [63]. However, other studies have shown no alteration or even a decrease in BCAA levels between obese children as compared to children with normal weight [55,64,79,84]. This implies that future studies must be carried out to clearly elucidate the association between BCAAs and insulin resistance in non-adult populations. Such discrepancies are not related to differences in the methodology used to gather the information about metabolites, despite that there is no uniformity in the terminology. Moreover, as stated above it is important to differentiate metabolomic profiles between boys and girls during childhood. A study by Newbern et al. [74] reported an increase in BCAA levels and BCAA by-products in boys compared to girls, together with an inverse relationship between adiponectin and BCAA in boys.

Most of the targeted metabolomics studies concerning insulin resistance and obesity in childhood have focused on amino acids and acylcarnitines. Nevertheless, measurement of the alterations in the lipid profile with a metabolomics approach, i.e., lipidomics, has demonstrated a strong correlation of one lysophosphophatidylcholine (LPC(14:1)) and one phosphatidylcholine (PC(16:0/2:0)) with cardiovascular disease risk factors in adolescents [86]. In addition, alterations in steroid hormone levels have been found in children with insulin resistance [53,56,59,76].

4.7. Combining Metabolomics Information in Obesity and Insulin Resistance

As no single technique can provide coverage of the whole metabolite, the samples must be analyzed by different techniques, and the information must be integrated. Metabolomics can supply a large amount of useful information, but other determinations are still necessary. For instance, Newgard et al. combined information of the so-called "conventional" metabolite determination (glucose, lactate, cholesterol, etc.) with the targeted MS/MS analysis of acylcarnitines and amino acids, plus the free and total fatty acids, and short-chain organic acids by GC/MS to characterize the metabolic signature that was different between lean and obese [69].

We analyzed possible correlations between the metabolites measured and other clinical determinations such as the HOMA index, total triglycerides, leptin and adiponectin [57]. In the ROC analysis, the combination of leptin and alanine showed a high IR discrimination value in the whole cohort (area under curve, AUCALL = 0.87), as well as in boys (AUCM = 0.84) and girls (AUCF = 0.91) when considered separately. However, the specific metabolite/adipokine combinations with highest sensitivity were different between the sexes. Therefore, combined sets of metabolic, adipokine and metabolomic parameters can identify pathophysiological relevant IR in a single fasting sample, suggesting a potential application of metabolomic analysis in clinical practice to better identify children at risk without using invasive protocols.

Based on our current understanding of this problem, more research is clearly needed to elucidate reliable biomarkers for future complications in childhood obesity, including employing different types of samples, such as feces. Diseases associated with lifestyle, as well as their complications, are complex and multifactorial in nature. Genetic heritage, dietary habits, and other environmental factors, as well as their interaction with the microbiome, conditioning gene expression and transcription and the subsequent regulation of protein translation and activity, all impact on the metabolic outcome. All these factors are molecularly related to the metabolome. Moreover, the role of factors such as the gut microbiome and low-grade inflammation in modulating the response to insulin and other hormones cannot be questioned but is exceedingly difficult to quantify.

Table 1. Metabolomic studies about obesity, insulin resistance or type 2 diabetes mellitus (T2DM) in children.

Methodology	Instrumental Analysis	Disease	Study Design	Sample	Findings	Ref.
Untargeted	LC-MS, CE-MS, GC-MS	Obesity and IR	Fingerprinting study: 60 prepubertal obese children. Boys (n = 30, 50% IR and 50% non-IR) Girls (n = 30, 50% IR and 50% non-IR) Validation study: 100 prepubertal obese children. Boys (n = 50, 50% IR and 50% non-IR) Girls (n = 50, 50% IR and 50% non-IR)	Serum	IR vs non-IR: • Inflammation, central carbon metabolism and gut microbiota are the most altered processes. • Increased BCAA, ArAAs, Ala, Pro, Pyr, taurodeoxycholate, glycodeoxycholate, piperidine, pyroglutamate. • In females, increased free carnitine, propionylcarnitine and butyrylcarnitine, but in males only propionylcarnitine.	[51]
Untargeted	LC-MS/MS	Metabolic Risk	Boys (n = 113) Girls (n = 125) (8–14 years)	Serum	Metabolic Risk: In girls: • Positive association of DG(16:0/16:0), 1,3-dielaidin, myo-inositol, and urate. • Inverse association of thymine, dodecenedioic acid, and n-acetylglycine with metabolic risks. In boys:: • Positive associations of BCAA, DG(16:0/16:0), tyrosine, and 5′-methylthioadenosine.	[54]
Untargeted	NMR	IR	Cross sectional study: 78 non diabetic adolescents (8–18 years) Longitudinal study: 16 subjects after a mean follow-up of 2.3 years	Plasma	Higher baseline 2-hydroxybutyrate and BCAA levels in insulin resistant youth and predict worsening of glycemic control Alterations of 2-hydroxybutyrate metabolism predict incipient deterioration of β-cell function and longitudinal worsening of glycemic tolerance.	[63]

Table 1. *Cont.*

Methodology	Instrumental Analysis	Disease	Study Design	Sample	Findings	Ref.
Untargeted	NMR	IR	170 healthy normal weight children (5, 14 and 16 years)	Serum	IR higher in girls than in boys. In healthy normal weight children IR was associated with reduced concentrations of BCAA, 2-ketobutyrate, citrate and 3-hydroxybutyrate, and higher concentrations of lactate and alanine.	[64]
Semi-targeted	LC-MS/MS, GC-MS	Obesity	Obese ($n = 84$) Overweight ($n = 28$) Normal-weight ($n = 150$) Median age 7.7 years 50% boys 50% girls	Plasma	OB vs NW: • Increased BCAA (Val, Leu, Ile) and androgen hormones (DHEA-S).	[53]
Semi-targeted	LC-MS/MS, GC-MS	Obesity and IR	Hispanic children Obese ($n = 450$) Non-obese ($n = 353$) Boys ($n = 405$) Girls ($n = 398$). (4–19 years). Mean age 11.1 years	Plasma	OB vs NOB: • Increased BCAA and acylcarnitine catabolism and changes in nucleotides, lysolipids, steroid derivatives and inflammation markers. • Reduced fatty acid catabolism. • BCAAs, ArAAs, aspartate, dipeptides, citrate, asparagine, glycine and serine is associated with risk factors for IR, hyperleptinemia, hypertriglyceridemia, hyperuricemia and inflammation.	[56]

Table 1. *Cont.*

Methodology	Instrumental Analysis	Disease	Study Design	Sample	Findings	Ref.
Semi-targeted	LC-MS/MS, GC-MS	Obesity	Longitudinal study for 5 years: Obese ($n = 68$) Overweight ($n = 23$) Normal weight ($n = 122$) 48.8% boys Median age 7.7 years	Plasma	BCAA is not associated with worsening metabolic health during early adolescence. Inverse association of the BCAA pattern with a change in fasting glucose in boys. Direct relation of BCAA pattern with a change in serum triglycerides in girls. Higher score for androgen hormone pattern at baseline corresponds with a decrease in leptin an increase in CRP in girls.	[55]
Targeted	LC-MS/MS, GC-MS	Obesity and IR	100 prepubertal obese children. Boys ($n = 50$, 50% IR and 50% non-IR) Girls ($n = 50$, 50% IR and 50% non-IR) 5–10 years	Serum	IR vs non-IR: • Higher ALT, GPT and TAG levels • Higher leptin and reduce leptin/adiponectin ratio • Increase BCAA, ArAAs (Phe, Tyr and Trp), and Ala • The most altered pathway is the urea cycle, alanine metabolism and the glucose-alanine cycle. • C12 acylcarnitine and methionine correlate with HOMA-IR exclusively in males	
Targeted	MS/MS	Obesity and T2D	Case-control: Obese ($n = 64$) Obese with TD2 ($n = 17$) Normal-weight ($n = 39$) 12–17 years	Plasma	T2D vs OB/NW: • Decreased BCAA. T2D/OB vs NW: • No difference in long-chain AcylCN • Reduced short and medium-chain AcycCN • No defects in fatty acid or amino acid metabolism No differences in fasting FFA levels	[62] [87]

Table 1. *Cont.*

Methodology	Instrumental Analysis	Disease	Study Design	Sample	Findings	Ref.
Targeted	MS/MS	Obesity, IR and T2D	Case-control: Obese ($n = 57$) Obese prediabetes ($n = 27$) Obese T2D ($n = 17$) Normal-weight ($n = 38$) 13–14 years	Plasma	BCAA and BCAA intermediates correlated: positively with insulin sensitivity and DI	[79]
Targeted	LC-MS/MS	Obesity and IR	Cross sectional study: 69 healthy children and adolescents 8-18 years Longitudinal cohort study in subset: Subgroup of 17 participants 8-13 years	Plasma	OB *vs* NW: • Increased BCAA • Increased BCAA not associated with measures of insulin resistance at baseline. • Baseline BCAAs predicted HOMA-IR at 18 months. • Elevations in the concentrations of BCAAs were associated with reduced insulin sensitivity at 12 months.	[72]
Targeted	MS/MS	Obesity and IR	Cross-sectional study: Obese ($n = 82$) Boys ($n = 41$) Girls ($n = 41$) 12–18 years	Plasma	BCAA levels and by products of BCAA catabolism are higher in males than females with similar BMI. In males, HOMA-IR correlated: • Positively: BMI z-score, BCAA, uric acid, long-chain acyl-carnitines • Negatively: fatty-acid oxidation products In females, HOMA-IR correlated: • Positively: BMI z-score Adiponectin correlated inversely with BCAA and uric acid in males, but not females	[74]

Table 1. *Cont.*

Methodology	Instrumental Analysis	Disease	Study Design	Sample	Findings	Ref.
Targeted	LC-MS/MS	Obesity and IR	Identify biomarkers predictive of future disease risk- Obese ($n = 46$) Obese to normal weight ($n = 18$) Normal-weight ($n = 45$) 9–11 years	Plasma	Baseline BCAA concentration as a predictor of future risk of insulin resistance and metabolic syndrome OB vs NW: • Increased levels of BCAA, Tyr, Phe, 2-AAA and several acyl-carnitines • Lower levels of acyl-alkyl phosphatidylcholines	[85]
Targeted	MS/MS	Obesity and IR	Longitudinal study: 80 obese Caucasian children. 40 participate in one-year lifestyle interventions 8–15 years	Serum	Tyr was the only metabolite significantly associated with HOMA-IR at baseline and after 1-year intervention. No association between HOMA-IR and BCAA.	[84]
Targeted	MS/MS	Obesity and IR	430 control (13–15 years). 91 morbid obese (12–16 years)	Plasma	Accumulation of ADMA is associated with modulation of insulin signaling and insulin resistance. ADMA decreased after obesity intervention program	[68]
Targeted	MS/MS, LC-MS/MS	Obesity and IR	Meta-analysis 1020 pre-pubertal children from three European studies. 8–10 years	Plasma	• Positive association of SM (32:2) with BMI z-score. • SM 32:2 as a potential molecular marker for mechanistic alterations involved in the pathogenesis of obesity. • Ala and Tyr was associated positively with HOMA-IR. • Acylcarnitines and non-esterified fatty acids were negatively associated with HOMA.	[67]

Table 1. *Cont.*

Methodology	Instrumental Analysis	Disease	Study Design	Sample	Findings	Ref.
Targeted	GC-MS	Obesity and IR	20 obese with IR 67 obese without IR 8.5–17.9 years	Urine	The steroidal signature IR vs non-IR: • High adrenal androgens, glucocorticoids and mineralocorticoid metabolites • Higher 5α-reductase and 21-hidroxylase activity • Lower 11bHSD1 activity The authors suggest a vicious cycle model, whereby glucocorticoids induce IR.	[59]
Targeted	MS/MS	Obesity and Metabolic Risk	Non-OW/OB and low MetRisk ($n = 335$) Non-OW/OB and high MetRisk ($n = 29$) OW/OB and low MetRisk ($n = 58$) OW/OB and high MetRisk ($n = 102$) Girls 48.3% Boys 51.7% 11–16 years	Plasma	Lower levels of LCFA in non-OW/OB with high MetRisk and OW/OB with high MetRisk compared to mon-OW/OB with low MetRisk. Higher levels of BCAA metabolite pattern in OW/OB with high MetRisk compared to non-OW/OB with low MetRisk. Higher levels of DAG in OW/OB with high MetRisk vs non-OB/OW with low MetRisk. Higher score of androgen steroid hormones pattern in OW/OB with high MetRisk compared to Non-OW/OB with low MetRisk. Higher levels of AcylCN in non-OW/OB with high MetRisk compared to non-OW/OB with low MetRisk. Lower levels of AcylCN in OW/OB with high MetRisk compared to Non-OW/OB with low MetRisk.	[76]

Abbreviations: 11bHSD1: 11β-hydroxysteroid dehydrogenase type 1; 2-AAA: alpha amino adipic acid; AcylCN: acylcarnitines; ADMA: asymmetric dimethylarginine; Ala: alanine; ALT: alanine transaminase; ArAAs: aromatic amino acids; BCAA: branched chain amino acids; BMI: body mass index; CE-MS: capillary electrophoresis – mass spectrometry; CRP: C-reactive protein; DAG: diacylglycerides; DG: diglyceride; DI: disposition index; GC-MS: gas chromatography – mass spectrometry; GPT: gamma-glutamyltransferase; HOMA-IR: homeostatic model assessment – insulin resistance; IR: insulin resistance; LCFA: long-chain fatty acids; LC-MS: liquid chromatography – mass spectrometry; NMR: nuclear magnetic resonance; NOB: non obese; NW: normal weight; OB: obese; OW: overweight; Phe: phenylalanine; Pro: proline; Pyr: pyruvate; SM: sphingomyelin; T2D: type 2 diabetes; TAG: triacylglycerides; Trp: tryptophan; Tyr: tyrosine.

5. Conclusions

Not only is the concept of obesity in children and adolescents unclear, but the definition of insulin resistance continues to be controversial. In this regard, the combined analysis of adipokines (particularly leptin and adiponectin), growth factors, inflammatory markers, chemokines, metabolic and metabolomic markers could be useful to predict the existence of insulin resistance in children with obesity prior to overt glucose metabolism impairment.

The evolution of obesity and its comorbidities differ between children and adults and more studies are necessary in children to define insulin resistance, as well as metabolic syndrome, and determine its implications in further complications.

New and precise markers of the evolution of glucose metabolism in children and adolescents with obesity are necessary to provide a correct diagnosis and early intervention.

Metabolomics, untargeted, targeted and the combination of these, is a powerful new technology to understand metabolism and to highlight possible biomarkers with clinical relevance.

Metabolomics can provide valuable information from bench to bedside and backward, and the information gathered from large metabolomics studies can be applied to the pursuance of precision nutrition. Ideally, we will be able to relate the presence of some metabolites, at least to some extent, to characterize the individual needs in terms of nutrition.

More studies are clearly necessary to precisely determine the progression of alterations in glucose metabolism in young patients with obesity to identify clear biomarkers of risk of further complications. The metabolites that are most often found associated with obesity and/or insulin resistance (BCAA and acylcarnitines) still need to be studied in children and adolescents. Moreover, other biomarkers coming from untargeted studies (related to inflammation or the gut microflora) should be tested in the clinic.

The patient's sex must be taken into consideration even in prepubertal periods.

Funding: This work has been partially funded by the Ministerio de Ciencia, Innovación y Universidades of Spain (MICINN/FEDER RTI2018-095166-B-I00) and Spanish Ministry of Health (FIS-PI19/00166).

Conflicts of Interest: The authors declare no conflict of interest.

References

1. World Health Organization. Childhood Overweight and Obesity. Available online: https://www.who.int/dietphysicalactivity/childhood/en/ (accessed on 21 October 2020).
2. World Health Organization. Tenfold Increase in Childhood and Adolescent Obesity in Four Decades: New Study by Imperial College London and WHO. Available online: https://www.who.int/news/item/11-10-2017-tenfold-increase-in-childhood-and-adolescent-obesity-in-four-decades-new-study-by-imperial-college-london-and-who(accessed on 21 October 2020).
3. Freedman, D.S.; Wang, J.; Maynard, L.M.; Thornton, J.C.; Mei, Z.; Pierson, R.N.; Dietz, W.H.; Horlick, M. Relation of BMI to fat and fat-free mass among children and adolescents. *Int. J. Obes.* **2005**, *29*, 1–8. [CrossRef] [PubMed]
4. World Health Organization. Obesity. Available online: https://www.who.int/health-topics/obesity#tab=tab_1 (accessed on 21 October 2020).
5. CDC (Centers for Disease Control and prevention). Defining Adult Overweight and Obesity. Available online: https://www.cdc.gov/obesity/adult/defining.html (accessed on 21 October 2020).
6. Martos-Moreno, G.A.; Argente, J. Obesidades pediátricas: De la lactancia a la adolescencia. In *Anales de Pediatría*; Elsevier Doyma: Barcelona, Spain, 2011; Volume 75. [CrossRef]
7. De Onis, M.; Blössner, M. The World Health Organization Global Database on Child Growth and Malnutrition: Methodology and applications. *Int. J. Epidemiol.* **2003**, *32*, 518–526. [CrossRef] [PubMed]
8. Wright, N.; Wales, J. Assessment and management of severely obese children and adolescents. *Arch. Dis. Child.* **2016**, *101*, 1161–1167. [CrossRef]
9. Flegal, K.M.; Wei, R.; Ogden, C.L.; Freedman, D.S.; Johnson, C.L.; Curtin, L.R. Characterizing extreme values of body mass index-for-age by using the 2000 Centers for Disease Control and Prevention growth charts. *Am. J. Clin. Nutr.* **2009**, *90*, 1314–1320. [CrossRef]

10. Pacheco, L.S.; Blanco, E.; Burrows, R.; Reyes, M.; Lozoff, B.; Gahagan, S. Early onset obesity and risk of metabolic syndrome among Chilean adolescents. *Prev. Chronic Dis.* **2017**, *14*, 1–9. [CrossRef] [PubMed]

11. Fève, B. Adipogenesis: Cellular and molecular aspects. *Best Pract. Res. Clin. Endocrinol. Metab.* **2005**, *19*, 483–499. [CrossRef]

12. Cristancho, A.G.; Lazar, M.A. Forming Functional Fat: A Growing Understanding of Adipocyte Differentiation. *Nat. Rev. Mol. Cell Biol.* **2011**, *12*, 722–734. [CrossRef]

13. Majka, S.M.; Barak, Y.; Klemm, D.J. Adipocyte Origins Weighing the Possibilities. *Stem Cells.* **2011**, *29*, 1034–1040. [CrossRef]

14. Martos-Moreno, G.Á.; Barrios, V.; Chowen, J.A.; Argente, J. Adipokines in Childhood Obesity. In *Vitamins & Hormones*; Elsevier: Amsterdam, The Netherlands, 2013; Volume 91, ISBN 9780124077669.

15. Coppack, S.W. Adipose tissue changes in obesity. *Biochem. Soc. Trans.* **2005**, *33*, 1049–1052. [CrossRef]

16. Després, J.P.; Lemieux, I. Abdominal obesity and metabolic syndrome. *Nature* **2006**, *444*, 881–887. [CrossRef]

17. Wellen, K.E.; Hotamisligil, G.S. Obesity-induced inflammatory changes in adipose tissue. *J. Clin. Invest.* **2003**, *112*, 1785–1788. [CrossRef]

18. Charo, I.F.; Ransohoff, R.M. Mechanisms of disease: The many roles of chemokines and chemokine receptors in inflammation. *N. Engl. J. Med.* **2006**, *354*, 610–621. [CrossRef] [PubMed]

19. Lumeng, C.N.; DeYoung, S.M.; Bodzin, J.L.; Saltiel, A.R. Increased inflammatory properties of adipose tissue macrophages recruited during diet-induced obesity. *Diabetes* **2007**, *56*, 16–23. [CrossRef]

20. Fain, J.N. Release of Interleukins and Other Inflammatory Cytokines by Human Adipose Tissue Is Enhanced in Obesity and Primarily due to the Nonfat Cells. *Vitam. Horm.* **2006**, *74*, 443–477. [CrossRef] [PubMed]

21. Hausman, D.B.; DiGirolamo, M.; Bartness, T.J.; Hausman, G.J.; Martin, R.J. The biology of white adipocyte proliferation. *Obes. Rev. Off. J. Int. Assoc. Study Obes.* **2001**, *2*, 239–254. [CrossRef]

22. Tsuchida, A.; Yamauchi, T.; Kadowaki, T. Nuclear receptors as targets for drug development: Molecular mechanisms for regulation of obesity and insulin resistance by peroxisome proliferator-activated receptor gamma, CREB-binding protein, and adiponectin. *J. Pharmacol. Sci.* **2005**, *97*, 164–170. [CrossRef] [PubMed]

23. Skurk, T.; Alberti-Huber, C.; Herder, C.; Hauner, H. Relationship between adipocyte size and adipokine expression and secretion. *J. Clin. Endocrinol. Metab.* **2007**, *92*, 1023–1033. [CrossRef] [PubMed]

24. Spalding, K.L.; Arner, E.; Westermark, P.O.; Bernard, S.; Buchholz, B.A.; Bergmann, O.; Blomqvist, L.; Hoffstedt, J.; Näslund, E.; Britton, T.; et al. Dynamics of fat cell turnover in humans. *Nature* **2008**, *453*, 783–787. [CrossRef]

25. Martos-Moreno, G.A.; Barrios, V.; Martínez, G.; Hawkins, F.; Argente, J. Effect of weight loss on high-molecular weight adiponectin in obese children. *Obesity* **2010**, *18*, 2288–2294. [CrossRef]

26. Sone, M.; Osamura, R.Y. Leptin and the pituitary. *Pituitary* **2001**, *4*, 15–23. [CrossRef]

27. Farooqi, I.S.; Depaoli, A.M.; Rahilly, S.O.; Farooqi, I.S.; Matarese, G.; Lord, G.M.; Keogh, J.M.; Lawrence, E.; Agwu, C.; Sanna, V.; et al. Beneficial effects of leptin on obesity. *J. Clin. Investig.* **2002**, *110*, 1093–1103. [CrossRef] [PubMed]

28. Mantzoros, C.S.; Magkos, F.; Brinkoetter, M.; Sienkiewicz, E.; Dardeno, T.A.; Kim, S.Y.; Hamnvik, O.P.R.; Koniaris, A. Leptin in human physiology and pathophysiology. *Am. J. Physiol. Endocrinol. Metab.* **2011**, *301*. [CrossRef] [PubMed]

29. Martos-Moreno, G.Á.; Barrios, V.; De Pipaón, M.S.; Pozo, J.; Dorronsoro, I.; Martínez-Biarge, M.; Quero, J.; Argente, J. Influence of prematurity and growth restriction on the adipokine profile, IGF1, and ghrelin levels in cord blood: Relationship with glucose metabolism. *Eur. J. Endocrinol.* **2009**, *161*, 381–389. [CrossRef] [PubMed]

30. Martos-Moreno, G.Á.; Barrios, V.; Argente, J. Normative data for adiponectin, resistin, interleukin 6 and leptin/receptor ratio in a healthy Spanish pediatric population: Relationship with sex steroids. *Eur. J. Endocrinol.* **2006**, *155*, 429–434. [CrossRef]

31. Meier, U.; Gressner, A.M. Endocrine regulation of energy metabolism: Review of pathobiochemical and clinical chemical aspects of leptin, ghrelin, adiponectin, and resistin. *Clin. Chem.* **2004**, *50*, 1511–1525. [CrossRef]

32. Matsuzawa, Y. Adiponectin: A key player in obesity related disorders. *Curr. Pharm. Des.* **2010**, *16*, 1896–1901. [CrossRef]

33. Sowers, J.R. Endocrine Functions of Adipose Tissue: Focus on Adiponectin. *Clin. Cornerstone* **2008**, *9*, 32–40. [CrossRef]

34. Jeffery, A.N.; Murphy, M.J.; Metcalf, B.S.; Hosking, J.; Voss, L.D.; English, P.; Sattar, N.; Wilkin, T.J. Adiponectin in childhood. *Int. J. Pediatr. Obes.* **2008**, *3*, 130–140. [CrossRef]

35. Arita, Y.; Kihara, S.; Ouchi, N.; Takahashi, M.; Maeda, K.; Miyagawa, J.I.; Hotta, K.; Shimomura, I.; Nakamura, T.; Miyaoka, K.; et al. Paradoxical decrease of an adipose-specific protein, adiponectin, in obesity. *Biochem. Biophys. Res. Commun.* **2012**, *425*, 560–564. [CrossRef]

36. Yu, J.G.; Javorschi, S.; Hevener, A.L.; Kruszynska, Y.T.; Norman, R.A.; Sinha, M.; Olefsky, J.M. The effect of thiazolidinediones on plasma adiponectin levels in normal, obese, and type 2 diabetic subjects. *Diabetes* **2002**, *51*, 2968–2974. [CrossRef]

37. Nishimura, R.; Sano, H.; Matsudaira, T.; Miyashita, Y.; Morimoto, A.; Shirasawa, T.; Takahashi, E.; Kawaguchi, T.; Tajima, N. Childhood obesity and its relation to serum adiponectin and leptin: A report from a population-based study. *Diabetes Res. Clin. Pract.* **2007**, *76*, 245–250. [CrossRef]

38. Martos-Moreno, G.Á.; Martínez-Villanueva, J.; González-Leal, R.; Chowen, J.A.; Argente, J. Sex, puberty, and ethnicity have a strong influence on growth and metabolic comorbidities in children and adolescents with obesity: Report on 1300 patients (the Madrid Cohort). *Pediatric Obes.* **2019**, *14*, e12565. [CrossRef]

39. Eyzaguirre, F.; Mericq, V. Insulin resistance markers in children. *Horm. Res.* **2009**, *71*, 65–74. [CrossRef]

40. Gutt, M.; Davis, C.L.; Spitzer, S.B.; Llabre, M.M.; Kumar, M.; Czarnecki, E.M.; Schneiderman, N.; Skyler, J.S.; Marks, J.B. Validation of the insulin sensitivity index (ISI0,120): Comparison with other measures. *Diabetes Res. Clin. Pract.* **2000**, *47*, 177–184. [CrossRef]

41. Reaven, G.M. Role of insulin resistance in human disease. *Diabetes* **1988**, *37*, 1595–1607. [CrossRef] [PubMed]

42. Stefan, N. Causes, consequences, and treatment of metabolically unhealthy fat distribution. *Lancet Diabetes Endocrinol.* **2020**, *8*, 616–627. [CrossRef]

43. Zimmet, P.; Alberti, K.G.M.; Kaufman, F.; Tajima, N.; Silink, M.; Arslanian, S.; Wong, G.; Bennett, P.; Shaw, J.; Caprio, S. The metabolic syndrome in children and adolescents—An IDF consensus report. *Pediatric Diabetes* **2007**, *8*, 299–306. [CrossRef]

44. Damanhoury, S.; Newton, A.S.; Rashid, M.; Hartling, L.; Byrne, J.L.S.; Ball, G.D.C. Defining metabolically healthy obesity in children: A scoping review. *Obes. Rev. Off. J. Int. Assoc. Study Obes.* **2018**, *19*, 1476–1491. [CrossRef]

45. Prince, R.L.; Kuk, J.L.; Ambler, K.A.; Dhaliwal, J.; Ball, G.D.C. Predictors of metabolically healthy obesity in children. *Diabetes Care* **2014**, *37*, 1462–1468. [CrossRef] [PubMed]

46. Heinzle, S.; Ball, G.D.C.; Kuk, J.L. Variations in the prevalence and predictors of prevalent metabolically healthy obesity in adolescents. *Pediatric Obes.* **2016**, *11*, 425–433. [CrossRef]

47. Schwartz, M.W.; Niswender, K.D. Adiposity signaling and biological defense against weight gain: Absence of protection or central hormone resistance? *J. Clin. Endocrinol. Metab.* **2004**, *89*, 5889–5897. [CrossRef]

48. Rivera, P.; Martos-Moreno, G.Á.; Barrios, V.; Suárez, J.; Pavón, F.J.; Chowen, J.A.; Rodríguez de Fonseca, F.; Argente, J. A novel approach to childhood obesity: Circulating chemokines and growth factors as biomarkers of insulin resistance. *Pediatric Obes.* **2019**, *14*, e12473. [CrossRef]

49. Rivera, P.; Martos-Moreno, G.Á.; Barrios, V.; Suárez, J.; Pavón, F.J.; Chowen, J.A.; Rodríguez de Fonseca, F.; Argente, J. A combination of circulating chemokines as biomarkers of obesity-induced insulin resistance at puberty. *Pediatric Obes.* **2020**, e12711. [CrossRef]

50. González-Riano, C.; Dudzik, D.; Garcia, A.; Gil-De-La-Fuente, A.; Gradillas, A.; Godzien, J.; López-Gonzálvez, Á.; Rey-Stolle, F.; Rojo, D.; Ruperez, F.J.; et al. Recent Developments along the Analytical Process for Metabolomics Workflows. *Anal. Chem.* **2019**, *92*, 203–226. [CrossRef] [PubMed]

51. Mastrangelo, A.; Martos-Moreno, G.; García, A.; Barrios, V.; Rupérez, F.J.; Chowen, J.A.; Barbas, C.; Argente, J. Insulin resistance in prepubertal obese children correlates with sex-dependent early onset metabolomic alterations. *Int. J. Obes.* **2016**, *40*, 1494–1502. [CrossRef]

52. Alonso, A.; Marsal, S.; Julià, A. Analytical methods in untargeted metabolomics: State of the art in 2015. *Front. Bioeng. Biotechnol.* **2015**, *3*, 23. [CrossRef]

53. Perng, W.; Gillman, M.W.; Fleisch, A.F.; Michalek, R.D.; Watkins, S.M.; Isganaitis, E.; Patti, M.E.; Oken, E. Metabolomic profiles and childhood obesity. *Obesity* **2014**, *22*, 2570–2578. [CrossRef] [PubMed]

54. Perng, W.; Hector, E.C.; Song, P.X.K.; Tellez Rojo, M.M.; Raskind, S.; Kachman, M.; Cantoral, A.; Burant, C.F.; Peterson, K.E. Metabolomic Determinants of Metabolic Risk in Mexican Adolescents. *Obesity* **2017**, *25*, 1594–1602. [CrossRef] [PubMed]

55. Perng, W.; Rifas-Shiman, S.L.; Hivert, M.F.; Chavarro, J.E.; Oken, E. Branched Chain Amino Acids, Androgen Hormones, and Metabolic Risk Across Early Adolescence: A Prospective Study in Project Viva. *Obesity* **2018**, *26*, 916–926. [CrossRef]

56. Butte, N.F.; Liu, Y.; Zakeri, I.F.; Mohney, R.P.; Mehta, N.; Voruganti, V.S.; Göring, H.; Cole, S.A.; Comuzzie, A.G. Global metabolomic profiling targeting childhood obesity in the Hispanic population. *Am. J. Clin. Nutr.* **2015**, *102*, 256–267. [CrossRef]

57. Naz, S.; García, A.; Barbas, C. Multiplatform Analytical Methodology for Metabolic Fingerprinting of Lung Tissue. *Anal. Chem.* **2013**, *85*, 10941–10948. [CrossRef] [PubMed]

58. Zeki, Ö.C.; Eylem, C.C.; Reçber, T.; Kır, S.; Nemutlu, E. Integration of GC–MS and LC–MS for untargeted metabolomics profiling. *J. Pharm. Biomed. Anal.* **2020**, *190*, 113509. [CrossRef]

59. Gawlik, A.M.; Shmoish, M.; Hartmann, M.F.; Wudy, S.A.; Hochberg, Z. Steroid metabolomic signature of insulin resistance in childhood obesity. *Diabetes Care* **2020**, *43*, 405–410. [CrossRef] [PubMed]

60. Ramautar, R.; Somsen, G.W.; de Jong, G.J. CE-MS for metabolomics: Developments and applications in the period 2016–2018. *Electrophoresis* **2019**, *40*, 165–179. [CrossRef]

61. Gil-de-la-Fuente, A.; Godzien, J.; Saugar, S.; Garcia-Carmona, R.; Badran, H.; Wishart, D.S.; Barbas, C.; Otero, A. CEU Mass Mediator 3.0: A Metabolite Annotation Tool. *J. Proteome Res.* **2019**, *18*, 797–802. [CrossRef] [PubMed]

62. Martos-Moreno, G.; Mastrangelo, A.; Barrios, V.; Garciá, A.; Chowen, J.A.; Rupérez, F.J.; Barbas, C.; Argente, J. Metabolomics allows the discrimination of the pathophysiological relevance of hyperinsulinism in obese prepubertal children. *Int. J. Obes.* **2017**, *41*, 1473–1480. [CrossRef]

63. Tricò, D.; Prinsen, H.; Giannini, C.; De Graaf, R.; Juchem, C.; Li, F.; Caprio, S.; Santoro, N.; Herzog, R.I. Elevated a-hydroxybutyrate and branched-chain amino acid levels predict deterioration of glycemic control in adolescents. *J. Clin. Endocrinol. Metab.* **2017**, *102*, 2473–2481. [CrossRef]

64. Hosking, J.; Pinkney, J.; Jeffery, A.; Cominetti, O.; Da Silva, L.; Collino, S.; Kussmann, M.; Hager, J.; Martin, F.P. Insulin Resistance during normal child growth and development is associated with a distinct blood metabolic phenotype (Earlybird 72). *Pediatric Diabetes* **2019**, *20*, 832–841. [CrossRef]

65. Evans, A.M.; DeHaven, C.D.; Barrett, T.; Mitchell, M.; Milgram, E. Integrated, Nontargeted Ultrahigh Performance Liquid Chromatography/Electrospray Ionization Tandem Mass Spectrometry Platform for the Identification and Relative Quantification of the Small-Molecule Complement of Biological Systems. *Anal. Chem.* **2009**, *81*, 6656–6667. [CrossRef]

66. Shin, S.-Y.; Fauman, E.B.; Petersen, A.-K.; Krumsiek, J.; Santos, R.; Huang, J.; Arnold, M.; Erte, I.; Forgetta, V.; Yang, T.-P.; et al. An atlas of genetic influences on human blood metabolites. *Nat. Genet.* **2014**, *46*, 543–550. [CrossRef]

67. Hellmuth, C.; Kirchberg, F.F.; Brandt, S.; Moß, A.; Walter, V.; Rothenbacher, D.; Brenner, H.; Grote, V.; Gruszfeld, D.; Socha, P.; et al. An individual participant data meta-analysis on metabolomics profiles for obesity and insulin resistance in European children. *Sci. Rep.* **2019**, *9*, 1–14. [CrossRef] [PubMed]

68. Lee, W.; Lee, H.J.; Jang, H.B.; Kim, H.J.; Ban, H.J.; Kim, K.Y.; Nam, M.S.; Choi, J.S.; Lee, K.T.; Cho, S.B.; et al. Asymmetric dimethylarginine (ADMA) is identified as a potential biomarker of insulin resistance in skeletal muscle. *Sci. Rep.* **2018**, *8*, 1–13. [CrossRef] [PubMed]

69. Newgard, C.B.; An, J.; Bain, J.R.; Muehlbauer, M.J.; Stevens, R.D.; Lien, L.F.; Haqq, A.M.; Shah, S.H.; Arlotto, M.; Slentz, C.A.; et al. A BCAA Related Metabolic Signature that differentiates obese and lean. *Cell Metab.* **2009**, *9*, 311–326. [CrossRef]

70. Newgard, C.B. Interplay between lipids and branched-chain amino acids in development of insulin resistance. *Cell Metab.* **2012**, *15*, 606–614. [CrossRef] [PubMed]

71. Balikcioglu, P.G.; Newgard, C.B. Metabolomic signatures and metabolic complications in childhood obesity. *Contemp. Endocrinol.* **2018**, 343–361. [CrossRef]

72. Mccormack, S.E.; Shaham, O.; Mccarthy, M.A.; Deik, A.A.; Wang, T.J.; Gerszten, R.E.; Clish, C.B.; Mootha, V.K.; Grinspoon, S.K.; Fleischman, A. Circulating branched-chain amino acid concentrations are associated with obesity and future insulin resistance in children and adolescents. *Pediatric Obes.* **2013**, *8*, 52–61. [CrossRef]

73. Gkourogianni, A.; Kosteria, I.; Telonis, A.G.; Margeli, A.; Mantzou, E.; Konsta, M.; Loutradis, D.; Mastorakos, G.; Papassotiriou, I.; Klapa, M.I.; et al. Plasma metabolomic profiling suggests early indications for predisposition to latent insulin resistance in children conceived by ICSI. *PLoS ONE* **2014**, *9*, e94001. [CrossRef]

74. Newbern, D.; Balikcioglu, P.G.; Balikcioglu, M.; Bain, J.; Muehlbauer, M.; Stevens, R.; Ilkayeva, O.; Dolinsky, D.; Armstrong, S.; Irizarry, K.; et al. Sex differences in biomarkers associated with insulin resistance in obese adolescents: Metabolomic profiling and principal components analysis. *J. Clin. Endocrinol. Metab.* **2014**, *99*, 4730–4739. [CrossRef]

75. Zhao, X.; Gang, X.; Liu, Y.; Sun, C.; Han, Q.; Wang, G. Using Metabolomic Profiles as Biomarkers for Insulin Resistance in Childhood Obesity: A Systematic Review. *J. Diabetes Res.* **2016**, *2016*. [CrossRef]

76. Perng, W.; Rifas-Shiman, S.L.; Sordillo, J.; Hivert, M.F.; Oken, E. Metabolomic Profiles of Overweight/Obesity Phenotypes During Adolescence: A Cross-Sectional Study in Project Viva. *Obesity* **2020**, *28*, 379–387. [CrossRef]

77. Bagheri, M.; Djazayery, A.; Qi, L.; Yekaninejad, M.S.; Chamari, M.; Naderi, M.; Ebrahimi, Z.; Koletzko, B.; Uhl, O.; Farzadfar, F. Effectiveness of vitamin D therapy in improving metabolomic biomarkers in obesity phenotypes: Two randomized clinical trials. *Int. J. Obes.* **2018**, *42*, 1782–1796. [CrossRef] [PubMed]

78. Menni, C.; Zhai, G.; MacGregor, A.; Prehn, C.; Römisch-Margl, W.; Suhre, K.; Adamski, J.; Cassidy, A.; Illig, T.; Spector, T.D.; et al. Targeted metabolomics profiles are strongly correlated with nutritional patterns in women. *Metabolomics* **2013**, *9*, 506–514. [CrossRef] [PubMed]

79. Michaliszyn, S.F.; Sjaarda, L.A.; Mihalik, S.J.; Lee, S.J.; Bacha, F.; Chace, D.H.; De Jesus, V.R.; Vockley, J.; Arslanian, S.A. Metabolomic profiling of amino acids and β-cell function relative to insulin sensitivity in youth. *J. Clin. Endocrinol. Metab.* **2012**, *97*, 2119–2124. [CrossRef] [PubMed]

80. Benítez-Páez, A.; Gómez del Pugar, E.M.; López-Almela, I.; Moya-Pérez, Á.; Codoñer-Franch, P.; Sanz, Y. Depletion of Blautia Species in the Microbiota of Obese Children Relates to Intestinal Inflammation and Metabolic Phenotype Worsening. *Msystems* **2020**, *5*, 1–13. [CrossRef]

81. Shah, R.; Murthy, V.; Pacold, M.; Danielson, K.; Tanriverdi, K.; Larson, M.G.; Hanspers, K.; Pico, A.; Mick, E.; Reis, J.; et al. Extracellular RNAs are associated with insulin resistance and metabolic phenotypes. *Diabetes Care* **2017**, *40*, 546–553. [CrossRef]

82. Park, S.E.; Park, C.Y.; Sweeney, G. Biomarkers of insulin sensitivity and insulin resistance: Past, present and future. *Crit. Rev. Clin. Lab. Sci.* **2015**, *52*, 180–190. [CrossRef]

83. Rauschert, S.; Uhl, O.; Koletzko, B.; Hellmuth, C. Metabolomic biomarkers for obesity in humans: A short review. *Ann. Nutr. Metab.* **2014**, *64*, 314–324. [CrossRef]

84. Hellmuth, C.; Kirchberg, F.F.; Lass, N.; Harder, U.; Peissner, W.; Koletzko, B.; Reinehr, T. Tyrosine Is Associated with Insulin Resistance in Longitudinal Metabolomic Profiling of Obese Children. *J. Diabetes Res.* **2016**, *2016*. [CrossRef]

85. Lee, A.; Jang, H.B.; Ra, M.; Choi, Y.; Lee, H.J.; Park, J.Y.; Kang, J.H.; Park, K.H.; Park, S.I.; Song, J. Prediction of future risk of insulin resistance and metabolic syndrome based on Korean boy's metabolite profiling. *Obes. Res. Clin. Pract.* **2015**, *9*, 336–345. [CrossRef]

86. Syme, C.; Czajkowski, S.; Shin, J.; Abrahamowicz, M.; Leonard, G.; Perron, M.; Richer, L.; Veillette, S.; Gaudet, D.; Strug, L.; et al. Glycerophosphocholine Metabolites and Cardiovascular Disease Risk Factors in Adolescents: A Cohort Study. *Circulation* **2016**, *134*, 1629–1636. [CrossRef]

87. Mihalik, S.J.; Michaliszyn, S.F.; De Las Heras, J.; Bacha, F.; Lee, S.J.; Chace, D.H.; DeJesus, V.R.; Vockley, J.; Arslanian, S.A. Metabolomic profiling of fatty acid and amino acid metabolism in youth with obesity and type 2 diabetes: Evidence for enhanced mitochondrial oxidation. *Diabetes Care* **2012**, *35*, 605–611. [CrossRef] [PubMed]

Publisher's Note: MDPI stays neutral with regard to jurisdictional claims in published maps and institutional affiliations.

Article

Influence of Single Nucleotide Polymorphisms of ELOVL on Biomarkers of Metabolic Alterations in the Mexican Population

María Luisa Maycotte-Cervantes [1], Adriana Aguilar-Galarza [1], Miriam Aracely Anaya-Loyola [1], Ma. de Lourdes Anzures-Cortes [2], Lorenza Haddad-Talancón [2], Akram Sharim Méndez-Rangel [2], Teresa García-Gasca [1], Víctor Manuel Rodríguez-García [3,*] and Ulisses Moreno-Celis [1,*]

[1] Facultad de Ciencias Naturales, Universidad Autónoma de Querétaro, Juriquilla, Querétaro CP 76230, Mexico; maru.maycotte@gmail.com (M.L.M.-C.); adrianaag.ga@gmail.com (A.A.-G.); aracely.anaya@uaq.mx (M.A.A.-L.); tggasca@uaq.edu.mx (T.G.-G.)

[2] Código 46 SA de CV, Cuernavaca, Morelos 62498, Mexico; lourdes@codigo46.com.mx (M.d.L.A.-C.); lorenza@codigo46.com.mx (L.H.-T.); akramsharim@gmail.com (A.S.M.-R.)

[3] Tecnologico de Monterrey, Escuela de Ingeniería y Ciencias, San Pablo, Querétaro 76130, Mexico

* Correspondence: vmrodrigg@tec.mx (V.M.R.-G.); ulisses.moreno@uaq.mx (U.M.-C.); Tel.: +52-442-238-3525 (V.M.R.-G.); +52-442-192-1200 (ext. 5367) (U.M.-C.)

Received: 30 September 2020; Accepted: 31 October 2020; Published: 4 November 2020

Abstract: The elongation of very long chain fatty acids (ELOVL) is a family of seven enzymes that have specific functions in the synthesis of fatty acids. Some have been shown to be related to insulin secretion (ELOVL2), and in the lipid profile (ELOVL6) and patients with various pathologies. The present work focused on the study of ELOVL polymorphs with clinical markers of non-communicable chronic diseases in the Mexican population. A sample of 1075 participants was obtained, who underwent clinical, biochemical, and nutritional evaluation, and a genetic evaluation of 91 genetic variants of ELOVL was considered (2–7). The results indicate a 33.16% prevalence of obesity by body mass index, 13.84% prevalence of insulin resistance by homeostatic model assessment (HOMA) index, 7.85% prevalence of high cholesterol, and 20.37% prevalence of hypercholesterolemia. The deprived alleles showed that there is no association between them and clinical disease risk markers, and the notable finding of the association studies is that the ELOVL2 variants are exclusive in men and ELVOL7 in women. There is also a strong association of ELOVL6 with various markers. The present study shows, for the first time, the association between the different ELOVLs and clinical markers of chronic non-communicable diseases.

Keywords: biomarkers; ELOVL; metabolic alterations; Mexican population; Single Nucleotide Polymorphisms

---

1. Introduction

The elongases of very long chain fatty acid (ELOVL) family of fatty acid elongases includes seven enzymes that catalyze elongation of the carbon chain of fatty acids (FAs) during their condensation phase. Other studies have reported that ELOVLs are specific to the type of FA that elongate; ELOVL1, ELOVL3, ELOVL6, and ELOVL7 preferably elongate saturated fatty acids (SFAs) and monounsaturated fatty acids (MFAs); whereas ELOVL2 and ELOVL5 are specific to polyunsaturated fatty acids (PUFAs), and ELOVL4 elongates PUFAs and SFAs of very long chains [1,2]. Given the important role of lipids in the metabolism, their associations with the development of various pathological processes, such as hepatic steatosis, obesity, insulin resistance and diabetes mellitus 2, and cancer, are currently being studied [2]. As an example, the importance of ELOVL2 in pancreatic insulin secretion has been demonstrated in animal models [3]. An ELOVL3 knockout model has also been shown to be related to adiponectin reduction, adipose tissue expansion, and diet-induced resistance to obesity, in addition to a reduction in liver lipids and a decrease in triglycerides (TGs) in blood [4]. By comparison, the determination of the metabolic function of ELOVL4 has shown some influence on ceramides and very long chain fatty acids (VLFAs), mainly in the skin, brain, and eyes of mice [5,6]. Furthermore, possible participation in the pathophysiology of some diseases has not yet been evaluated. Moreover, ELOVL5 has been identified as an intermediate in the signaling for the export of glucose receptors to the membrane and the increase in the activity that promotes insulin resistance, mainly suppressing the activity of the transcriptional factor mTorc1. Studies have also reported that ELOVL5 regulates the levels of triglycerides in the liver [7]. The activity of ELOVL6 is the best studied, in addition to its importance in adipogenesis as a target of the transcription factor SRBP1 [8] and its effect in preventing insulin resistance when it is knocked down or knocked out in transgenic mice [9]. ELOVL7 is the least studied of all fatty acid elongases, but it has been associated with pathologies such as cancer and Parkinson's [10].

Polymorphisms in the ELOVL enzymes affect the fatty acid composition of breast milk [11]. A study of a Costa Rican population found no association between the Single Nucleotide Polymorphisms (SNPs) of ELOVL2 (rs2295601, rs10498676, and rs3734397), ELOVL4 (rs17239120), and ELOVL5 (rs17544464, rs2115564, rs2294867, and rs761179) and the risk of acute myocardial infarction, but found an association with the metabolism of lipids [12]. Thus, the present work is designed to determine the relationship between 91 SNPs of ELOVL found in a Mexican population and the biomarkers of chronic non-communicable diseases.

2. Materials and Methods

A total of 1075 Mexican subjects participating in the SUSALUD-UAQ (University Health Program from the Autonomous University of Queretaro) program were sampled, comprising 563 women (52.3%) and 512 men (47.6%) of 18 to 30 years old. Subjects with previously diagnosed health problems were not included for analysis and those who did not have complete clinical, biochemical, and genetic information were excluded from the study. This study was approved by the Bioethics Committee (52FCN2017) of the Natural Science Department of the Autonomous University of Queretaro under the guidelines of the Declaration of Helsinki [13].

2.1. Evaluation of Nutritional Status and Body Composition

Anthropometric data was collected, including weight, height, and waist and hip circumferences, following the standard procedures of the World Health Organization, 2006 [14]. Height measurement was performed with a wireless transmission stadiometer (Brand SECA, Model 264; Hamburg, Germany). Waist circumference was measured by placing a tape measure on a line that is at the midpoint between the upper iliac crest and the border inferior costal, at the end of a normal expiration. Weight and body composition were determined using a multifrequency bioelectrical impedance device (SECA, model mBCA 515; Hamburg, Germany). Medical staff determined blood pressure of patients subject to the following conditions: without physical exercise and after resting for 10 to 15 min before measurement, back and arm supported, legs not crossed. The reference values of the Treatment of High Blood Cholesterol in Adults (ATPIII) 2005 revision (130/85 mmHg) were considered.

2.2. Evaluation of Biochemical Markers

Blood samples were collected from each subject to obtain plasma (BD Vacutainer Plus) and serum (BD Vacutainer SST II). Following collection of the blood samples, a complete blood count was performed in a Cell-dyn 1400 device (Abbot Mark, IL, USA). Plasma and serum were obtained by centrifugation of the whole blood sample at 2500 rpm for 10 min, and then aliquoted in 1.5 mL cryovials for subsequent analyses and stored at −70 °C (REVCO, Thermo Scientific, Waltham, MA, USA). The serum was used to analyze the concentrations of glucose, total cholesterol, triglycerides, and high-density lipoprotein (HDL), whereas the concentration of low-density lipoprotein (LDL) was calculated using the formula of Fridelwald (i.e., LDL = CT- (TG/5) + HDL)) in patients with TG < 400 mg/dL [15]. For those samples with TG > 400 mg/dL, the determination was made using colorimetric analysis. All biochemical determinations were carried out in duplicate in an automated Mindray BS 120 device (Medical International Limited, Shenzhen, China) using colorimetric enzymatic methods (SPINREACT S.A./S.A.U, Girona, Spain). Fasting glucose concentrations of less than 100 md/dL were considered normal [16].

The reference values from the National High Blood Pressure Education Program Working Group on High Blood Pressure in Children and Adolescents, 2004, were used for diagnostics of the lipid profile [17].

The plasma was used to analyze insulin levels. The insulin levels were analyzed according to the manufacturer's instructions using an ELISA kit (Insulin ELISA 80-INSHU-E01.1, ALPCO INMUNOASSAYS). The readings were compiled using a Multiskan ascent spectrophotometer (Thermo, electron corporation) at a wavelength of 450 nm.

The homeostatic model assessment (*HOMA*) index was calculated according to the following equation:

$$HOMA\ (IR) = [Insulin(\mu U)/mL) \times Glucose(mmol/L)]/22.5 \tag{1}$$

The cut-off values for insulin and *HOMA* index were based on those suggested by Munguía-Romero et al. (2013) [18].

2.3. DNA Purification and Genotyping

Genomic DNA was extracted using the Agencourt DNAdvance kit according to the manufacturer's instructions (Beckmann Coulter, CA, USA). The extracted DNA was quantified and analyzed for its quality using the Spectrophotometer 190 (Molecular Devices, CA, USA) device. As quality criteria, a concentration higher than 40 ng/μl and an absorbance ratio 260/280 of 1.8–2 were used. The integrity of the DNA was also taken into account by electrophoretic running. Next, the Infinium HTS Automated Protocol (Illumina) [19] was followed with the objective of genotyping the samples. For this, the beadchip Global Screening Array-24 + V1.0/HTS CODIGO46_2017_01 was used. Genotypes described in Table 1 were determined using Genome Studio software.

Table 1. Genetic markers analyzed.

Gene	Genetic Variant	Alleles	Functional Consequence
ELOVL2	rs8523	(A/G)	genic downstream transcript variant, 3' UTR variant.
ELOVL2	rs3734396	(A/G)	genic downstream transcript variant, 3' UTR variant.
ELOVL2	rs17606561	(A/G)	genic downstream transcript variant, 3' UTR variant.
ELOVL2	rs3734398	(T/C)	genic downstream transcript variant, 3' UTR variant.
ELOVL2	rs2281591	(A/G)	intron variant, genic downstream transcript variant.
ELOVL2	rs2236212	(G/C)	intron variant, genic downstream transcript variant.
ELOVL2	rs3798713	(G/C)	intron variant.
ELOVL2	rs7765206	(A/C)	intron variant, genic upstream transcript variant.
ELOVL2	rs116279801	(T/C)	intron variant, genic upstream transcript variant.
ELOVL2	rs9295757	(T/G)	intron variant, genic upstream transcript variant.
ELOVL2	rs3798721	(A/C)	intron variant, genic upstream transcript variant.
ELOVL2	rs16870899	(A/G)	intron variant, genic upstream transcript variant.
ELOVL2	rs3798722	(A/G)	intron variant, genic upstream transcript variant.
ELOVL2	rs9393903	(A/G)	intron variant, upstream transcript variant, genic upstream transcript variant.
ELOVL2	rs4532436	(G/C)	genic downstream transcript variant, 3' UTR variant.
ELOVL2	rs12195587	(A/G)	downstream gene transcription variant, synonym variant, coding sequence variant.

Table 1. *Cont.*

Gene	Genetic Variant	Alleles	Functional Consequence
ELOVL3	rs10748816	(A/G)	intron variant.
ELOVL3	rs36103207	(A/G)	nonsense variant, coding sequence variant.
ELOVL4	rs3812153	(T/C)	coding sequence variant, nonsense variant.
ELOVL4	6:80628844	(A/G)	
ELOVL4	rs117891930	(T/C)	intron variant.
ELOVL4	rs144198896	(A/G)	intron variant.
ELOVL4	rs80246554	(T/C)	intron variant.
ELOVL4	rs12196014	(A/G)	intron variant.
ELOVL4	rs9448863	(A/G)	intron variant.
ELOVL4	rs16891339	(A/G)	intron variant.
ELOVL5	rs41273878	(A/C)	coding sequence variant, synonym variant, downstream gene transcription variant.
ELOVL5	rs41273880	(T/C)	coding sequence variant, nonsense variant, synonym variant, downstream gene transcription variant.
ELOVL5	rs72938776	(A/G)	intron variant, genic downstream transcript variant.
ELOVL5	rs182937551	(T/C)	downstream gene transcription variant, intron variant
ELOVL5	rs36054518	(A/G)	intron variant.
ELOVL5	rs209487	(T/G)	intron variant.
ELOVL5	rs115397424	(A/G)	intron variant.
ELOVL5	rs13208390	(T/G)	intron variant.
ELOVL5	rs72940713	(T/C)	intron variant.
ELOVL5	rs114271869	(T/G)	intron variant.
ELOVL5	rs2073040	(A/G)	intron variant, genic downstream transcript variant.
ELOVL5	rs9370194	(C/T)	intron variant.
ELOVL6	rs11098065	(A/G)	intron variant.
ELOVL6	rs17041284	(T/C)	intron variant.
ELOVL6	rs7662161	(T/C)	intron variant.
ELOVL6	rs77958351	(A/G)	intron variant.
ELOVL6	rs77808755	(A/G)	intron variant.
ELOVL6	rs59634436	(A/C)	intron variant.
ELOVL6	rs78160528	(T/C)	intron variant.
ELOVL6	rs16997129	(T/C)	intron variant.
ELOVL6	rs3813827	(A/G)	intron variant.
ELOVL6	rs11737840	(T/C)	intron variant.
ELOVL6	rs10033691	(T/C)	intron variant.
ELOVL6	rs2005701	(T/C)	intron variant.
ELOVL6	rs76145164	(T/C)	intron variant.
ELOVL6	rs76338299	(T/G)	intron variant.
ELOVL6	rs6533491	(T/C)	intron variant.
ELOVL6	rs72679222	(A/G)	intron variant.
ELOVL6	rs11937052	(A/G)	intron variant.
ELOVL6	rs11098070	(A/G)	intron variant.
ELOVL6	rs80343897	(T/C)	intron variant.
ELOVL6	rs373773495	(T/C)	intron variant.
ELOVL6	rs114422025	(T/C)	intron variant.
ELOVL6	rs6533495	(A/G)	intron variant.
ELOVL6	rs28722886	(T/C)	intron variant.
ELOVL6	rs6533497	(T/C)	intron variant.
ELOVL6	rs77504516	(A/G)	intron variant.
ELOVL6	rs6815102	(T/C)	intron variant.
ELOVL6	rs4326075	(A/C)	intron variant.
ELOVL6	rs116418972	(A/G)	intron variant.
ELOVL6	rs11729740	(T/C)	intron variant.
ELOVL6	rs2035415	(T/C)	intron variant.
ELOVL6	rs17041402	(A/C)	intron variant.
ELOVL6	rs59111930	(A/G)	intron variant.
ELOVL6	rs74874270	(A/G)	intron variant.

Table 1. *Cont.*

Gene	Genetic Variant	Alleles	Functional Consequence
ELOVL6	rs1384331	(T/G)	intron variant.
ELOVL6	rs72679246	(A/C)	intron variant.
ELOVL6	rs78563565	(T/C)	intron variant.
ELOVL6	rs6533498	(A/C)	intron variant.
ELOVL6	rs9997926	(C/T)	intron variant.
ELOVL6	rs6824447	(A/G)	upstream transcription variant.
ELOVL6	rs17041272	(C/G)	3′ UTR region variant.
ELOVL7	rs75621404	(A/G)	intron variant, genic downstream transcript variant.
ELOVL7	rs143990657	(A/G)	downstream gene transcription variant, intron variant.
ELOVL7	rs115862620	(T/C)	intron variant, genic downstream transcript variant.
ELOVL7	rs1563517	(T/G)	intron variant, genic downstream transcript variant.
ELOVL7	rs12188996	(A/C)	downstream gene transcription variant, intron variant.
ELOVL7	rs60258111	(T/C)	intron variant, genic downstream transcript variant.
ELOVL7	rs16878426	(T/C)	downstream gene transcription variant, intron variant.
ELOVL7	rs6872863	(A/G)	intron variant, genic downstream transcript variant.
ELOVL7	rs76641655	(T/G)	intron variant, genic downstream transcript variant, coding sequence variant, synonym variant.
ELOVL7	rs145299240	(T/C)	downstream gene transcription variant, intron variant, genic upstream transcript variant.
ELOVL7	rs114011218	(T/C)	upstream gene transcription variant, intron variant, genic downstream transcript variant.
ELOVL7	rs115159664	(T/C)	upstream gene transcription variant, intron variant, genic downstream transcript variant.
ELOVL7	rs4700398	(A/G)	upstream gene transcription variant, intron variant.

2.4. Genetic and Statistical Analyses

Allelic and genotypic frequencies were calculated using GenAlEx. Null alleles were removed from the dataset for further analysis and the remaining markers were tested in reference to the Hardy–Weinberg equilibrium (HWE). Private alleles were also identified and quantified for the analyzed data. SNPs were examined for associations with the studied biomarkers, therefore means and their standard deviations were analyzed with Student's t-tests. We also tested the statistical homogeneity of the effects on body mass index (BMI) in the corresponding regression model between clinical biomarkers. The strength of association between variables was measured by calculating the odds ratio (OR) and 95% confidence intervals using logistic regressions performed with SPSS (ver. 9.6) [20] statistical software; the regression coefficients were tested for significance and the *p*-value used to reject the null hypothesis was 0.05. The multivariate logistic regression model used to calculate the risk associations was controlled for potential confounders such as sex and age. The recessive genotypes were compared to both dominant and heterozygous genotypes, and tested for statistical significance (*p*-value = 0.05).

3. Results

3.1. Characteristics of the Subjects

From the original sample of 1075, 476 participants were eliminated due to having incomplete data for the study, so a final sample of 599 subjects was used. Table 2 shows the general characteristics of the population, in which it can be seen that 311 of the subjects (51.9%) were women and 288 (48.1%) were men, with an average age of 19.1 ± 1.9 years. A comparison of means was carried out in both populations to establish that there were no significant differences in the variables independent of the sex. Significant statistical differences occurred in variables typically classified for both of the sexes.

Table 2. General characteristics of the population.

Biomarkers	Total (N = 599)		Women (N = 311)		Men (N = 288)		
	Mean	S.D.	Mean	S.D.	Mean	S.D.	*p*-Value
Age (years)	19.18	1.95	19.07	1.80	19.29	2.09	0.162
Weight (kg)	65.07	13.70	59.88	12.06	70.65	13.18	0.000
BMI (kg/m^2)	23.76	4.30	23.50	4.44	20.04	4.14	0.127
Waist circumference (cm)	80.98	11.83	78.24	11.62	83.92	11.36	0.000
Waist–Hip Ratio	0.83	0.07	0.81	0.07	0.86	0.06	0.000
Waist–height Ratio	0.49	0.07	0.49	0.07	0.49	0.07	0.970
Body Fat (%)	26.41	0.56	31.39	7.34	21.08	8.06	0.000
Glucose (mg/dL)	83.53	9.04	82.25	9.03	84.91	8.86	0.000
Insulin (µg/mL)	7.87	5.65	7.99	0.07	7.73	5.14	0.581
HOMA-IR Index	1.63	1.20	1.63	1.24	1.63	1.15	0.999
Triglycerides (mg/dL)	105.18	64.11	95.57	53.58	114.45	72.76	0.001
Total Cholesterol (mg/dL)	157.43	30.19	157.44	27.71	157.41	32.71	0.989
HDL (mg/dL)	50.72	12.55	53.25	13.33	48.02	11.07	0.000
LDL (mg/dL)	85.51	23.71	84.63	22.50	85.47	24.95	0.346

S.D.: Standard deviation; BMI: Body mass index; HDL: high-density lipoprotein; LDL: low-density lipoprotein. Student's t-test of statistical significance, *p*-value < 0.05.

3.2. Nutritional Alterations of Subjects

Metabolic alterations according to anthropometric and body composition variables (Figure 1A) indicate the evaluated population has a prevalence of overweight and obesity according to the BMI of 33.16%, which was 31.46% for women and 34.98% for men. Similarly, with respect to the High Waist–Hip Index (H-WHI), 42.35% of the population was above the recommended level, and was significantly higher in men (57.84%) than in women (27.92. Furthermore, the alteration of body composition with the highest prevalence in all of the population (n = 599) was due to the percentage of high body fat, which was 49.04% in the total population (46.10% in women and 52.17% in men).

Significant alterations were also found in biochemical variables (Figure 1B). According to glucose (H-Gluc) values, the prevalence of high levels was low (2.84%), and was 2.57% for women and 3.12% for men. High insulin (H-INS) levels prevailed in 18.2% of the population. It is worth noting that men had higher prevalence (23.12%) than women (13.84%) in this marker. This was similar to the case of the high HOMA index (H-HOMA), which was found be more prevalent in men (16.67%) than in women (11.15%). Regarding lipid metabolism markers, high total cholesterol (H-TC) had a prevalence of 7.85%, and was higher in men (10.07%) than in women (5.79%). Similarly, the high levels of low-density cholesterol (H-cLDL) had a prevalence of 4.51%, with values of 3.33% for women and 5.90% for men. The opposite case was observed in low levels of high-density cholesterol (L-cHDL), the overall prevalence of which was observed to 36.0%, and was higher in women (46.30%) than in men (25.0%). The prevalence of high triglycerides (H-TG) was observed to be 20.37%, and was higher in men (25.0%) than in women (16.08%).

Figure 1C shows the nutritional status of the study subjects according to their BMI, in which values of 6.46% for low weight (LW) were obtained, 7.62% for women and 5.30% for men; 60.34% for normal weight (NW), 59.72% for women and 60.93% for men; and for overweight (OW) the total prevalence was 25.47%, and slightly higher in men (27.21%) than in women (23.84%). For obesity (OB), the prevalence of the population was 7.69%, with 7.62% for women and 7.77% for men. Nutritional status was determined according to the percentage of body fat (Figure 1D), in which a prevalence of low body fat of 4.9% was observed, with values of 2.71% for women and 7.75% for men. By comparison, the total prevalence for normal fat percentage was 46.09%, with values of 51.19% for women and 40.58% for men. Regarding the percentage of high fat, a prevalence of 49.04% was found in the total population, with a higher proportion of men (52.17%) than women (46.10%) presenting this condition.

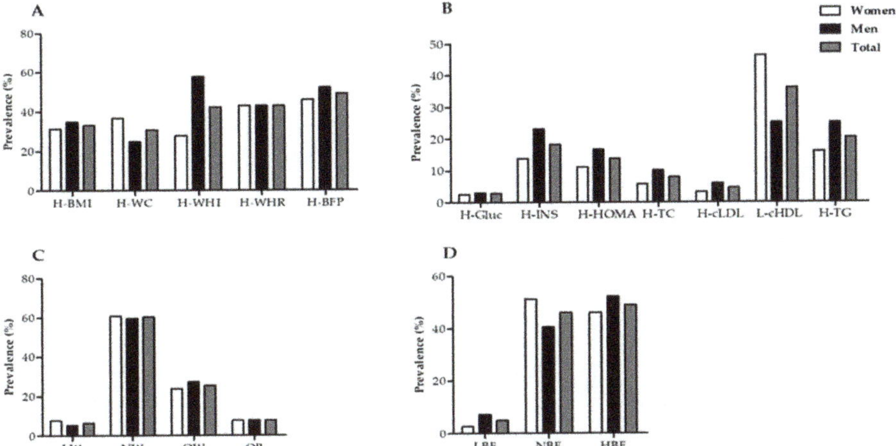

Figure 1. Distribution of nutritional status (**A**) Prevalence of anthropometric and body composition alterations. H-BMI: body mass index > 25.0 kg/m^2; H-WC: waist circumference (women > 0.80 cm and men > 90 cm); H-WHI: waist–hip index (women > 0.85 cm and men > 95 cm); H-WHR: waist–height ratio > 0.50; H-BFP (women > 25% and men > 20%). (**B**) Prevalence of biochemical alterations. H-Gluc: high glucose; >100 mg/dL; H-INS: high insulin, >14 μU/mL for women and >11 μU/mL for men; H-HOMA: high HOMA index: >2.9 for women and >2.3 for men; H-TC: high total cholesterol, >200 mg/dL; H-cLDL: elevated low-density lipoproteins, >130 mg/dL. L-cHDL: low high-density lipoproteins, ≤50 mg/dL for women and ≤40 mg/dL for men; H-TG: high triglycerides > 150 mg/dL. (**C**) Nutritional status according to BMI. LW: low weight: BMI ≤ 18.5 kg/m^2; NW: Standard weight: BMI 18.5–24.9 kg/m^2; OW: Overweight: BMI 25–29.9 kg/m^2; OB: obesity: BMI ≥ 30 kg/m^2. (**D**) Nutritional status according to the percentage of body fat for men: LBF: low body fat < 8%; NBF: body fat 8.1–20%; HBF: high body fat 20.1–25%; for women: low body fat < 15%; NBF: body fat 15–35%; HBF: high body fat > 35% [21].

3.3. Allelic Frequencies, HWE Analysis, and Private Alleles

From the 91 analyzed SNP polymorphisms of ELOVL, only 77 were used for further analysis after null alleles were excluded from the dataset. The allelic and genotypic frequencies analyzed in the sample (n = 599) are shown in Table S1 in the Supplementary Material. ELOVL2 and ELOVL6 were highly diverse with a total of 15 analyzed markers found for the former and 37 for the latter. Most markers showed a significant deviation from the HWE ($p > 0.05$) after sequential Bonferroni correction.

Private alleles were identified after the population was analyzed independently by males and females, and once separated, both subpopulations were analyzed against all biomarkers. This was because some biomarkers behave differentially among sexes. Most of the private alleles were found to have a positive input on the biomarkers (Table 3), with the exception of two SNPs from ELOVL5, rs72938776 (allele A, freq. 0.019) and rs72940713 (allele G, freq. 0.019), which were significantly ($p > 0.05$) associated with the waist–height ratio (WHR) and percentage of body fat (%BF).

Table 3. Single Nucleotide Polymorphisms private alleles with significant associations with biomarkers.

Sex	Gene	Biomarker	Clinical Diagnosis	Locus	Allele	Frequency
Males	ELOVL2	%BF	Low body fat	rs116279801	A	0.025
		Glucose	Normal levels	rs7765206	A	0.028
		HOMA	No insulin resistance	rs7765206	A	0.031
	ELOVL4	WHR	Without cardiovascular risk	rs16891339	G	0.023
			Without cardiovascular risk	rs117891930	A	0.020
		LDL	Without cardiovascular risk	rs80246554	G	0.037
			Without cardiovascular risk	rs12196014	A	0.095
	ELOVL5	WHR	Without cardiovascular risk	rs36054518	G	0.021
		TG	Normal levels	rs36054518	G	0.021
	ELOVL6	Glucose	Normal levels	rs76145164	A	0.021
			Normal levels	rs76338299	A	0.053
		HOMA	No insulin resistance	rs77958351	A	0.021
		Cholesterol	Normal levels	rs76145164	A	0.021
			Normal levels	rs74874270	G	0.031
		LDL	Without cardiovascular risk	rs76145164	A	0.021
			Without cardiovascular risk	rs74874270	G	0.030
	ELOVL7	HOMA	No insulin resistance	rs115159664	G	0.024
		TG	Normal levels	rs115159664	G	0.023
Females	ELOVL2	Cholesterol	Normal levels	rs7765206	A	0.031
			Normal levels	rs16870899	G	0.022
		LDL	Without cardiovascular risk	587/ELOVL2	A	0.071
	ELOVL5	WHR	Without cardiovascular risk	rs72938776	A	0.019
		%BF	High levels	rs72938776	A	0.018
			High levels	rs72940713	G	0.019
	ELOVL6	Glucose	Normal levels	rs76145164	A	0.020
			Normal levels	rs74874270	G	0.029
		HOMA	No insulin resistance	rs17041402	C	0.023
		Cholesterol	Normal levels	rs17041402	C	0.022
			Normal levels	rs74874270	G	0.027
			Normal levels	rs78563565	G	0.024
		TG	Normal levels	rs17041402	C	0.025
		LDL	Without cardiovascular risk	rs11729740	A	0.062
			Without cardiovascular risk	rs17041402	C	0.022
			Without cardiovascular risk	rs74874270	G	0.027
			Without cardiovascular risk	rs78563565	G	0.023
	ELOVL7	WHI	Normal distribution	rs76641655	C	0.022

3.4. Association of SNPs with Clinical Markers of Risk to Chronic Non-Communicable Diseases

To facilitate the analysis of the results, only those whose associations were statistically significant are presented in the following tables and all of the results of the association analyses and their respective *p*-values are shown in the Supplementary Material (S2). For the general population, the associations resulted in five variants associated with risk for ELOVL2, one for ELOVL4, two for ELOVL5, nine for ELOVL6, and three for ELOVL7, giving a total of 20 variants associated with risk markers (Table 4). By comparison, according to the results, 12 genetic variants were found to be protective factors for the risk markers for chronic non-communicable diseases, of which three are for ELOVL2, one for ELOVL3, one for ELOVL5, and seven for ELOVL6 (Table 5). The highest levels of risk identified by the analysis for the total population were found in ELOVL5 and ELOVL6.

Table 4. Association between SNP of elongases of very long chain fatty acids (ELOVL) and clinical markers of chronic non-communicable diseases (risk factors).

Gene	SNP	Clinical Marker	OR	95% CI		*p*-Value
ELOVL2	rs8523	H-Insulin	2.048	1.238	3.388	0.005
		H-HOMA	1.847	1.132	3.013	0.013
		H-Cholesterol	2.628	1.400	4.935	0.002
	rs3734398	H-BMI	1.441	1.020	2.036	0.038
		H-Insulin	1.894	1.146	3.130	0.012
		H-HOMA	1.780	1.088	2.912	0.020
		H-Cholesterol	2.341	1.251	4.378	0.006
	rs2236212	H-Insulin	1.922	1.135	3.254	0.014
		H-HOMA	1.995	1.184	3.361	0.009
		H-Cholesterol	2.432	1.236	4.785	0.008
		H-LDL	2.856	1.136	7.181	0.020
	rs3798713	H-HOMA	1.765	1.058	2.944	0.028
		H-Cholesterol	2.485	1.263	4.888	0.007
		H-LDL	2.915	1.159	7.330	0.018
	rs4532436	H-Insulin	2.248	1.357	3.722	0.001
		H-HOMA	2.025	1.237	3.315	0.004
		H-Cholesterol	2.654	1.418	4.966	0.002
ELOVL4	rs80246554	H-HOMA	2.622	1.164	5.908	0.016
ELOVL5	rs72938776	H-BMI	12.447	1.488	104.127	0.003
		H-Insulin	6.973	1.145	42.469	0.015
	rs72940713	H-BMI	7.32	1.506	35.579	0.004
		H-%BF	8.561	1.064	68.899	0.016
		H-Insulin	6.953	1.142	42.346	0.015
ELOVL6	rs59634436	H-BMI	2.129	1.256	3.608	0.004
		H-%BF	1.879	1.091	3.238	0.021
		H-Insulin	2.237	1.168	4.284	0.013
		H-Glucose	3.597	1.225	10.561	0.013
		H-HOMA	2.023	1.052	3.888	0.032
	rs78160528	H-LDL	4.642	1.258	17.12	0.012
		H-BMI	1.844	1.14	2.983	0.012
	rs10033691	H-%BF	2.27	1.371	3.759	0.001
		H-Insulin	2.094	1.132	3.874	0.017
		H-Glucose	3.687	1.325	10.265	0.008
	rs76145164	H-Triglycerides	3.073	1.264	7.471	0.009
	rs11937052	H-BMI	1.797	1.113	2.901	0.015
		H-Waist	1.776	1.104	2.858	0.017
	rs114422025	H-HOMA	4.678	1.445	15.141	0.005
	rs72679246	H-Glucose	6.144	1.263	29.902	0.011
	rs9997926	H-WHR	1.975	1.098	3.553	0.021
		H-BMI	1.569	1	2.46	0.049
	rs17041272	H-%BF	1.794	1.133	2.839	0.012
		H-Insulin	2.147	1.198	3.847	0.009
		H-HOMA	1.956	1.096	3.491	0.021
ELOVL7	rs1563517	H- Waist	1.517	1.049	2.194	0.027
		H-Insulin	1.844	1.107	3.074	0.018
	rs115159664	H-LDL	3.847	1.06	13.959	0.028
		H-BMI	1.57	1.109	2.221	0.011
	rs4700398	H-WHI	1.438	1.036	1.996	0.03
		H-%BF	1.473	1.058	2.051	0.022

H-BMI: high body mass index > 25.0 kg/m^2; H-Waist: high waist circumference (women > 0.80 cm and men > 90 cm); H-WHI: high waist–hip Index (women > 0.85 cm and men > 95 cm); H-WHR: high waist–height ratio > 0.50; H-%BF: high body fat percent (women > 35% and men > 20%). H-Glucose: high glucose; >100 mg/dL; H-INS: high insulin (>14 µU/mL for women and >11µU/mL for men); H-HOMA: high HOMA index (>2.9 for women and >2.3 for men); H-Cholesterol: high total cholesterol (>200 mg/dL); H-LDL: elevated low-density lipoproteins (>130 mg/dL); L-HDL: low high-density lipoproteins (≤50 mg/dL for women and ≤40 mg/dL for men); H-Triglycerides: high triglycerides (>150 mg/dL). The statistical analysis applied to this dataset was a multinominal regression ($p \leq 0.05$).

Table 5. Association between SNP of ELOVL and clinical markers of chronic non-communicable diseases (protective factors).

Gene	SNP	Clinical Marker	OR	95% CI		*p*-Value
ELOVL2	rs17606561	H-HOMA	0.521	0.298	0.914	0.021
		H-Triglycerides	0.625	0.404	0.968	0.034
	rs2281591	H-Waist	0.641	0.445	0.924	0.017
		H-Insulin	0.517	0.297	0.899	0.018
		H-HOMA	0.479	0.276	0.831	0.008
	rs9393903	H-HOMA	0.560	0.322	0.972	0.038
ELOVL3	rs10748816	H-WHR	0.578	0.401	0.832	0.003
ELOVL5	rs2073040	L-HDL	0.686	0.475	0.992	0.045
ELOVL6	rs11098065	H-Insulin	0.499	0.289	0.861	0.011
		L-HDL	0.686	0.487	0.966	0.031
	rs7662161	H-LDL	0.331	0.123	0.89	0.022
	rs6533491	H-Waist	0.676	0.476	0.958	0.028
		H-WHR	0.704	0.508	0.977	0.035
	rs80343897	H-Glucose	0.202	0.045	0.895	0.02
		L-HDL	0.654	0.462	0.924	0.016
	rs6815102	H-Insulin	0.575	0.349	0.949	0.029
	rs4326075	L-HDL	0.552	0.312	0.977	0.042
	rs6824447	H-Cholesterol	0.447	0.242	0.825	0.008
		H-LDL	0.435	0.196	0.968	0.036

H-BMI: high body mass index > 25.0 kg/m^2; H-Waist: high waist circumference (women > 0.80 cm and men > 90 cm); H-WHI: high waist–hip Index (women > 0.85 cm and men > 95 cm); H-WHR: high waist–height ratio > 0.50; H-%BF: high body fat percent (women > 35% and men > 20%). H-Glucose: high glucose; >100 mg/dL; H-INS: high insulin (>14 μU/mL for women and >11μU/mL for men); H-HOMA: high HOMA index (>2.9 for women and >2.3 for men); H-Cholesterol: high total cholesterol (>200 mg/dL); H-LDL: elevated low-density lipoproteins (>130 mg/dL); L-HDL: low high-density lipoproteins (≤50 mg/dL for women and ≤40 mg/dL for men); H-Triglycerides: high triglycerides (>150 mg/dL). The statistical analysis applied to this dataset was a multinominal regression ($p \leq 0.05$).

When the population was divided by sex for posterior analyses, the results showed 15 SNPs of ELOVL for women were related to clinical markers of chronic non-communicable diseases as risk factors (two for ELOVL5, eight for ELOVL6, and five for ELOVL7) in which the variant rs72938776 of ELOVL5 obtained an OR of 11.37 for high LDL, whereas the rs9370194 variant of the same ELOVL was found with an OR of 2.92 for high total cholesterol. Regarding ELOVL6, the variants rs59634436 (OR = 2.761), rs10033691 (OR = 2.102), rs2005701 (OR = 1.717), and rs11937052 (OR = 2.0) were associated with high BMI; similarly, rs2005701 (OR = 2.163) was associated with elevated waist circumference. Regarding the association with biochemical markers, it was observed that rs76145164 was associated with elevated triglycerides (OR = 5.667) and rs72679246 with high glucose (OR = 19.2), and high total cholesterol had an association with rs17041272 (OR = 3.1), whereas rs10033691 (OR = 4.479) and rs78160528 (OR = 3.605) were associated with high LDL. The SNP of ELOVL7 rs1563517 was associated with elevated waist circumference (OR = 1.716) and high waist–height index (OR = 1.806) and rs4700398 showed associations with high BMI (OR 1.964). Regarding biochemical markers, rs1563517 was also associated with elevated insulin (OR = 3.126) and rs76641655 with high triglycerides (OR = 4.452), whereas rs115159664 was associated with high total cholesterol (OR = 7.125) and high LDL (OR = 8.111), and rs12188996 was associated with high LDL with an OR of 10.963 (Table 6).

Table 6. Association of SNPs with clinical markers of chronic non-communicable diseases in women (risk factors).

Gene	SNP	Clinical Marker	OR	95% CI		*p*-Value
ELOVL5	rs72938776	H-LDL	11.037	1.044	116.696	0.013
	rs9370194	H-Cholesterol	2.926	1.085	7.892	0.027
ELOVL6	rs59634436	H-BMI	2.761	1.314	5.802	0.006
	rs10033691	H-BMI	2.102	1.076	4.106	0.027
		H-LDL	4.479	1.21	16.584	0.015
	rs78160528	H-LDL	3.605	0.412	31.569	0.218
	rs2005701	H-BMI	1.717	1.004	2.937	0.047
		H-Waist	2.163	1.287	3.634	0.003
	rs76145164	H-Triglycerides	5.667	1.576	20.37	0.003
	rs11937052	H-BMI	2	1.03	3.884	0.038
	rs72679246	H-Glucose	19.2	3.085	119.485	0.001
	rs17041272	H-Cholesterol	3.1	1.101	8.73	0.025
ELOVL7	rs1563517	H-Waist	1.716	1.051	2.801	0.03
		H-WHR	1.806	1.075	3.033	0.025
		H-Insulin	3.126	1.441	6.778	0.003
	rs12188996	H-LDL	10.963	1.037	115.915	0.013
	rs76641655	H-Triglycerides	4.452	1.152	17.204	0.019
	rs115159664	H-Cholesterol	7.125	1.714	29.621	0.002
		H-LDL	8.111	1.503	43.76	0.004
	rs4700398	H-BMI	1.964	1.196	3.225	0.007

H-BMI: high body mass index > 25.0 kg/m^2; H-Waist: high waist circumference (women > 0.80 cm and men > 90 cm); H-WHI: high waist–hip Index (women > 0.85 cm and men > 95 cm); H-WHR: high waist–height ratio > 0.50; H-%BF: high body fat percent (women > 35% and men > 20%). H-Glucose: high glucose; >100 mg/dL; H-INS: high insulin (>14 µU/mL for women and >11µU/mL for men); H-HOMA: high HOMA index (>2.9 for women and >2.3 for men); H-Cholesterol: high total cholesterol (>200 mg/dL); H-LDL: elevated low-density lipoproteins (>130 mg/dL); L-HDL: low high-density lipoproteins (≤50 mg/dL for women and ≤40 mg/dL for men); H-Triglycerides: high triglycerides (>150 mg/dL). The statistical analysis applied to this dataset was a multinominal regression ($p \leq 0.05$).

Interestingly, in women, protective factors were found only in ELOVL2 and ELOVL6 variants. For ELOVL2, rs2281591 is associated with protection for a high waist circumference (OR = 0.610). The two SNPs of ELOVL6 are rs6533491, which is a protective factor for high waist circumference (OR = 0.526) and high waist–height radius (OR = 0.559); and rs11098065, which shows protection against low HDL (OR = 0.582) (Table 7).

Table 7. Association of SNPs with clinical markers of chronic non-communicable diseases in women (protective factors).

Gene	SNP	Clinical Marker	OR	95% CI		*p*-Value
ELOVL2	rs2281591	H-Waist	0.610	0.377	0.988	0.044
ELOVL6	rs6533491	H-Waist	0.526	0.329	0.841	0.007
		H-WHR	0.559	0.338	0.925	0.023
	rs11098065	L-HDL	0.582	0.369	0.919	0.02

H-BMI: high body mass index > 25.0 kg/m^2; H-Waist: high waist circumference (women > 0.80 cm and men > 90 cm); H-WHI: high waist–hip Index (women > 0.85 cm and men > 95 cm); H-WHR: high waist–height ratio > 0.50; H-%BF: high body fat percent (women > 35% and men > 20%). H-Glucose: high glucose; > 100 mg/dL; H-INS: high insulin (>14 µU/mL for women and >11µU/mL for men); H-HOMA: high HOMA index (>2.9 for women and >2.3 for men); H-Cholesterol: high total cholesterol (>200 mg/dL); H-LDL: elevated low-density lipoproteins (>130 mg/dL); L-HDL: low high-density lipoproteins (≤50 mg/dL for women and ≤40 mg/dL for men); H-Triglycerides: high triglycerides (>150 mg/dL). The statistical analysis applied to this dataset was a multinominal regression ($p \leq 0.05$).

In men, the results indicate that 15 SNPs are associated with clinical markers of risk for chronic non-communicable diseases (five for ELOVL2, one for ELOVL4, two for ELOVL5, and seven for ELOVL6). Regarding ELOVL2, the results surprisingly showed that the five SNPs with significant risk associations—rs8523, rs3734398, rs2236212, rs3798713, and rs4532436—are related to high levels of insulin and cholesterol, in addition to a high HOMA index, with an OR ranging from 2.080 to 3.154, thus showing an important relationship. Similarly, for ELOVL4 it was found that rs12196014 is associated with high triglycerides (OR = 1.978). For ELOVL5, rs41273878 and rs114271869 were associated with high triglycerides (OR = 4.176) and low HDL (OR = 5.274), respectively. Regarding the SNPs of ELOVL6, it was observed that rs114422025 is associated with a high HOMA index (OR = 4.929); rs59634436 is associated with a high percentage of body fat (OR = 2.333), with elevated levels of insulin (OR = 2.87) and blood glucose (OR = 7.407), and with a high HOMA index (OR = 3.635). In addition, rs10033691 was found to have associations with ORs for the same markers of 3.299, 2.824, 5.765, and 2.957, respectively. rs78160528 is associated with low HDL (OR = 5.299) and high LDL (OR = 5.889). rs11937052 has associations with elevated waist circumference (OR = 2.441), elevated body fat percentage (OR = 2.101), elevated glucose (OR = 5.577), and low HDL levels (OR = 2.105). rs9997926 was found to be a risk factor for elevated insulin (OR = 5.714), high BMI (OR = 2.068), and elevated waist circumference (OR = 2.131). Finally, for ELOVL6, rs17041272 risk was associated with elevated body fat percentage with an OR = 1.933 (Table 8).

The consistency in the results of ELOVL2 is surprising, because in the risk factors (Table 8) in the associations that indicate protection, the six variants that were found to have significant associations—rs17606561, rs2281591, rs9295757, rs3798721, rs3798722, and rs9393903—are only associated with elevated insulin levels and an elevated HOMA index, with an OR ranging from 0.204 to 0.422. We also found protective associations in ELOVL3, for which rs10748816 was associated with the radius waist height (OR = 5.08); and in ELOVL5, for which rs9370194 was associated with a high percentage of fat (OR = 0.537). By comparison, the results of protective associations of ELOVL6 show that rs11098065 is associated with high levels of insulin (OR = 0.436), as is rs6815102 (OR = 0.478); rs766216 also shows protective associations with elevated total cholesterol (OR = 0.34) and elevated LDL levels (OR = 0.177). rs4326075 is associated with H-LDL (OR = 0.552), whereas rs59111930 was found to be associated with protection against high triglyceride levels (OR = 0.5) and low HDL levels (OR = 0.582). Finally, rs6824447 was found to be a protective factor for high BMI (OR = 0.516), high waist circumference (OR = 0.551), high total cholesterol (OR = 0.268), and high LDL levels (OR = 0.227) (Table 9). It is interesting to note that ELOVL2 SNPs are present in the male population whereas ELOVL7 SNPs are only present in women.

Table 8. Association of SNPs with clinical markers of chronic non-communicable diseases in men (risk factors).

Gene	SNP	Clinical Marker	OR	95% CI		*p*-Value
ELOVL2	rs8523	H-Insulin	2.305	1.172	4.532	0.014
		H-HOMA	2.099	1.094	4.027	0.024
		H-Cholesterol	3.154	1.408	7.061	0.004
	rs3734398	H-Insulin	2.222	1.133	4.360	0.019
		H-HOMA	2.080	1.079	4.009	0.027
		H-Cholesterol	3.010	1.320	6.864	0.006
	rs2236212	H-Insulin	2.198	1.088	4.441	0.026
		H-HOMA	2.415	1.200	4.859	0.012
		H-Cholesterol	2.842	1.173	6.886	0.016
	rs3798713	H-Cholesterol	2.885	1.191	6.990	0.015
	rs4532436	H-Insulin	2.924	1.476	5.792	0.002
		H-HOMA	2.710	1.395	5.267	0.003
		H-Cholesterol	2.935	1.311	6.568	0.007
ELOVL4	rs12196014	H-Triglycerides	1.978	1.011	3.871	0.044
ELOVL5	rs41273878	H-Triglycerides	4.176	0.912	19.127	0.047
	rs114271869	L-HDL	5.274	1.228	22.651	0.013

Table 8. *Cont.*

Gene	SNP	Clinical Marker	OR	95% CI		*p*-Value
ELOVL6	rs114422025	H-HOMA	4.929	0.961	25.264	0.036
		H-%BF	2.333	1.027	5.299	0.039
	rs59634436	H-Insulin	2.87	1.212	6.797	0.013
		H-Glucose	7.407	1.876	29.251	0.001
		H-HOMA	3.635	1.567	8.433	0.002
	rs78160528	L-HDL	5.299	1.234	22.757	0.013
		H-LDL	5.889	1.095	31.683	0.02
		H-%BF	3.299	1.493	7.288	0.002
	rs10033691	H-Insulin	2.824	1.232	6.469	0.011
		H-Glucose	5.765	1.475	22.524	0.005
		H-HOMA	2.957	1.342	6.514	0.005
		H-Waist	2.441	1.206	4.94	0.011
	rs11937052	H-%BF	2.101	1.009	4.375	0.044
		H-Glucose	5.577	1.429	21.766	0.006
		L-HDL	2.105	1.035	4.28	0.037
		H-Insulin	5.714	2.035	16.048	0.001
	rs9997926	H-BMI	2.068	1.118	3.826	0.019
		H-Waist	2.131	1.116	4.069	0.02
	rs17041272	H-%BF	1.933	1.016	3.68	0.043

H-BMI: high body mass index > 25.0 kg/m^2; H-Waist: high waist circumference (women > 0.80 cm and men > 90 cm); H-WHI: high waist–hip Index (women > 0.85 cm and men > 95 cm); H-WHR: high waist–height ratio > 0.50; H-%BF: high body fat percent (women > 35% and men > 20%). H-Glucose: high glucose; >100 mg/dL; H-INS: high insulin (>14 μU/mL for women and >11μU/mL for men); H-HOMA: high HOMA index (>2.9 for women and >2.3 for men); H-Cholesterol: high total cholesterol (>200 mg/dL); H-LDL: elevated low-density lipoproteins (>130 mg/dL); L-HDL: low high-density lipoproteins (≤50 mg/dL for women and ≤40 mg/dL for men); H-Triglycerides: high triglycerides (>150 mg/dL). The statistical analysis applied to this dataset was a multinominal regression ($p \leq 0.05$).

Table 9. Association of SNPs with clinical markers of chronic non-communicable diseases in men (protective factors).

Gene	SNP	Clinical marker	OR	95% CI		*p*-Value
ELOVL2	rs17606561	H-Insulin	0.291	0.122	0.693	0.004
		H-HOMA	0.204	0.077	0.539	0.001
	rs2281591	H-Insulin	0.240	0.101	0.572	0.001
		H-HOMA	0.204	0.083	0.503	0.001
	rs9295757	H-HOMA	0.403	0.184	0.881	0.020
	rs3798721	H-HOMA	0.381	0.174	0.833	0.013
	rs3798722	H-Insulin	0.422	0.199	0.894	0.022
		H-HOMA	0.357	0.168	0.759	0.006
	rs9393903	H-Insulin	0.299	0.125	0.713	0.004
		H-HOMA	0.204	0.077	0.539	0.001
ELOVL3	rs10748816	H-WHR	0.508	0.298	0.866	0.012
ELOVL5	rs9370194	H-%BF	0.537	0.319	0.907	0.019
ELOVL6	rs11098065	H-Insulin	0.436	0.21	0.906	0.024
	rs7662161	H-Cholesterol	0.34	0.134	0.863	0.018
		H-LDL	0.177	0.04	0.789	0.011
	rs6815102	H-Insulin	0.478	0.244	0.939	0.03
	rs4326075	H-LDL	0.552	0.312	0.977	0.04
	rs59111930	H-Triglycerides	0.5	0.291	0.859	0.011
		L-HDL	0.582	0.34	0.995	0.047
		H-BFI	0.516	0.313	0.85	0.009
	rs6824447	H-Waist	0.551	0.319	0.952	0.031
		H-Cholesterol	0.268	0.114	0.628	0.001
		H-LDL	0.227	0.072	0.714	0.006

H-BMI: high body mass index > 25.0 kg/m^2; H-Waist: high waist circumference (women > 0.80 cm and men > 90 cm); H-WHI: high waist–hip Index (women > 0.85 cm and men > 95 cm); H-WHR: high waist–height ratio > 0.50; H-%BF: high body fat percent (women > 35% and men > 20%). H-Glucose: high glucose; >100 mg/dL; H-INS: high insulin (>14 μU/mL for women and >11μU/mL for men); H-HOMA: high HOMA index (>2.9 for women and >2.3 for men); H-Cholesterol: high total cholesterol (>200 mg/dL); H-LDL: elevated low-density lipoproteins (>130 mg/dL); L-HDL: low high-density lipoproteins (≤50 mg/dL for women and ≤40 mg/dL for men); H-Triglycerides: high triglycerides (>150 mg/dL). The statistical analysis applied to this dataset was a multinominal regression ($p \leq 0.05$).

4. Discussion

The current work presents, for the first time, a global study in human populations of the association between genetic variants of ELOVL and clinical markers of chronic non-communicable diseases.

Mexico ranks first in the world in childhood obesity and second in adult obesity, according to data from the National Health and Nutrition Survey (ENSANUT, for its acronym in Spanish) 2016. The combined prevalence of overweight and obesity in adolescents between 12 and 19 years was 36.3%, which is 1.4 percentage points higher than the prevalence in 2012 (34.9%). In adults over 20 years of age, the combined prevalence of overweight and obesity was 71.2% in 2012 and 72.5% in 2016, representing a non-significant difference of 1.3 percentage points [22,23]. This shows a clear trend of increase in the prevalence of overweight and obesity. Furthermore, this trend is reflected in our results, which show a 33% prevalence of overweight and obesity in the population with an average age of 19 years, but a marked difference when the percentage of body fat is analyzed, which shows a prevalence of high fat of 49% in the total population evaluated.

Similar studies in the Mexican population have reported the cumulative prevalence of overweight and obesity as 32% [24,25]. With respect to the WHR, 42.35% of the population was found to be above the recommended level, with this ratio considerably higher in men than in women [26]. Similarly, the body composition alteration with the highest prevalence in the population was the elevated body fat percentage, with 49.04% of young people experiencing the alteration (Figure 1A). This may be worrying because one of the main causes of obesity comorbidities is lipotoxicity or ectopic fat accumulation [27–29]. According to Murguía-Romero (2013) [17], the prevalence of elevated insulin found in a sample of young Mexicans was 9.5% in women and 12.4% in men, and in the case of high HOMA-IR, it was found to be 12.1% in women and 15.3% in men. Our results suggest that the prevalence of obesity and insulin resistance have increased considerably [30]. The results obtained are similar to those reported in other studies, both nationally, 35% [31], and at the state level, 33% [32]. It is known that the BMI underestimates the real prevalence of obesity due to its low specificity in the diagnosis, so the nutritional status of the sample was evaluated according to the %BF.

As mentioned previously, few studies have reported certain correlations between the SNPs found in other human populations and the metabolism of PUFA. The current work found the presence of private alleles of ELOVL5, which could be directly related to the %BF and the risk of cardiovascular diseases in females. Although these and other alleles of the ELOVLs have a low frequency represented within the population, this finding might also reveal the evolutionary process that was undergone by these markers. The frequencies of these alleles within the sampled population, and their HWE deviations and negative health impacts, suggest a possible negative selection process of these makers within the studied population [33].

There was a difference in the associations of SNPs with clinical markers between men and women, shown by the difference in ELOVL2 associated with clinical markers only in men, and ELOVL7 only in women. Although there is no scientific evidence for genetic variants of ELOVL5 related to chronic non-communicable diseases, we observe that ELOVL5 increases the risk for elevated blood levels of LDL (rs72938776) and total cholesterol (rs9370194) by 11.0 and 2.9 times, respectively. For the same gene in men, the associations were for SNP rs41273878 with high triglycerides (OR = 4.1) and rs114271869 with low LDL (OR = 5.2). ELOVL5 in experimental mice has been shown to regulate the degradation of hepatic triglycerides, which is related to an increase in lipase levels in adipocytes [34]. Similarly, epigenetic association studies have shown promising results in the analysis of the gene regulation of ELOVL in relation to type 2 diabetes mellitus [35].

For ELOVL6 it was observed that, of the eight SNPs of ELOVL6, four are related to a high body mass index in women (rs59634436, rs10033691, rs2005701, and rs11937052). Four different SMPs are related to %BF in men (rs59634436, rs10033691, rs11937052, and rs17041272), of which only rs17041272 has been studied for its participation in insulin sensitivity in the Spanish population; results show this SNP to be at risk for insulin resistance. This provides a starting point for the specific investigation of this genetic variant [36]. Similarly, it was observed that mice deficient in ELOVL6 and fed a high fat diet developed obesity but not hyperinsulinemia, hyperglycemia, or insulin resistance [37]. Our findings suggest a difference in ELOVL6 activity in men compared to women, because although the association of SNPs in women was skewed more towards anthropometric markers than in men, an important correlation was observed between elevated insulin levels, and the index of elevated HOMA and elevated glucose.

Regarding the genetic variants of ELOVL7 only associated with risk markers in women, until now no scientific information has existed to support or rule out the role of this enzyme in the physiopathogenesis of any chronic non-communicable disease. The one exception is the relationship attributed to prostate cancer, in which it was observed that enzyme's overexpression occurred in tumor cells. It was found that eliminating the enzyme significantly decreased the growth of these cells, leading the authors to propose a relationship with the androgenic pathway of ELOVL7 [38].

ELOVL2 is clearly one of the most-studied members of the ELOVL family due to its effects on different pathological processes. It has been observed that changes in the methylation of ELOVL2 are related to the aging process [39]. Similarly, protective effects have been attributed to ELOVL2 in pancreatic beta cells of both rodents and humans, through an increase in the oxidation of mitochondrial palmitate, thus avoiding cell death induced by glycolipoxidation [40]. In another study conducted in experimental animals, it was discovered that ELOVL2 is necessary for glucose-mediated insulin secretion, which depends on the endogenous production of docosahexaenoic acid in which ELOVL2 is actively involved [3]. In a study carried out in a population of Tunisia, in which two SNPs of desaturases (FAD1 and FAD2) and one of ELOVL2 (rs3756963) were analyzed, the authors observed that the minor allele C is related to low triglyceride levels and low body mass index, which confers a protective factor to the population with this variant [41]. Similarly, in another study carried out in Italian children and adolescents, a similar behavior was observed in the association of genetic variants of ELOVL2 with indicators of obesity and insulin resistance, in addition to alterations in the blood lipid profile [42]. Like the aforementioned authors, our results show an important association of ELOVL2 with markers of insulin resistance and elevated total cholesterol, which indicates the value of a larger study to specifically identify the role of this enzyme in the pathophysiology of chronic non-communicable diseases.

5. Conclusions

In the present study, a tendency to increase the prevalence of clinical markers of chronic non-communicable diseases in the Mexican population was observed. Although these markers are subject to an environmental influence, we also observed the genetic component and, in particular, the association of ELOVL2, ELOVL5, ELOVL6, and ELOVL7. In addition, results highlighted the differences in the associations between men and women, such as in the cases of ELOVL2 and ELOVL7. These findings indicate the value of a more in-depth study of these genetic variants and their metabolic and physiological functions.

Supplementary Materials: The following are available online at http://www.mdpi.com/2072-6643/12/11/3389/s1, S1: Table S1: Allele and genotype frequencies for all SNP markers. S2: Table S2: Association of SNPs with clinical markers of chronic non-communicable diseases.

Author Contributions: M.L.M.-C.: Performed the total analysis of the data; A.A.-G.: Participated in obtaining clinical data; M.A.A.-L.: Participated in obtaining clinical data; M.d.L.A.-C.: Participated in the genetic analysis of the samples; L.H.-T.: Participated in the genetic analysis of the samples, A.S.M.-R.: Participated in the genetic analysis of the samples and data curation; V.M.R.-G.: Participated in the genetic analysis of the samples and data curation; T.G.-G.: Participated in obtaining data and samples and in data analysis, V.M.R.-G. and U.M.-C.: Directed the entire work. All authors participated in the writing, discussion and structuring of the manuscript. All authors have read and agreed to the published version of the manuscript.

Funding: This research was partially funded by the Fund for Strengthening Research of the Autonomous University of Queretaro, 2018 (FOFI-UAQ 2018). Likewise, for the contribution of the genetic analysis of the samples by CÓDIGO 46 S.A. de C.V. and CONACyT for the Masters scholarship to María Luisa Maycotte Cervantes, from the 2017–2019 National Scholarship call.

Acknowledgments: To the university health system (SUSALUD-UAQ); to the Nutrition Clinic "Carlos Alcocer Cuarón" FCN-UAQ, Human Nutrition Laboratory (FCN-UAQ), the Academic Body Biomedical Research and Functional Foods (UAQ-CA-140) and to CÓDIGO 46 S.A. de C.V., for their support with the facilities for the development of this project.

Conflicts of Interest: The authors declare no conflict of interest.

References

1. Green, C.D.; Ozguden-Akkoc, C.G.; Wang, Y.; Jump, D.B.; Olson, L.K. Role of fatty acid elongases in determination of de novo synthesized monounsaturated fatty acid species. *J. Lipid Res.* **2010**, *51*, 1871–1877. [CrossRef] [PubMed]

2. Jakobsson, A.; Westerberg, R.; Jacobsson, A. Fatty acid elongases in mammals: Their regulation and roles in metabolism. *Prog. Lipid Res.* **2006**, *45*, 237–249. [CrossRef] [PubMed]

3. Cruciani-Guglielmacci, C.; Bellini, L.; Denom, J.; Oshima, M.; Fernandez, N.; Normandie-Levi, P.; Berney, X.P.; Kassis, N.; Rouch, C.; Dairou, J.; et al. Molecular phenotyping of multiple mouse strains under metabolic challenge uncovers a role for Elovl2 in glucose-induced insulin secretion. *Mol. Metab.* **2017**, *6*, 340–351. [CrossRef] [PubMed]

4. Zadravec, D.; Brolinson, A.; Fisher, R.M.; Carneheim, C.; Csikasz, R.I.; Bertrand-Michel, J.; Borén, J.; Guillou, H.; Rudling, M.; Jacobsson, A. Ablation of the very-long-chain fatty acid elongase ELOVL3 in mice leads to constrained lipid storage and resistance to diet-induced obesity. *FASEB J.* **2010**, *24*, 4366–4377.

5. Conn, J.R.; Catchpoole, E.M.; Runnegar, N.; Mapp, S.J.; Markey, K.A. Low rates of antibiotic resistance and infectious mortality in a cohort of high-risk hematology patients: A single center, retrospective analysis of blood stream infection. *PLoS ONE* **2017**, *12*, e0178059. [CrossRef]

6. Vasireddy, V.; Sharon, M.; Salem, N.; Ayyagari, R. Role of ELOVL4 in Fatty Acid Metabolism. In *Advances in Experimental Medicine and Biology*; Anderson, R.E., LaVail, M.M., Hollyfield, J.G., Eds.; Springer: New York, NY, USA, 2008; Volume 613, pp. 283–290. ISBN 9780387749020.

7. Tripathy, S.; Jump, D.B. Elovl5 regulates the mTORC2-Akt-FOXO1 pathway by controlling hepatic cis-vaccenic acid synthesis in diet-induced obese mice. *J. Lipid Res.* **2013**, *54*, 71–84. [CrossRef]

8. Kumadaki, S.; Matsuzaka, T.; Kato, T.; Yahagi, N.; Yamamoto, T.; Okada, S.; Kobayashi, K.; Takahashi, A.; Yatoh, S.; Suzuki, H.; et al. Mouse Elovl-6 promoter is an SREBP target. *Biochem. Biophys. Res. Commun.* **2008**, *368*, 261–266. [CrossRef]

9. Matsuzaka, T.; Shimano, H. Elovl6: A new player in fatty acid metabolism and insulin sensitivity. *J. Mol. Med.* **2009**, *87*, 379–384. [CrossRef] [PubMed]

10. Li, G.; Cui, S.; Du, J.; Liu, J.; Zhang, P.; Fu, Y.; He, Y.; Zhou, H.; Ma, J.; Chen, S. Association of GALC, ZNF184, IL1R2 and ELOVL7 With Parkinson's Disease in Southern Chinese. *Front. Aging Neurosci.* **2018**, *10*, 1–6. [CrossRef]

11. Morales, E.; Bustamante, M.; Gonzalez, J.R.; Guxens, M.; Torrent, M.; Mendez, M.; Garcia-Esteban, R.; Julvez, J.; Forns, J.; Vrijheid, M.; et al. Genetic variants of the FADS gene cluster and ELOVL gene family, colostrums LC-PUFA levels, breastfeeding, and child cognition. *PLoS ONE* **2011**, *6*, e17181. [CrossRef] [PubMed]

12. Aslibekyan, S.; Jensen, M.K.; Campos, H.; Linkletter, C.D.; Loucks, E.B.; Ordovas, J.M.; Deka, R.; Rimm, E.B.; Baylin, A. Genetic variation in fatty acid elongases is not associated with intermediate cardiovascular phenotypes or myocardial infarction. *Eur. J. Clin. Nutr.* **2012**, *66*, 353–359. [CrossRef]

13. World Medical Association. World Medical Association Declaration of Helsinki. Ethical principles for medical research involving human subjects. *Bull. World Health Organ.* **2001**, *79*, 373.

14. WHO. WHO World Health Organization. Available online: http://www.scopus.com/inward/record.url?eid=2-s2.0-84894477459&partnerID=tZOtx3y1 (accessed on 25 March 2020).

15. Parra, I.; Jonguitud, V. La fórmula de Friedewald no debe ser utilizada para el cálculo de colesterol de baja densidad en pacientes con triglicéridos elevados. *Rev Mex Patol Clin* **2007**, *54*, 112–115.

16. Alberti, K.G.M.M.; Eckel, R.H.; Grundy, S.M.; Zimmet, P.Z.; Cleeman, J.I.; Donato, K.A.; Fruchart, J.-C.; James, W.P.T.; Loria, C.M.; Smith, S.C. Harmonizing the Metabolic Syndrome. *Circulation* **2009**, *120*, 1640–1645. [CrossRef]

17. Roccella, E.J. Update on the 1987 Task Force Report on High Blood Pressure in Children and Adolescents: A working group report from the National High Blood Pressure Education Program. National High Blood Pressure Education Program Working Group on Hypertension Control i. *Pediatrics* **1996**, *98*, 649–658.

18. Murguía-Romero, M.; Jiménez-Flores, J.R.; Sigrist-Flores, S.C.; Espinoza-Camacho, M.A.; Jiménez-Morales, M.; Piña, E.; Méndez-Cruz, A.R.; Villalobos-Molina, R.; Reaven, G.M. Plasma triglyceride/HDL-cholesterol ratio, insulin resistance, and cardiometabolic risk in young adults. *J. Lipid Res.* **2013**, *54*, 2795–2799. [CrossRef]

19. *Ilumina Inc Infinium HTS Assay: Reference Guide*; Ilumina Inc.: San Diego, CA, USA, 2019; Volume 2, pp. 1–83, ISBN 1000000074604.

20. IBM Corp IBM SPSS Statistics for Windows, Armonk, NY, USA. 2015. Available online: https://www.ibm.com/support/home/ (accessed on 1 April 2020).

21. Gallagher, D.; Heymsfield, S.B.; Heo, M.; Jebb, S.A.; Murgatroyd, P.R.; Sakamoto, Y. Healthy percentage body fat ranges: An approach for developing guidelines based on body mass index. *Am. J. Clin. Nutr.* **2000**, *72*, 694–701. [CrossRef]

22. Shamah, T.; Cuevas, L.; Gaona, E.B.; Gómez, L.M.; Morales, M.C.; Hernández, M.; Rivera, J.Á. Sobrepeso y obesidad en niños y adolescentes en México, actualización de la Encuesta Nacional de Salud y Nutrición de Medio Camino 2016. *Salud Publica Mex.* **2018**, *60*, 244–253. [CrossRef]

23. Shamah-Levy, T.; Romero-Martínez, M.; Cuevas-Nasu, L.; Gómez-Humaran, I.M.; Avila-Arcos, M.A.; Rivera-Dommarco, J.A. The Mexican national health and nutrition survey as a basis for public policy planning: Overweight and obesity. *Nutrients* **2019**, *11*, 1727. [CrossRef]

24. González Sandoval, C.E.; Díaz Burke, Y.; Mendizabal-Ruiz, A.P.; Medina Díaz, E.; Alejandro Morales, J. Prevalencia de obesidad y perfil lipídico alterado en jóvenes universitarios. *Nutr. Hosp.* **2014**, *29*, 315–321.

25. Secretaría de Salud (SSA) Sobrepeso y Obesidad, Factores de Riesgos Para Desarrollar Diabetes|Secretaría de Salud|Gobierno|Gob.mx. Available online: https://www.gob.mx/salud/articulos/sobrepeso-y-obesidad-factores-de-riesgos-para-desarrollar-diabetes (accessed on 17 April 2020).

26. González Jiménez, E.; Aguilar Cordero, M.J.; Padilla López, C.A.; García García, I. Obesidad monogénica humana: Papel del sistema leptina-melanocortina en la regulación de la ingesta de alimentos y el peso corporal en humanos. *An. Sist. Sanit. Navar.* **2012**, *35*, 285–293. [CrossRef]

27. Zhou, Y.-T.; Grayburn, P.; Karim, A.; Shimabukuro, M.; Higa, M.; Baetens, D.; Orci, L.; Unger, R.H. Lipotoxic heart disease in obese rats: Implications for human obesity. *Proc. Natl. Acad. Sci. USA* **2000**, *97*, 1784–1789. [CrossRef]

28. Izquierdo, A.; Medina-Gómez, G. Papel de la lipotoxicidad en el desarrollo de la lesión renal en el síndrome metabólico y el envejecimiento. *Diálisis y Traspl.* **2012**, *33*, 89–96. [CrossRef]

29. Cazanave, S.C.; Sanyal, A.J. Molecular mechanisms of lipotoxity in nonalcoholic fatty liver disease. *Hepatic Novo Lipogenes. Regul. Metab.* **2015**, *65*, 101–129.

30. Murguia-Romero, M.; Jimenez-Flores, J.R.; Sigrist-Flores, S.C.; Tapia-Pancardo, D.C.; Jimenez-Ramos, A.; Mendez-Cruz, A.R.; Villalobos-Molina, R. Prevalence of Metabolic Syndrome in Young Mexicans: A Sensitivity Analysis on Its Components. *Nutr. Hosp.* **2015**, *32*, 189–195.

31. Sugawara, E.; Nikaido, H. Properties of AdeABC and AdeIJK Efflux Systems of Acinetobacter baumannii Compared with Those of the AcrAB-TolC System of Escherichia coli. *Antimicrob. Agents Chemother.* **2014**, *58*, 7250–7257. [CrossRef] [PubMed]

32. Campos-Nonato, I.; Hernández-Barrera, L.; Rojas-Martínez, R. Hipertensión arterial: Prevalencia, diagnóstico oportuno, control y tendencias en adultos mexicanos. *Salud Publica Mex.* **2013**, *55*, 144. [CrossRef]

33. Christine, A. Andrews Natural Selection, Genetic Drift, and Gene Flow Do Not Act in Isolation in Natural Populations. *Nat. Educ. Knowl.* **2010**, *3*, 5.

34. Peakall, R.; Smouse, P.E. GenAlEx 6.5: Genetic analysis in Excel. Population genetic software for teaching and research–an update. *Bioinformatics* **2012**, *28*, 2537–2539. [CrossRef]

35. Hwang, J.-Y.; Lee, H.J.; Go, M.J.; Jang, H.B.; Choi, N.-H.; Bae, J.B.; Castillo-Fernandez, J.E.; Bell, J.T.; Spector, T.D.; Lee, H.-J.; et al. Genome-wide methylation analysis identifies ELOVL5 as an epigenetic biomarker for the risk of type 2 diabetes mellitus. *Sci. Rep.* **2018**, *8*, 14862. [CrossRef]

36. Morcillo, S.; Martín-Núñez, G.M.; Rojo-Martínez, G.; Almaraz, M.C.; García-Escobar, E.; Mansego, M.L.; de Marco, G.; Chaves, F.J.; Soriguer, F. ELOVL6 Genetic Variation Is Related to Insulin Sensitivity: A New Candidate Gene in Energy Metabolism. *PLoS ONE* **2011**, *6*, e21198. [CrossRef] [PubMed]

37. Matsuzaka, T.; Shimano, H.; Yahagi, N.; Kato, T.; Atsumi, A.; Yamamoto, T.; Inoue, N.; Ishikawa, M.; Okada, S.; Ishigaki, N.; et al. Crucial role of a long-chain fatty acid elongase, Elovl6, in obesity-induced insulin resistance. *Nat. Med.* **2007**, *13*, 1193–1202. [CrossRef]
38. Tamura, K.; Makino, A.; Hullin-Matsuda, F.; Kobayashi, T.; Furihata, M.; Chung, S.; Ashida, S.; Miki, T.; Fujioka, T.; Shuin, T.; et al. Novel Lipogenic Enzyme ELOVL7 Is Involved in Prostate Cancer Growth through Saturated Long-Chain Fatty Acid Metabolism. *Cancer Res.* **2009**, *69*, 8133–8140. [CrossRef]
39. Mansego, M.; Milagro, F.; Zulet, M.; Moreno-Aliaga, M.; Martínez, J. Differential DNA Methylation in Relation to Age and Health Risks of Obesity. *Int. J. Mol. Sci.* **2015**, *16*, 16816–16832. [CrossRef]
40. Bellini, L.; Campana, M.; Rouch, C.; Chacinska, M.; Bugliani, M.; Meneyrol, K.; Hainault, I.; Lenoir, V.; Denom, J.; Véret, J.; et al. Protective role of the ELOVL2/docosahexaenoic acid axis in glucolipotoxicity-induced apoptosis in rodent beta cells and human islets. *Diabetologia* **2018**, *61*, 1780–1793. [CrossRef]
41. Khamlaoui, W.; Mehri, S.; Hammami, S.; Hammouda, S.; Chraeif, I.; Elosua, R.; Hammami, M. Association Between Genetic Variants in FADS1-FADS2 and ELOVL2 and Obesity, Lipid Traits, and Fatty Acids in Tunisian Population. *Clin. Appl. Thromb.* **2020**, *26*, 107602962091528. [CrossRef]
42. Maguolo, A.; Zusi, C.; Giontella, A.; Miraglia Del Giudice, E.; Tagetti, A.; Fava, C.; Morandi, A.; Maffeis, C. Influence of genetic variants in FADS2 and ELOVL2 genes on BMI and PUFAs homeostasis in children and adolescents with obesity. *Int. J. Obes.* **2020**, *613*, 283–290. [CrossRef]

Publisher's Note: MDPI stays neutral with regard to jurisdictional claims in published maps and institutional affiliations.

Review

Innovations in Infant Feeding: Future Challenges and Opportunities in Obesity and Cardiometabolic Disease

Julio Alvarez-Pitti [1,2,3,*], **Ana de Blas** [1] and **Empar Lurbe** [1,2,3]

1   Department of Pediatrics, Consorcio Hospital General, University of Valencia, 46014 Valencia, Spain; adeblaszapata@gmail.com (A.d.B.); Empar.Lurbe@uv.es (E.L.)
2   CIBER Fisiopatología Obesidad y Nutrición (CB06/03), Instituto de Salud Carlos III, 28029 Madrid, Spain
3   INCLIVA Biomedical Research Institute, Hospital Clínico, University of Valencia, 46010 Valencia, Spain
*   Correspondence: alvarez_jul@gva.es; Tel.: +34-96-1820772

Received: 16 October 2020; Accepted: 11 November 2020; Published: 14 November 2020

Abstract: The field of nutrition in early life, as an effective tool to prevent and treat chronic diseases, has attracted a large amount of interest over recent years. The vital roles of food products and nutrients on the body's molecular mechanisms have been demonstrated. The knowledge of the mechanisms and the possibility of controlling them via what we eat has opened up the field of precision nutrition, which aims to set dietary strategies in order to improve health with the greatest effectiveness. However, this objective is achieved only if the genetic profile of individuals and their living conditions are also considered. The relevance of this topic is strengthened considering the importance of nutrition during childhood and the impact on the development of obesity. In fact, the prevalence of global childhood obesity has increased substantially from 1990 and has now reached epidemic proportions. The current narrative review presents recent research on precision nutrition and its role on the prevention and treatment of obesity during pediatric years, a novel and promising area of research.

Keywords: pediatrics; obesity; cardiometabolic risk factors; precision nutrition; eating behavior; nutrigenetics; nutrigenomics; metabolomics; microbiota

---

1. Introduction

Nutrition is known to play one of the key roles in the prevention and treatment of non-communicable chronic diseases. Many of the molecular mechanisms through which nutrients affect the functioning of our bodies are now known. Promoting the correct functioning of these mechanisms through what we eat is the basis of precision nutrition. The goal now is to find personalized dietary strategies that improve people's health.

Cardiometabolic diseases (CMDs) are the leading global cause of death, among which, obesity is the most frequent in childhood and adolescence [1]. Obesity, even during childhood, is a chronic disease with multifactorial etiology. Genetics, lifestyle factors, an unhealthy diet, sedentarism, and poor sleeping habits are some of its main causes, and all of them play an important role in its progression and the development of its comorbidities [2,3].

Physiological functions in the body can be modulated by nutrients because of their ability to interact with molecular mechanisms: "Nutritional genomics focuses on the interaction between bioactive food components and the genome, which includes nutrigenetics and nutrigenomics. The influence of nutrients on gene expression is called nutrigenomics, while the heterogeneous response of gene variants to nutrients, dietary components, and developing nutraceuticals is called nutrigenetics" [4] (Figure 1).

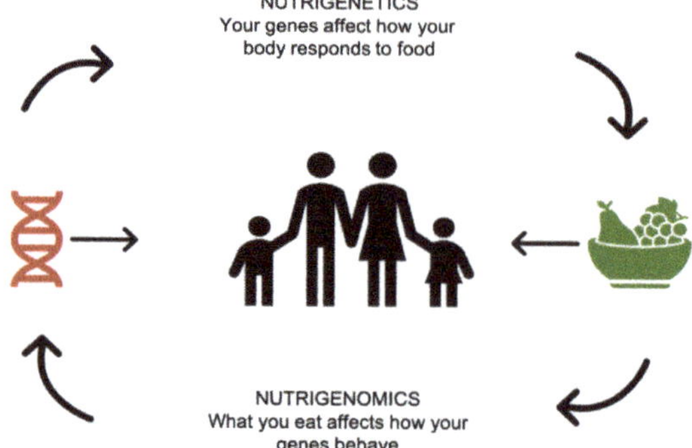

Figure 1. Nutrigenetics vs. nutrigenomics.

Herein, we highlight three of the hypotheses underpinning nutrigenetics and nutrigenomics [5]:

- "The health effects of nutrients and nutriomes (nutrient combinations) depend on inherited genetic variants that alter the uptake and metabolism of nutrients and/or the molecular interaction of enzymes with their nutrient cofactor and hence the activity of biochemical reactions."
- "Nutrition may exert its impact on health outcomes by directly affecting the expression of genes in critical metabolic pathways and/or indirectly by affecting the incidence of genetic mutation at the base sequence or chromosomal level which, in turn, causes alterations in gene dosage and gene expression."
- "Better health outcomes can be achieved if nutritional requirements are customized for each individual taking into consideration both his/her inherited and acquired genetic characteristics depending on life stage, dietary preferences, and health status."

The objective pursued by precision nutrition is to design personalized feeding recommendations that allow for the treatment or prevention of CMDs [6]. For this purpose, it is not only based on genetic information, but also other components such as dietary habits, physical activity, the microbiota, and the metabolome.

Purpose of Review

Nutrition plays a critical role during childhood and adolescence. Understanding the impact of nutrition in early life is essential for the development of future intervention strategies in order to modulate both immediate adult, offspring, grand offspring, and further phenotypes. Information on the potential of prescribing personalized nutrition is, however, scarce. Therefore, the aim of this narrative review is to provide an update on the evidence of the results of precision nutrition interventions during the early period of life, focusing primarily on its impact on obesity and cardiometabolic health. We focused our review on interventions (non-randomized pre-post intervention studies, clinical trials) or meta-analysis of interventions, mainly based on a personalized nutrition approach, in which their objective was to prevent or to treat obesity in infants and adolescents. We also included interventions during pregnancy due to the impact of a mother's nutrition on their offspring. In order to structure the review, we followed the classification of precision nutrition interventions levels, proposed by Ordovás et al. [7].

2. Precision Nutrition Levels

Ordovás et al., propose two levels of personalization of nutrition advice. The first level incorporates a characterization of the subject's behaviors and phenotype (such as adiposity) in order to develop a personalized nutritional advice. Interventions based in this level of personalization, were reviewed throughout the different pediatric stages (e.g., pregnancy, childhood, and adolescence). The second level of personalization builds on the first layer, while also considering different responses to foods and/or nutrients that are conditioned by genotypic or other biological characteristics (Table 1).

Table 1. Levels of personalization of nutrition-based interventions in the prevention and treatment of obesity throughout the different pediatric stages.

Level of Precision Nutrition		Pediatric Stage		
1. Behavioral Level		1.1. Pregnancy [8–18]	1.2. Lactation period [19–50]	1.3. Childhood and adolescence [51–69]
		First 1000 days		
2. Biological Levels	2.1. Biomarkers [70–77]			
	2.2. Genetics [78–87]			
	2.3. Metabolomics [88–91]			
	2.4. Microbiota [92–106]			

[xx] References of articles included in each subheading.

2.1. Behavioral Level of Personalized Nutrition

The collection of an individual's eating habits, behaviors, and phenotypic characteristics is the first level of personalization [7]. All this information combined is used to provide a dietary plan personalized to these characteristics and adapted to each period along the first stages of life.

2.1.1. Pregnancy

Development during the intrauterine period determines life-lasting characteristics. Growing evidence suggests that childhood obesity and predisposition to metabolic disorders might be acquired as a result of erratic intrauterine attributes. In this aspect, the Developmental Origins of Health and Disease (DOHaD) hypothesis outlines the significant effects that the exposure to certain environmental stimuli since fetus development might have on an individual's health. Even at the early stages of conception, the fetus, as a result of alterations that happen in its environment, is able to develop a series of predictive adaptations and adjustments in the homeostatic systems which will prepare it for a possible unfavorable postnatal setting. These adaptations, however, could have a negative impact as a result of an erratic interpretation of environmental changes. In this scenario, adaptations could be responsible for the development of diseases that might become chronic and even spread to future generations [8].

Therefore, pregnancy malnutrition impacts offspring adiposity and obesity, and its effects have been proven to spread to adult life independent of other lifestyle factors. Among the most referred to examples of undernutrition and its implications is the infamous Dutch famine case, in which these effects were observed in the offspring of women who had experienced severe caloric restriction during pregnancy [9]. In contrast, excess weight gain during pregnancy has also been related to metabolic and weight alterations in offspring. Additionally, the chances of developing pregnancy comorbidities such as gestational diabetes, preeclampsia, or cesarean birth are also enhanced [10].

Pregnancy weight control is, therefore, a key element in the health surveillance of mothers and their offspring. Numerous diet interventions have been shown to effectively control pregnancy weight gain, but their beneficial effects on offspring in the long term remain inconclusive according to a systematic review by Tanentsapf et al. [11].

Apparently, it is not only a matter of high energy intake, but also of the type of nutrients that women consume during pregnancy. Long-standing research suggests that what a woman eats during pregnancy not only influences children's taste preferences across their lifetime [12,13], but may also influence children's body composition and appetite. Few human research studies exist on this topic, although higher correlations have been shown between children's protein and fat consumption at 10 years of age with their mother's intake of these macronutrients during pregnancy than with her postnatal intake [14]. The relationship between saturated fat, and especially sugar dietary intake during pregnancy and offspring weight status was described by Murrin et al. [15].

A relationship between diet quality not only with an increased risk of developing obesity in offspring, but also with high blood pressure (BP) has been found. Data from two cohort studies in Scotland established in the 1950s [16,17] reported associations between a high-protein diet during pregnancy and high blood pressure in offspring in adulthood. Since then, many studies have shown similar evidence about the risk of a high-protein diet. A study performed among 965 women from the Danish Fetal Origins Cohort and their offspring after a 20-year follow-up showed an association between a higher mother´s intake of dietary protein during pregnancy with a slightly higher offspring BP [18].

Generating awareness of the risks associated with excess pregnancy adiposity could be an important factor in the establishment of healthy nutritional habits among parents that would prevent harmful events in offspring. Developing precision nutrition strategies for pregnant women could, therefore, be a high-impact preventive intervention.

2.1.2. Lactation Period

Breastfeeding offers children and mothers unrivaled health benefits described since antiquity. The World Health Organization (WHO) suggests that breast milk is the "perfect food for the newborn and recommends all infants be exclusively breastfed up to six months of age, with continued breastfeeding along with appropriate complementary foods up to two years of age or beyond" [19].

Breastfeeding is shown to be a key component in the development of neurological, digestive and immune systems, and cellular health [20]. It is widely known that breastfeeding also delivers a protective effect against overweight [21]. A meta-analysis of observational studies suggested that this protective effect is dose-dependent, being more relevant the longer the duration of breastfeeding [22]. Another large meta-analysis demonstrated that the probability of developing obesity in children who had never been breastfed or who had been so during a period of less than six months was higher than for those having been breastfed for six months or longer [23]. In this regard, as demonstrated by previous studies [24], overweight or obese women during pregnancy are at a higher risk of not being able to breastfeed or having to interrupt it at an early stage. Therefore, the offspring become more vulnerable given the combination of an unfavorable fetal environment and the increased risk of developing cardiometabolic disease as a consequence of not benefiting from the properties of breastfeeding.

However, the physiopathological mechanism behind this protective effect is not fully understood, despite the large body of evidence supporting these benefits. Some hypotheses point toward the micronutrient and bioactive composition of breast milk [25,26]. On the one hand, it seems that the type of dietary constituent that the mother takes in while breastfeeding is an important part in modeling the future health of her offspring. Tahir et al. described how there was an inverse relationship between a higher maternal diet quality from pregnancy through three months postpartum with relative infant weight and adiposity from birth to six months and breastfeeding percentage at six months. Similarly, higher maternal diet quality during lactation was inversely related to infant fat mass at six months [27]. The exposure of infants to different flavors through breast milk can condition or shape their food preferences, both in the process of weaning and in later life [12]. In addition, some studies have demonstrated a lower likelihood of emptying a bottle or cup when compared to bottle-fed children [28], and that breastfeeding is associated with better self-regulation of eating by children [29], including better satiety responsiveness [30,31]. On the other hand, it is widely known that breast milk contains many bioactive hormones and peptides, with leptin being one of the most relevant. It is well documented that leptin is directly involved in the regulation of energy balance and food intake, and is crucial in the programming of metabolic pathways and infant appetite. This could explain that breastfeeding has a protective effect against childhood obesity compared to formula feeding [32], probably due to the lack of leptin in the later. Different studies have shown a clear association between leptin concentration in infants and in breast milk and infant body mass index (BMI) at two years of age [33]. They have established that moderate amounts of leptin supplied through breast milk seem to moderately protect infants from excess weight gain [34]. This benefit seems to last the longer the span of breastfeeding.

The lower protein content of breast milk is also hypothesized to be a protective factor when it is being compared to formula milk [35,36]. Research with formula trials has demonstrated that enriched formulas lead to an increase in fat mass between five and eight years of age [37], and high protein content formulas have also been shown to pose a greater risk of obesity at school age [38–40]. The lower the amount of protein in the formula, down to certain limits, the greater the benefit. Comparisons between hydrolyzed and cow milk formulas have indicated faster weight gain and a doubled incidence in early rapid weight in the cow formula-fed group [41]. A modified formula with lower protein content has also been shown to have a protective effect on rapid weight gain [42].

Genome analysis has become a useful tool for the detection of individuals at a higher risk of developing obesity. Breastfeeding might have a direct impact on these patients. A recent study conducted by Wu et al. demonstrated that offering up to five months exclusive breastfeeding to children with a higher genetic risk obtained through an obesity-specific genetic risk score implied a substantial decrease in BMI [19]. This study proposed that breastfeeding is a key candidate intervention to reduce overweight risks in predisposed patients, since it demonstrated direct impact from the first moments of life. Detecting those people who carry alleles related to obesity, and then carrying out interventions for the promotion of breastfeeding in this group of people, should become a priority.

Lastly, during the lactation period, weaning is crucial, since caregiver feeding behaviors could be an obesogenic influence on child eating behaviors [43]. Non-directive strategies such as repeatedly offering foods [44] and having caregivers portray eating the food with enjoyment have been demonstrated to increase the consumption of a given food and to support children's liking for a wider variety of healthy foods, and may help maintain responsivity in the feeding environment [45–47]. However, pressure to eat has been associated with an impaired ability to self-regulate eating behaviors in preschool and poorer energy compensation in childhood [48]. Taken together, this evidence suggests a bidirectionality between child eating behaviors and adiposity, as well as caregiver feeding behaviors [49,50]. Therefore, interventions should also focus on how parents should feed their children and maintain healthy habits in the family from birth.

Taking all of this scientific evidence into account, it seems that there is no better precision nutrition during the lactation period than breastfeeding, without ignoring the effect that caregivers have on children's relationship with food.

2.1.3. Childhood and Adolescence

The treatment of obesity is based on changes in lifestyle, mainly eating habits and physical activity. A meta-analysis including 3436 adults in 16 randomized controlled trials concluded that the Mediterranean diet, especially if the diet is energy-reduced and accompanied by physical activity, favors greater weight loss than a control diet [51]. According to the evidence obtained in this and other studies, vegetable-based diets such as the Mediterranean diet are the most appropriate for treating obesity [52]. However, there are also studies that show greater weight loss in protein-rich diets [53]. Another aspect is the quality of fats. Increasing the consumption of omega-3 or -6 polyunsaturated fatty acids happens to improve plasma lipid levels and on the other hand lowering the risk of cardiovascular events [54,55]. However, it appears that losing weight does not depend on macronutrient composition when comparing the effect of different low-calorie diets in adults, as shown in a review of 48 studies [56].

There are also discrepancies in the consensus on the best dietary strategy for treating obesity in children and adolescents [57]. It has been shown that low-carb or low-glycemic diets to have the same effect as standard portion-controlled diets [58]. Nevertheless, children's adherence to modified carbohydrate diets in the long term may be low. In children and adolescents, semi-structured dietary approaches are best used to support children and their families to select healthier food groups and decrease portion sizes [59].

In a systematic review including 14 clinical trials conducted in children between 6 and 18 years old, the efficacy of seven interventions based on low-fat diets was compared against two isocaloric and five ad libitum low-carb diets [60]. This study showed that the weight status of participants improved by implementing a reduced-energy diet, irrespectively of macronutrient distribution. Moreover, the authors pointed out that the distribution of macronutrients in the diet could be tailored in each case according to the presence or not of comorbidities. For example, in the case of obesity associated with diabetes or insulin resistance, low-carbohydrate diets were chosen. However, they reported that more studies are needed to better define these personalized dietary guidelines.

All of this research suggests that it is the energy content and not the macronutrient composition of a diet playing the key role in weight loss. However, is this a consequence of the fact that, considering the complexity of obesity, we are not personalizing the type of nutrition enough?

Recently, the American Association of Clinical Endocrinologists (AACE) and the European Association for the Study of Obesity (EASO) proposed the use of a new diagnostic term in adults, namely, "adiposity-based chronic disease" (ABCD) [61,62]. Obesity is considered "adiposity-based" because the disease is primarily a result of abnormalities in the mass, distribution, and/or function of adipose tissue. A new coding system was proposed addressing the pathophysiological heterogeneity of obesity [63]. This code is based on three dimensions. First, etiology, including two mechanistic categories: (1) multifactorial disease (the majority of patients) and (2) specific identifiable factor obesity. Second, the degree of adiposity. Third, health risk, taking into account the presence or absence of comorbidities. This proposal constitutes an advancement because obesity is recognized as a heterogeneous disease with complex pathophysiology, and it includes a dimension revealing its impact on health [61].

In order to have enough information to tailor an adequate treatment, the diagnostic procedures of an obese child may include family history, prenatal factors, feeding history, sleep duration and matters, exercise and the time spent looking at screens, when and where meals are taken, bullying or social isolation, motivation and the capacity to make modifications within a family, and the family's cultural expectations and financial limitations. Laboratory tests and studies are considered according to age, BMI percentile, and the presence of risk factors. Children presenting with growth deceleration, symptoms of hypo-hyperthyroidism or other endocrinopathies, sustained hypertension, symptoms of type 2 diabetes (T2DM), a family history of early cardiovascular disease, hirsutism, snoring, and/or daytime sleepiness need further work-up. Therefore, obesity is increasingly seen as a complex disease with different etiologies. As a consequence, the need for an individual nutritional approach instead of "one-size-fits-all" recommendations is growing.

Personalized nutrition-based interventions in obese children and adolescents are scarce in the literature. Recently, Lim et al. reported the results of a 16-week long lasting evidence-based customized nutritional intervention in 103 subjects [64]. For each patient, an intervention was implemented following a nutrition care process (NCP) model. Both at baseline and follow-up, the sociodemographic and anthropometric data, health and dietary behavior, and dietary intake of the participants were assessed. All participants engaged in 30-min nutritional sessions held each month. Four steps were used to implement the NCP: (1) Nutrition assessment, (2) diagnosis, (3) intervention, and (4) monitoring/evaluation. Nutritional diagnosis was performed on the basis of three areas: Intake (amount of food or nutrient), clinical factors (nutritional problems linked to medical or physical circumstances), and behavioral–environmental factors (attitudes, knowledge, beliefs, food access and safety). On the basis of this information, a personalized intervention was planned for each subject. The study has some limitations, such as not having a control group and the short duration of the follow-up. Moreover, although participants globally improved in body composition, macronutrient intake, and nutritional behavior, the results were not superior to those of other non-tailored interventions.

Another area of study is the relationship between obesity and eating disorders (EDs), because the presence of an eating disorder can suppose a radical change in the therapeutic approach of the patient. Overweight and obesity knowingly increases the risk of developing EDs [65]. However, the assessment and management of EDs caused by or being a consequence of obesity are not contemplated with sufficient caution in children and adolescents [66]. Developing effective tools to properly diagnose these disorders and designing specific therapies that combine nutritional, psychological, behavioral, and/or pharmacological treatment in children and adolescents can be a further step in the development of personalized therapeutic strategies [67].

A systematic review conducted following preferred reporting items for systematic reviews and meta-analyses (PRISMA) guidelines tried to answer the question: "Is multidisciplinary treatment (MT) effective on eating disorder symptoms in children with obesity?" [68]. Concerning psychology, all studies except one included support group therapy. The review showed that MT not only decreased the BMI Z-score (zBMI), but also positively influenced external eating, disinhibition of control, and emotional eating. They concluded that MT might be influenced by eating behavior, and participants learned to face emotional stress and external stimuli. Nevertheless, a recently published study identified four patterns of ED pathology [69]: only loss of control, shape and weight concerns (SWCs), low ED pathology, and high ED pathology. After the intervention, all groups shared an important reduction of their zBMI; however, no clinically significant weight loss was reached by those with the highest ED pathology. The authors concluded that further research is recommended in order to identify strategies modifying treatment for children who suffer of both obesity and high ED pathology to improve weight loss success.

More research is needed in this field in order to be able to offer obese children and adolescents an intervention that is adjusted to the etiology of the disease, the degree of obesity, and the comorbidity that it presents, in addition to the fact that it must be viable considering the patient's sociocultural environment. Behavioral level interventions included in the review are summarized in Table 2.

Table 2. Summary of behavioral level interventions included in the review.

Authors/Reference	Type of Study	Subjects/Studies	Clinical Condition	Type of Intervention	Outcomes: Primary (1st) and Secondary (2nd)	Key Findings
Pregnancy						
Tanentsapf et al.; BMCpregnancy and Child-birth. 2011 [11].	Mct	N: 1434 13: RCT (10) quasi-RCT (3) pregnant women	H, Ow, Ob > 18 years	D or DI vs. UC ($n = 9$) D vs. DI vs. UC ($n = 1$) D + placebo vs. UC + placebo ($n = 1$). D vs. UC ($n = 1$) DI vs. US ($n = 1$)	1st: % of women excess GWG 2nd: maternal (GD, PRE, CS, WG) & infant (BW, PTB, GA)	Dietary advice in pregnancy can decrease GWG and postpartum weight retention
Lactation period						
Singhal et al., Am. J. Clin. Nutr. 2010 [37].	Mct	Study 1 ($n = 299$) Study 2 ($n = 246$) Follow up from birth to 5–8 years	>37 weeks Study 1: BW < 10th pct. Study 2: BW < 20th pct	STF vs. NEF	1st: childhood adiposity (fat mass)	NEF increased body fat mass in later childhood independent of other confounding factors
Weber et al., Am. J. Clin. Nutr. 2014 [39].	RCT	$n = 755$ (HPF 256, LPF 262, BF 237) 6 y follow-up	H term infants	Intervention group: HPF vs. LPFReference group BF	1st: BMI2nd: weight, height, and obesity	HPF implies higher risk of obesity in childhood
Mennella et al., Pediatrics 2011 [41].	RCT	$n = 56$ (CMF 32, PHF 24) 7 mo follow up	H term infants	CMF vs. PHF	1st: trajectories growth measures2nd: F acceptance	CMF greater weight gain velocity
Kouwenhoven, Am. J. Clin. Nutr. 2020 [42].	RCT	$n = 235$ (mLPF 88, CTF 82, BF 65) 6 mo follow up	H term infants	Intervention group: mLPF vs. CTFReference group BF	1st: daily WG2nd: F intake, growth, BC, BT, AEs	mLPF has protective effect on rapid weight gain

Table 2. *Cont.*

Authors/Reference	Type of Study	Subjects/Studies	Clinical Condition	Type of Intervention	Outcomes: Primary (1st) and Secondary (2nd)	Key Findings
Childhood and adolescence						
Kirk et al., J Pediatr. 2012 [58].	RCT	102 (7–12 years) 43% male	Ob BMI Z-score range (1.60–2.65)	3-month intervention randomly assigned to LC, RGL, or PC diet+exerc.	1st: BMI z-score, WC, %BF and diet adherence 2nd: clinical metabolic parameters	LC and RGL as effective as PC at at 3, 6 and 12 months, but significantly lower adherence to LC
Go et al. Nutr Rev. 2014 [60].	Mct	14: RCT (12) quasi-RCT (2) <18 years 8 weeks-2 years follow up	Ow + Ob	LC vs. PC ($n = 7$) IP vs. SP ($n = 6$) IF vs. SF ($n = 1$)	1st: BMI, BMI z-score, %BF 2nd: CMB parameters	Achieved improvements in weight status irrespective of macronutrient distribution of a reduced-energy diet
Kim et al. Nutr Res Pract. 2020 [64].	RCT	103 (7–12 years) 61% male	Ob 67% moderate 33% severe	16-week nutritional intervention tailored for each subject based subject's specific nutrition diagnosis.	1st: BMI z-score, WC, %BF 2nd: stage of change and diet adherence	Intervention improved body composition and decreased calorie intake in adherent subjects
De Giuseppe et al. Front. Pediatr. 2019 [68].	Mct	9: RCT (3), non-RCT (1), ITS (5) 6 to 18 years	Ow/Ob + ED	Heterogeneous MT All: support group therapy + CBT	1st: ED symptoms and MT features 2nd: BMI	MTs reduced BMI positive short- and long-term impact of MTs on ED symptoms

AEs: adverse events, BC: body composition, BF: breastfed, %BF: body fat percentage, BMI: body mass index, BW: birth weight, BT: blood test, CBT: cognitive behavioral therapy, CMB: cardiometabolic, CMF: cow's milk formula, CS: cesarean section, CTF: control formula, D: diet, DI: diet and intervention, ED: eating disorder, F: formula, GA: gestational age, GD: gestational diabetes, GWG: gestational weight gain, H: healthy, HPF: high-protein formula, IF: increased fat diet, IP: increased protein diet, ITS: interrupted time series without comparison group, LC: low-carbohydrate diet, LPF: low-protein formula, Mct: meta-analysis of clinical trials, mLPF: modified low-protein formula, MT: multidisciplinary treatment, NEF: nutrient-enriched formula, Ob: obese, Ow: overweight, PHF: protein hydrolyzed formula, PC: standard portion-controlled diet, PRE: preeclampsia, PTB: preterm birth, RCT: randomized clinical trial, RGL: reduced glycemic load diet, SF: standard-fat diet, SP: standard protein diet, STF: standard-term formula, UC: usual care, WG: weight gain.

2.2. Biological Levels of Personalized Nutrition

With the information obtained from the regular diagnostic work with an obese patient, it does not seem that we can identify the type of nutritional intervention that is most appropriate for each patient [70]. Therefore, the identification of biomarkers that would allow us to predict the response of an individual to a given diet or lifestyle advice is of enormous value. This is the current challenge in the field of precision nutrition [71].

2.2.1. Predictive Biomarkers

There is enormous variability in the response of obese children and adolescents to dietary interventions. Currently, research continues in order to develop pretreatment prognostic markers that can identify patients who may benefit from a particular dietary weight loss intervention. However, some factors, such as the differences in dietary adherence, psychological, socioeconomic factors, and other unknown ones (e.g., genes, epigenetics, and pollutants), make the identification of these biomarkers a scientific challenge.

However, recent discoveries have identified that alterations in the mechanisms linking glucose metabolism to appetite control could be involved in the big variability in weight loss observed in obese patients prescribed the same diet. In adults, studies conducted by Astrup and Hjorth [72] suggested that "individual weight loss responsiveness to diets depends on glucose metabolic traits that, clinically, can be characterized by fasting glucose and insulin levels for each patient before initiation of treatment. Based on these measurements, an optimal diet composition can be tailored to enhance satiety, adherence, and weight loss." They hypothesized that carbohydrate-rich meals may be very satiating in insulin-sensitive overweight (type A), less so in more insulin-resistant obese prediabetics (type B), and even less so in obese individuals with T2DM (type C). This hypothesis is based on reanalysis by these authors of some of the largest clinical trials (e.g., PREDIMED, the NUGENOB study, and the Diet, Obesity, and Genes (DiOGenes) trial) that seemed to indicate that the macronutrient composition of diets does not really matter. They showed that, when reclassifying each participant into one of these conditions (types A, B, or C), weight loss responsiveness and best-fit diet are completely different [73,74]. These studies showed that prescribing personalized weight-control diets on the basis of pretreatment glycemic (and insulinemic) status is a promising field of research.

A review conducted by Lister et al. [75] identified specific dietary strategies that can reduce the risk of T2DM development in obese youths. The authors concluded that, in addition to weight loss, a reduced carbohydrate diet may optimize improvements in T2DM risk factors, including hepatic steatosis, insulin resistance, and hyperglycemia. The DiOGenes study, included in this review is, to date, the largest study aimed at examining the effect of modifying the glycemic index (GI) and protein content of a diet on weight status and cardiometabolic outcomes in children. Participants were randomized as a family unit to one of five ad libitum diets: low-protein and low-GI, low-protein and high-GI, high-protein and low-GI, high-protein and high-GI, and control diet (medium protein content and no instructions on GI) [76]. The results of this study showed that, among children, neither GI nor protein ad libitum diets had an isolated effect on body composition. However, the low-protein and high-GI combination increased body fat, while the high-protein and low-GI combination was protective against obesity [77]. Although these types of diet have some inconveniences, such as worse adherence, fatigue-inducing physical activity, and the risk of reducing the intake of fiber and phytochemicals if vegetable intake is not increased, they could be considered as an option to tailor nutrition in obese children and adolescents at greater risk of developing T2DM.

2.2.2. Genetics

It is estimated that monogenic diseases with Mendelian inheritance represent approximately 5% of non-syndromic cases of obesity, including mutations in the leptin receptor and *BDNF*, *MC4R*, *MC3R*, *PCSK1*, *PCSK2*, *POMC*, *PPARG*, *SIM1*, and *TRKB* genes, among others. However, obesity is generally considered to be a multifactorial disease with high heritability (50–75%), probably higher in cases of early onset. In children who develop severe obesity before 5 y.o., genetic causes should be taken into account. These children may present clinical features such as short stature, developmental delay, hyperphagia, or dysmorphic facies. Work-up in these patients involves DNA methylation studies, exome sequencing, and karyotyping for syndromes such as Albright's hereditary osteodystrophy, Alstrom, Bardet–Biedl, Cohen syndrome, Fragile X, Prader Willi, congenital leptin deficiency, *POMC* deficiency and *MC4R* deficiency. Nevertheless, other genetic causes of obesity not associated with described syndromes could also be contributors [78].

The association of body weight and genetic loci has been identified in hypothesis-driven candidate gene studies [79]. In these studies, the investigated genes were mainly chosen for their role in regulating food intake. For example, the transmembrane protein 18 (*TMEM18*) gene and melanocortin-4 receptor (*MC4R*) have a function in regulating food intake; fatty acid-binding protein 2 (*FABP2*) is involved in lipid metabolism; peroxisome proliferator-activated receptor-gamma (*PPARG*) and the fat mass- and obesity-associated (*FTO*) locus have a role in adipocyte differentiation. Specifically, almost 100 loci were identified, accounting for approximately 2.7% of the variation in BMI, of which the *FTO*, *MC4R*, and *TMEM18* genes showed the strongest associations. To date, the gene with the strongest relation on body fat is the *FTO* locus; however, the biological functions of most of these loci remain unclear [80]. Understanding the genetic mechanism behind obesity will likely help to develop future tailored interventions for the prevention and treatment of childhood obesity.

On the basis of these facts, it could be hypothesized that subjects with these non-favorable mutations respond worse to weight-loss interventions. Nevertheless, the results of a systematic review and meta-analysis including 9563 adults based on the association between the *FTO* locus gene and weight loss [81] showed that after dietary intervention, carriers of the *FTO* locus-risk allele of single-nucleotide polymorphisms (SNP) rs9939609 achieved similar weight loss compared to that of non-risk allele carriers. Recently, the results of the randomized controlled trial "Diet Intervention Examining the Factors Interacting with Treatment Success" (DIETFITS) have been published. These results again showed, in 609 overweight adults, that weight reduction is independent of genotypes [82].

Data on children are scarce. One of the studies with a larger number of recruited subjects (684; 280 boys and 404 girls) assessed how the *NYD-SP18* and *TMEM18* variants affected the efficacy of an intensive one-month inpatient diet and physical activity weight-reduction program [83]. Neither *NYD-SP18* nor the *TMEM18* variant was associated with changes in anthropometric measurements after the lifestyle intervention. Few studies on similar topics, including *TMEM18* polymorphism, have been published. Hiney et al. [84] analyzed 282 obese children and found no evidence for the effects of this gene on weight loss or regain one year after an intervention, and similar results were observed in an analysis of 400 children/adolescents [85]. However, a Spanish study analyzed 168 overweight/obese adolescents for the contribution of nine obesity-related polymorphisms, as well as genetic predisposition scores (GPSs) on the changes in body composition and cardiometabolic risk factors in a three-month intervention, on the basis of a personalized diet and a physical activity program [86]. They included *FTO*, *MC4R*, *PPARG*, and *TMEM18* SNPs. The authors detected, after adjusting for baseline BMI standard deviation score that subjects with a higher GPS had smaller improvements in metabolic profile and a worse response to physical activity compared to those subjects with a lower GPS. The limitations of this study were that the sample was small and that it included just three months of follow-up. Nevertheless, these findings were not confirmed by a larger intervention including 920 overweight and obese subjects from the Danish Childhood Obesity Biobank [87]. A genetic risk score (GRS) comprising 15 SNPs associated with childhood obesity was assessed. Authors demonstrate that these genetic variants did not predict the response to lifestyle changes. It could be concluded that a high genetic predisposition to overweight during childhood in fact had no influence on whether the children reacted to lifestyle intervention compared to children with low genetic predisposition to overweight, so these genetic risk scores have not demonstrated being a useful tool to tailor nutritional interventions.

2.2.3. Metabolomics

Metabolomics studies are essential to understand the effect of food on an individual's health. The identification of food-derived biomarkers makes it possible to know how different subjects distinctly metabolize the same food, and how such food metabolites may influence health outcomes. In this regard, obtaining reference values for food metabolites is necessary. Recently, a study established a reference dataset of mainly healthy subjects [88] and its results allowed for differentiating normal metabolomes between different groups (e.g., men/women and elderly/young).

Another useful field of metabolomics is its potential in determining the pattern of food consumption of an individual [89]. A major challenge of precision nutrition remains the objective measurement of adherence to a dietary pattern. In this sense, the spectroscopic profiling of urine was validated for the objective measurement of an overall dietary pattern [90]. This technique is being used to characterize eating patterns in an ongoing study in which our group participated, with the main aim of exploring the individual eating/behavioral patterns, parental styles, life events, cognitive styles, and biological-endocrine factors associated with the development of abnormal eating behaviors, EDs, or obesity (i.e., the EAT4HEALTHYLIFE Project) [91].

2.2.4. Microbiota

Microbiotainteract directly with the host's cells through a bidirectional equilibrium relationship, and impact multiple facets of its physiological and biochemical pathways involved in immunity and energy homeostasis. Microbiota are considered one of the most densely populated ecosystems with a vast gene pool, sometimes referred to as our "second genome," with the capacity of being malleable and significantly varying from host to host.

As mentioned, variations in our genome have a direct impact on the metabolism of nutrients between individuals. The gut microbiome also appears to significantly impact these variations. In this regard, precision microbiomics can be described as "the use of the gut microbiome as a biomarker to predict the effect of specific dietary components on host health and the use of these data to design precision diets and interventions that ensure optimal health" [92]. This approach sets a much more complex scenario when compared to the one-fit-for-all diet approaches that appear to perform poorly [93].

The understanding of the paths in which the microbiome is related to the substantial changes in inter-individual responses to lifestyle and diet interventions is a key element for the development of new tools that help enhance the potential of individual microbiomes as a source of human variation modulating dietary responses [94]. Diet shapes microbiota composition [95]; however, microbiota vary their interaction within the same diet depending on their composition and host characteristics [94]. Data on how a diet directly shapes the microbiome and how this affects body composition are complex and not completely understood.

In a review by Biesiekierski et al., microbiota are described as a predictive tool regarding host behavior toward diet interventions [96]. Although there are data that demonstrate that gut microbiota composition and response to diet are linked, this review pointed that only a relatively small number of interventions show health benefits, and current data are inconsistent to establish whether certain strains can predict the response that each individual has to a certain dietary intervention.

However, some authors have explained that this analysis between the response of a population group as described in randomized trials and a specific dietary pattern may result in an oversimplified vision of the correlations between diet and gut microbiota [97] without considering the diversity within an individual. When analyzing the response to a certain intervention, it seems important to know the status of the basal microbiome of each individual, since it determines the response. Studies have shown that the initial microbiota of the host at the time of initiating a dietary intervention influence the response to it and the possible changes that may occur both in the host and in their own microbiota [98].

It is now becoming apparent that long-term healthy dietary patterns—in particular, habitual fiber intake—are crucial for highly diverse gut microbiota responsiveness to specific interventions.

Low-diversity gut microbiota can benefit from specific dietary interventions if they contain even minimal amounts of responding and effector microorganisms [99]. Patients whose poor dietary practices or very low dietary fiber intake are prolonged during long periods of time could experience the loss of beneficial microbial lineages which would, in turn, make any dietary intervention fail [100]. There are some studies focused on promoting favorable changes in microbiota of obese children and adolescents using prebiotic supplementation [101] or through food (mainly whole grain) [102].

Understanding the relevance of these interventions on gut microbiota and their impact on health is in an early phase, and evidence-based scientific validation in children is scarce, but the use of the microbiome as a biomarker of response to certain interventions is encouraging. In the aforementioned review by Biesiekierski et al. [103], the results may not have been significant because the immense variability among different study participants was not considered. Methodological consistency between studies is also improvable. As far as we know, no such interventions have been performed on obese children or adolescents. Whole microbiome next-generation sequencing performs differently when compared to selective sequencing combined with specimen cultures.

Given the vast amount of information that can be obtained from a whole microbiome, its understanding and, therefore, the knowledge that can be applied to dietary decision making are still limited. Big data analysis is also experimenting with exponential improvements, and modifies and impacts obtained conclusions from previous data pools. There are already studies that use microbiome information for better approaches to machine learning and developing personally tailored diet interventions [104]. The first was carried out by Zeevi et al. [105], who developed an innovative algorithm to predict postprandial glucose responses according to anthropometry, blood test results, and the microbiome composition of 1000 people. In a study by Korem et al., the impact of different types of bread on health benefits was found to be determined by the gut microbiome as opposed to its ingredients [106]. Biological level interventions included in the review are summarized in Table 3.

Table 3. Summary of pediatric biological level interventions included in the review.

Authors/Reference	Type of Study	Subjects/Studies	Clinical Condition	Type of Intervention	Outcomes: Primary (1st) and Secondary (2nd)	Key Findings
Predictive Biomarkers						
Papadaki et al., Peds, 2010 [71].	RCT	Baseline 827 (5–18 years) Completers, 26 weeks: 465 (43% male)	Healthy children from selected families	Ad libitum diets 1–5: LP/LGI *n* = 102.Do: 37% LP/HGI *n* = 87. Do: 48% HP/HGI *n* = 92. Do: 42% HP/HGI *n* = 96. Do: 39% PC *n* = 88. Do: 42%	1st: BMI z-score, %BF 2nd: %OW-Ob and waist/hip ratio.	No effect on body composition was observed by GI nor protein isolated. However, LP/HGI increased body fat, while HP/LGI protected against obesity
Genetics						
Zlatohlaveketal, Med Sci Monit, 2018 [78].	Pre-post	684 (12–14 years) 41% male	Ob + Ow	One-month inpatient intensive lifestyle intervention: nutritional + physical activity *TMEM*-18 Vs *NYD-SP18* gene variants	1st: BMI z-score, waist, hip, abdominal skinfold 2nd: genotype	No significant differences in BMI changes among gene variants
Hiney et al. J Pediatr Endocr Met 2013, [84].	Pre-post	282 (8–13 years) 47% male	Ob + Ow 14% over-weight, 86% obese	1y. lifestyle intervention. Reevaluation 1y. after end intervention. Evaluate regain weight taking into account genes: *NEGR1*, *TNKS*, *SDCCAG8*, *FTO*, *MC4R*, *TMEM18*, *PTER*, *MTCH2*, *SH2B1*, *MAF*, *NPC1*, and *KCTD15*	1st: BMI SDS2nd: genotype	None of the SNPs including were related to weight regain after 1y.
Moleres et al. Jpeds. 2012, [86].	Pre-post	168 (12–16 years)	Ob + Ow 49% over-weight, 51% obese	6–24 months lifestyle intervention Subjects were genotyped for 9 obesity-related SNPs in the *FTO*, *MC4R*, *TMEM18*, *IL6*, *PPARG*, and *ADIPQ* genes and a GPS was calculated	1st: BMI, BMI z-score, %BF 2nd: genotype, CMB parameters	The GPS had relationship with BMI-SDS and fat mass both at baseline and after a 3-month intervention.
Hollensted et al. Obesity (2018) [87].	Pre-post	1674 (9–15 years)	Control group *n* = 754/ Ob + Ow group *n* = 920	3 months lifestyle intervention Subjects were genotyped for 15 obesity-related SNPs and a GPS was calculated	1st: BMI, BMI z-score, %BF 2nd: genotype, CMB parameters	The GPS had relationship with BMI-SDS at baseline in both groups, but did not influence response to intervention.

Table 3. *Cont.*

Metabolomics						
Fernandez Aranda et al., (study in progress) 2018-20 [91].	Pre-post	50 (8–10 years -prepubertal- and 12–16 years -post pubertal-)	Ob + -EDs	12-month lifestyle intervention	1st: BMI, BMI z-score, %B, EDs 2nd: Endocannabinoid profile	Hypothesis: relation among obesity, EDs and Biochemical, hormonal and Endocannabinoids blood levels. Metabolomic-based diet pattern
Microbiota						
Nicolucci et al. J.gastro. 2017. [101].	RCT	42 (7–12 years) 57% male	Ob + Ow 67%	16-week once daily supplementation OI group n = 22 vs. M group n = 20	1st: BMI z-score, %BF 2nd: effect of prebiotic supplementation on gut microbiota, FBAs, CMB parameters	OI significant decrease of BMI, BF, IL6, TGs vs. M. OI increases *Bifidobacterium* spp. and decreases *Bacteroides vulgatum*
Zhang a,1,et al. J.ebiom. 2015 [102].	Pre-post	40 (3–16 years)	PWS children n = 19 Ob children n = 21	Nutritional in-hospital intervention (ob 30 days, PWS 90 days) WTP diet.	1st: weight loss 2ng: changes gut microbiota	MTs reduced BMI positive short- and long-term impact of MTs on ED symptoms

ADIPQ: Human adiponectin gene, EDs: eating disorders, Do: drop-out rate, FBAs: fecal bile acids, FTO: fat mass and obesity-associated gene, GI: glycemic index, GPS: genetic predisposition score, HGI: high glycemic index diet, HP: high protein diet, IL6: Interleukin 6 gene , *KCTD15: potassium channel tetramerization domain containing 15 genes* LGI: low glycemic index diet, LP: low-protein diet, M: maltodextrin placebo, *MAF: v-maf musculoaponeurotic fibrosarcoma oncogene homolog (avian) gene*, MC4R: melanocortin 4 receptor gene, MTCH2: mitochondrial carrier 2 gene, NEGR1: Neuronal growth regulator 1 gene, *NPC1: Niemann-Pick disease, type C1, gene*; OI: oligofructose-enriched inulin, PC: standard portion-controlled diet, PPARG: Peroxisome proliferator-activated receptor gamma gene, PTER: phosphotriesterase related gene, PWS: Prader–Willi Syndrome, Pre-post: non-randomized pre-post intervention studies, SDS: standard deviation score, SDCCAG8: serologically defined colon cancer antigen 8 gene, SH2B1, SH2B: adaptor protein 1 gene, SNPs: single nucleotide polymorphisms, TMEM18: transmembrane protein 18 gene, TNKS: tankyrase gene, WTP: diet based on whole grains, traditional Chinese medicinal foods and prebiotics.

3. Future Directions

The prevalence of obesity and the associated cardiometabolic risk factors in children and adolescents has been globally increasing since 1990. Nowadays, there is growing evidence from basic nutritional science about the importance of dietary advice, and it is considered one of the main challenges of clinical nutrition. Moreover, tailored nutrition represents a promising approach to prevent and manage obesity. A concerted effort between clinical and basic science researchers is needed in order to establish a comprehensive framework to allow the implementation of these new findings to adequately apply novel and personalized dietary advice.

Author Contributions: J.A.-P. and A.d.B. wrote the manuscript. E.L. critically reviewed the manuscript. All authors approved the final version of the manuscript.

Funding: This work has been financially supported by: Ministerio de Ciencia, Innovación y Universidades (Grant Number PI17/01517), Instituto de Salud Carlos III and cofinanced by the European Regional Development Fund (to E.L.).

Conflicts of Interest: The authors declare no conflict of interest.

References

1. Farpour-Lambert, N.J.; Baker, J.L.; Hassapidou, M.; Holm, J.C.; Nowicka, P.; O'Malley, G.; Weiss, R. Childhood Obesity Is a Chronic Disease Demanding Specific Health Care—A Position Statement from the Childhood Obesity Task Force (COTF) of the European Association for the Study of Obesity (EASO). *Obes. Facts* **2015**, *8*, 342–349. [CrossRef] [PubMed]
2. Styne, D.M.; Arslanian, S.A.; Connor, E.L.; Farooqi, I.S.; Murad, M.H.; Silverstein, J.H.; Yanovski, J.A. Pediatric Obesity—Assessment, Treatment, and Prevention: An Endocrine Society Clinical Practice Guideline. *J. Clin. Endocrinol. Metab.* **2017**, *102*, 709–757. [CrossRef] [PubMed]
3. Goodarzi, M.O. Genetics of obesity: What genetic association studies have taught us about the biology of obesity and its complications. *Lancet Diabetes Endocrinol.* **2018**, *6*, 223–236. [CrossRef]
4. Sharma, P.; Dwivedi, S. Nutrigenomics and Nutrigenetics: New Insight in Disease Prevention and Cure. *Indian J. Clin. Biochem.* **2017**, *32*, 371–373. [CrossRef] [PubMed]
5. Fenech, M.; El-Sohemy, A.; Cahill, L.; Ferguson, L.R.; French, T.A.C.; Tai, E.S.; Milner, J.; Koh, W.P.; Xie, L.; Zucker, M.; et al. Nutrigenetics and nutrigenomics: Viewpoints on the current status and applications in nutrition research and practice. *J. Nutr. Nutr.* **2011**, *4*, 69–89. [CrossRef]
6. Betts, J.A.; Gonzalez, J.T. Personalised nutrition: What makes you so special? *Nutr. Bull.* **2016**, *41*, 353–359. [CrossRef]
7. Ordovas, J.M.; Ferguson, L.R.; Tai, E.S.; Mathers, J.C. Personalised nutrition and health. *BMJ* **2018**, 361. [CrossRef]
8. Barker, D.J.P. The origins of the developmental origins theory. *J. Intern. Med.* **2007**, *261*, 412–417. [CrossRef]
9. Ravelli, A.C.; van Der Meulen, J.H.; Osmond, C.; Barker, D.J.; Bleker, O.P. Obesity at the age of 50 y in men and women exposed to famine prenatally. *Am. J. Clin. Nutr.* **1999**, *70*, 811–816. [CrossRef]
10. Ramachenderan, J.; Bradford, J.; McLean, M. Maternal obesity and pregnancy complications: A review. *Aust. N. Z. J. Obstet. Gynaecol.* **2008**, *48*, 228–235. [CrossRef]
11. Tanentsapf, I.; Heitmann, B.L.; Adegboye, A.R. Systematic review of clinical trials on dietary interventions to prevent excessive weight gain during pregnancy among normal weight, overweight and obese women. *BMC Pregnancy Childbirth* **2011**, *11*, 81. [CrossRef] [PubMed]
12. Mennella, J.A.; Jagnow, C.P.; Beauchamp, G.K. Prenatal and postnatal flavor learning by human infants. *Pediatrics* **2001**, *107*, E88. [CrossRef] [PubMed]
13. Teegarden, S.L.; Scott, A.N.; Bale, T.L. Early life exposure to a high fat diet promotes long-term changes in dietary preferences and central reward signaling. *Neuroscience* **2009**, *162*, 924–932. [CrossRef] [PubMed]
14. Brion, M.J.; Ness, A.R.; Rogers, I.; Emmett, P.; Cribb, V.; Davey Smith, G.; Lawlor, D.A. Maternal macronutrient and energy intakes in pregnancy and offspring intake at 10 y: Exploring parental comparisons and prenatal effects. *Am. J. Clin. Nutr.* **2010**, *91*, 748–756. [CrossRef] [PubMed]

15. Murrin, C.; Shrivastava, A.; Kelleher, C.C. Maternal macronutrient intake during pregnancy and 5 years postpartum and associations with child weight status aged five. *Eur. J. Clin. Nutr.* **2013**, *67*, 670–679. [CrossRef] [PubMed]

16. Campbell, D.M.; Hall, M.H.; Barker, D.J.; Cross, J.; Shiell, A.W.; Godfrey, K.M. Diet in pregnancy and the offspring's blood pressure 40 years later. *Br. J. Obstet. Gynaecol.* **1996**, *103*, 273–280. [CrossRef] [PubMed]

17. Shiell, A.W.; Campbell-Brown, M.; Haselden, S.; Robinson, S.; Godfrey, K.M.; Barker, D.J. High-meat, low-carbohydrate diet in pregnancy: Relation to adult blood pressure in the offspring. *Hypertension* **2001**, *38*, 1282–1288. [CrossRef]

18. Hrolfsdottir, L.; Halldorsson, T.I.; Rytter, D.; Bech, B.H.; Birgisdottir, B.E.; Gunnarsdottir, I.; Granstrom, C.; Henriksen, T.B.; Olsen, S.F.; Maslova, E. Maternal Macronutrient Intake and Offspring Blood Pressure 20 Years Later. *J. Am. Heart Assoc.* **2017**, *6*, e005808. [CrossRef]

19. Wu, Y.; Lye, S.; Dennis, C.L.; Briollais, L. Exclusive breastfeeding can attenuate body-mass-index increase among genetically susceptible children: A longitudinal study from the ALSPAC cohort. *PLoS Genet.* **2020**, *16*, e1008790. [CrossRef]

20. Jackson, K.M.; Nazar, A.M. Breastfeeding, the immune response, and long-term health. *J. Am. Osteopath. Assoc.* **2006**, *106*, 203–207.

21. Khera, A.V.; Chaffin, M.; Wade, K.H.; Zahid, S.; Brancale, J.; Xia, R.; Distefano, M.; Senol-Cosar, O.; Haas, M.E.; Bick, A.; et al. Polygenic Prediction of Weight and Obesity Trajectories from Birth to Adulthood. *Cell* **2019**, *177*, 587–596.e9. [CrossRef] [PubMed]

22. Horta, B.L.; Loret de Mola, C.; Victora, C.G. Long-term consequences of breastfeeding on cholesterol, obesity, systolic blood pressure and type 2 diabetes: A systematic review and meta-analysis. *Acta Paediatr.* **2015**, *104*, 30–37. [CrossRef] [PubMed]

23. Yan, J.; Liu, L.; Zhu, Y.; Huang, G.; Wang, P.P. The association between breastfeeding and childhood obesity: A meta-analysis. *BMC Public Health* **2014**, *14*, 1267. [CrossRef] [PubMed]

24. Flores, T.R.; Mielke, G.I.; Wendt, A.; Nunes, B.P.; Bertoldi, A.D. Prepregnancy weight excess and cessation of exclusive breastfeeding: A systematic review and meta-analysis. *Eur. J. Clin. Nutr.* **2018**, *72*, 480–488. [CrossRef]

25. Rito, A.I.; Buoncristiano, M.; Spinelli, A.; Salanave, B.; Kunesova, M.; Hejgaard, T.; Garcia Solano, M.; Fijalkowska, A.; Sturua, L.; Hyska, J.; et al. Association between Characteristics at Birth, Breastfeeding and Obesity in 22 Countries: The WHO European Childhood Obesity Surveillance Initiative—COSI 2015/2017. *Obes. Facts* **2019**, *12*, 226–243. [CrossRef]

26. Nozhenko, Y.; Asnani-Kishnani, M.; Rodriguez, A.M.; Palou, A. Milk Leptin Surge and Biological Rhythms of Leptin and Other Regulatory Proteins in Breastmilk. *PLoS ONE* **2015**, *10*, e0145376. [CrossRef]

27. Tahir, M.J.; Haapala, J.L.; Foster, L.P.; Duncan, K.M.; Teague, A.M.; Kharbanda, E.O.; McGovern, P.M.; Whitaker, K.M.; Rasmussen, K.M.; Fields, D.A.; et al. Higher Maternal Diet Quality during Pregnancy and Lactation Is Associated with Lower Infant Weight-For-Length, Body Fat Percent, and Fat Mass in Early Postnatal Life. *Nutrients* **2019**, *11*, 632. [CrossRef]

28. Li, R.; Fein, S.B.; Grummer-Strawn, L.M. Do infants fed from bottles lack self-regulation of milk intake compared with directly breastfed infants? *Pediatrics* **2010**, *125*, 1386. [CrossRef]

29. Disantis, K.I.; Collins, B.N.; Fisher, J.O.; Davey, A. Do infants fed directly from the breast have improved appetite regulation and slower growth during early childhood compared with infants fed from a bottle? *Int. J. Behav. Nutr. Phys. Act.* **2011**, *8*, 89. [CrossRef]

30. Brown, A.; Lee, M.D. Early influences on child satiety-responsiveness: The role of weaning style. *Pediatr. Obes.* **2015**, *10*, 57–66. [CrossRef]

31. Brown, A.; Lee, M. Breastfeeding during the first year promotes satiety responsiveness in children aged 18–24 months. *Pediatr. Obes.* **2012**, *7*, 382–390. [CrossRef] [PubMed]

32. Obermann-Borst, S.A.; Eilers, P.H.; Tobi, E.W.; de Jong, F.H.; Slagboom, P.E.; Heijmans, B.T.; Steegers-Theunissen, R.P. Duration of breastfeeding and gender are associated with methylation of the LEPTIN gene in very young children. *Pediatr. Res.* **2013**, *74*, 344–349. [CrossRef] [PubMed]

33. Miralles, O.; Sanchez, J.; Palou, A.; Pico, C. A physiological role of breast milk leptin in body weight control in developing infants. *Obesity* **2006**, *14*, 1371–1377. [CrossRef] [PubMed]

34. Palou, A.; Pico, C. Leptin intake during lactation prevents obesity and affects food intake and food preferences in later life. *Appetite* **2009**, *52*, 249–252. [CrossRef]

35. Heinig, M.J.; Nommsen, L.A.; Peerson, J.M.; Lonnerdal, B.; Dewey, K.G. Energy and protein intakes of breast-fed and formula-fed infants during the first year of life and their association with growth velocity: The DARLING Study. *Am. J. Clin. Nutr.* **1993**, *58*, 152–161. [CrossRef]

36. Trabulsi, J.C.; Smethers, A.D.; Eosso, J.R.; Papas, M.A.; Stallings, V.A.; Mennella, J.A. Impact of early rapid weight gain on odds for overweight at one year differs between breastfed and formula-fed infants. *Pediatr. Obes.* **2020**, *15*, e12688. [CrossRef]

37. Singhal, A.; Kennedy, K.; Lanigan, J.; Fewtrell, M.; Cole, T.J.; Stephenson, T.; Elias-Jones, A.; Weaver, L.T.; Ibhanesebhor, S.; MacDonald, P.D.; et al. Nutrition in infancy and long-term risk of obesity: Evidence from 2 randomized controlled trials. *Am. J. Clin. Nutr.* **2010**, *92*, 1133–1144. [CrossRef]

38. Escribano, J.; Luque, V.; Ferre, N.; Mendez-Riera, G.; Koletzko, B.; Grote, V.; Demmelmair, H.; Bluck, L.; Wright, A.; Closa-Monasterolo, R. European Childhood Obesity Trial Study Group Effect of protein intake and weight gain velocity on body fat mass at 6 months of age: The EU Childhood Obesity Programme. *Int. J. Obes. (Lond.)* **2012**, *36*, 548–553. [CrossRef]

39. Weber, M.; Grote, V.; Closa-Monasterolo, R.; Escribano, J.; Langhendries, J.P.; Dain, E.; Giovannini, M.; Verduci, E.; Gruszfeld, D.; Socha, P.; et al. European Childhood Obesity Trial Study Group Lower protein content in infant formula reduces BMI and obesity risk at school age: Follow-up of a randomized trial. *Am. J. Clin. Nutr.* **2014**, *99*, 1041–1051. [CrossRef]

40. Koletzko, B.; von Kries, R.; Closa, R.; Escribano, J.; Scaglioni, S.; Giovannini, M.; Beyer, J.; Demmelmair, H.; Gruszfeld, D.; Dobrzanska, A. Lower protein in infant formula is associated with lower weight up to age 2 y: A randomized clinical trial. *Am. J. Clin. Nutr.* **2009**, *89*, 1836–1845. [CrossRef]

41. Mennella, J.A.; Ventura, A.K.; Beauchamp, G.K. Differential growth patterns among healthy infants fed protein hydrolysate or cow-milk formulas. *Pediatrics* **2011**, *127*, 110–118. [CrossRef] [PubMed]

42. Kouwenhoven, S.M.P.; Antl, N.; Finken, M.J.J.; Twisk, J.W.R.; van der Beek, E M.; Abrahamse-Berkeveld, M.; van de Heijning, B.J.M.; Schierbeek, H.; Holdt, L.M.; van Goudoever, J.B.; et al. A modified low-protein infant formula supports adequate growth in healthy, term infants: A randomized, double-blind, equivalence trial. *Am. J. Clin. Nutr.* **2020**, *111*, 962–974. [CrossRef] [PubMed]

43. Wood, A.C.; Blissett, J.M.; Brunstrom, J.M.; Carnell, S.; Faith, M.S.; Fisher, J.O.; Hayman, L.L.; Khalsa, A.S.; Hughes, S.O.; Miller, A.L.; et al. Caregiver Influences on Eating Behaviors in Young Children: A Scientific Statement From the American Heart Association. *J. Am. Heart Assoc.* **2020**, *9*, e014520. [CrossRef] [PubMed]

44. Carruth, B.R.; Ziegler, P.J.; Gordon, A.; Barr, S.I. Prevalence of picky eaters among infants and toddlers and their caregivers' decisions about offering a new food. *J. Am. Diet. Assoc.* **2004**, *104*, 57. [CrossRef] [PubMed]

45. Momin, S.R.; Hughes, S.O.; Elias, C.; Papaioannou, M.A.; Phan, M.; Vides, D.; Wood, A.C. Observations of Toddlers' sensory-based exploratory behaviors with a novel food. *Appetite* **2018**, *131*, 108–116. [CrossRef]

46. Luchini, V.; Musaad, S.; Lee, S.Y.; Donovan, S.M. Observed differences in child picky eating behavior between home and childcare locations. *Appetite* **2017**, *116*, 123–131. [CrossRef]

47. Horne, P.J.; Tapper, K.; Lowe, C.F.; Hardman, C.A.; Jackson, M.C.; Woolner, J. Increasing children's fruit and vegetable consumption: A peer-modelling and rewards-based intervention. *Eur. J. Clin. Nutr.* **2004**, *58*, 1649–1660. [CrossRef]

48. Carper, J.L.; Orlet Fisher, J.; Birch, L.L. Young girls' emerging dietary restraint and disinhibition are related to parental control in child feeding. *Appetite* **2000**, *35*, 121–129. [CrossRef]

49. Campbell, K.; Andrianopoulos, N.; Hesketh, K.; Ball, K.; Crawford, D.; Brennan, L.; Corsini, N.; Timperio, A. Parental use of restrictive feeding practices and child BMI z-score. A 3-year prospective cohort study. *Appetite* **2010**, *55*, 84–88. [CrossRef]

50. Jansen, P.W.; Tharner, A.; van der Ende, J.; Wake, M.; Raat, H.; Hofman, A.; Verhulst, F.C.; van Ijzendoorn, M.H.; Jaddoe, V.W.; Tiemeier, H. Feeding practices and child weight: Is the association bidirectional in preschool children? *Am. J. Clin. Nutr.* **2014**, *100*, 1329–1336. [CrossRef]

51. Esposito, K.; Kastorini, C.; Panagiotakos, D.B.; Giugliano, D. Mediterranean diet and weight loss: Meta-analysis of randomized controlled trials. *Metab. Syndr. Relat. Disord.* **2011**, *9*, 1–12. [CrossRef] [PubMed]

52. Shai, I.; Schwarzfuchs, D.; Henkin, Y.; Shahar, D.R.; Witkow, S.; Greenberg, I.; Golan, R.; Fraser, D.; Bolotin, A.; Vardi, H.; et al. Weight loss with a low-carbohydrate, Mediterranean, or low-fat diet. *N. Engl. J. Med.* **2008**, *359*, 229–241. [CrossRef] [PubMed]

53. Jenkins, D.J.A.; Wong, J.M.W.; Kendall, C.W.C.; Esfahani, A.; Ng, V.W.Y.; Leong, T.C.K.; Faulkner, D.A.; Vidgen, E.; Paul, G.; Mukherjea, R.; et al. Effect of a 6-month vegan low-carbohydrate ('Eco-Atkins') diet on cardiovascular risk factors and body weight in hyperlipidaemic adults: A randomised controlled trial. *BMJ Open* **2014**, *4*, e003505. [CrossRef] [PubMed]

54. Harris, W.S.; Mozaffarian, D.; Rimm, E.; Kris-Etherton, P.; Rudel, L.L.; Appel, L.J.; Engler, M.M.; Engler, M.B.; Sacks, F. Omega-6 Fatty Acids and Risk for Cardiovascular Disease: A Science Advisory From the American Heart Association Nutrition Subcommittee of the Council on Nutrition, Physical Activity, and Metabolism; Council on Cardiovascular Nursing; and Council on Epidemiology and Prevention. *Circulation* **2009**, *119*, 902–907. [CrossRef]

55. Bamberger, C.; Rossmeier, A.; Lechner, K.; Wu, L.; Waldmann, E.; Stark, R.; Altenhofer, J.; Henze, K.; Parhofer, K. A Walnut-Enriched Diet Reduces Lipids in Healthy Caucasian Subjects, Independent of Recommended Macronutrient Replacement and Time Point of Consumption: A Prospective, Randomized, Controlled Trial. *Nutrients* **2017**, *9*, 1097. [CrossRef]

56. Johnston, B.C.; Kanters, S.; Bandayrel, K.; Wu, P.; Naji, F.; Siemieniuk, R.A.; Ball, G.D.C.; Busse, J.W.; Thorlund, K.; Guyatt, G.; et al. Comparison of weight loss among named diet programs in overweight and obese adults: A meta-analysis. *JAMA* **2014**, *312*, 923–933. [CrossRef]

57. Kumar, S.; Kelly, A.S. Review of Childhood Obesity: From Epidemiology, Etiology, and Comorbidities to Clinical Assessment and Treatment. *Mayo Clin. Proc.* **2017**, *92*, 251–265. [CrossRef]

58. Kirk, S.; Brehm, B.; Saelens, B.E.; Woo, J.G.; Kissel, E.; D'Alessio, D.; Bolling, C.; Daniels, S.R. Role of Carbohydrate Modification in Weight Management among Obese Children: A Randomized Clinical Trial. *J. Pediatr.* **2012**, *161*, 320–327.e1. [CrossRef]

59. Daniels, S.R.; Hassink, S.G. The Role of the Pediatrician in Primary Prevention of Obesity. *Pediatrics* **2015**, *136*, 275. [CrossRef]

60. Gow, M.L.; Ho, M.; Burrows, T.L.; Baur, L.A.; Stewart, L.; Hutchesson, M.J.; Cowell, C.T.; Collins, C.E.; Garnett, S.P. Impact of dietary macronutrient distribution on BMI and cardiometabolic outcomes in overweight and obese children and adolescents: A systematic review. *Nutr. Rev.* **2014**, *72*, 453–470. [CrossRef]

61. Garvey, W.T.; Mechanick, J.I. Proposal for a Scientifically-Correct and Medically-Actionable Disease Classification System (ICD) for Obesity. *Obesity* **2020**, *28*, 484–492. [CrossRef] [PubMed]

62. Frühbeck, G.; Busetto, L.; Dicker, D.; Yumuk, V.; Goossens, G.H.; Hebebrand, J.; Halford, J.G.C.; Farpour-Lambert, N.J.; Blaak, E.E.; Woodward, E.; et al. The ABCD of Obesity: An EASO Position Statement on a Diagnostic Term with Clinical and Scientific Implications. *Obes. Facts* **2019**, *12*, 131–136. [CrossRef] [PubMed]

63. Hebebrand, J.; Holm, J.; Woodward, E.; Baker, J.L.; Blaak, E.; Schutz, D.D.; Farpour-Lambert, N.J.; Frühbeck, G.; Halford, J.G.C.; Lissner, L.; et al. A Proposal of the European Association for the Study of Obesity to Improve the ICD-11 Diagnostic Criteria for Obesity Based on the Three Dimensions Etiology, Degree of Adiposity and Health Risk. *OFA* **2017**, *10*, 284–307. [CrossRef] [PubMed]

64. Kim, J.; Kim, Y.; Seo, Y.; Park, K.; Jang, H.B.; Lee, H.; Park, S.I.; Lim, H. Evidence-based customized nutritional intervention improves body composition and nutritional factors for highly-adherent children and adolescents with moderate to severe obesity. *Nutr. Res. Pr.* **2020**, *14*, 262–275. [CrossRef]

65. Rancourt, D.; McCullough, M. Overlap in Eating Disorders and Obesity in Adolescence. *Curr. Diabetes Rep.* **2015**, *15*, 1–9. [CrossRef]

66. Cena, H.; Stanford, F.C.; Ochner, L.; Fonte, M.L.; Biino, G.; Giuseppe, R.D.; Taveras, E.; Misra, M. Association of a history of childhood-onset obesity and dieting with eating disorders. *Eat. Disord.* **2017**, *25*, 216–229. [CrossRef]

67. Cebolla, A.; Perpiñá, C.; Lurbe, E.; Alvarez-Pitti, J.; Botella, C. Prevalence of binge eating disorder among a clinical sample of obese children. *An. Pediatr.* **2012**, *77*, 98–102. [CrossRef]

68. De Giuseppe, R.; Di Napoli, I.; Porri, D.; Cena, H. Pediatric Obesity and Eating Disorders Symptoms: The Role of the Multidisciplinary Treatment. A Systematic Review. *Front. Pediatr.* **2019**, *7*. [CrossRef]

69. Balantekin, K.N.; Hayes, J.F.; Sheinbein, D.H.; Kolko, R.P.; Stein, R.I.; Saelens, B.E.; Hurst, K.T.; Welch, R.R.; Perri, M.G.; Schechtman, K.B.; et al. Patterns of Eating Disorder Pathology are Associated with Weight Change in Family-Based Behavioral Obesity Treatment. *Obesity* **2017**, *25*, 2115–2122. [CrossRef]

70. Tracy, R.P. 'Deep phenotyping': Characterizing populations in the era of genomics and systems biology. *Curr. Opin. Lipidol.* **2008**, *19*, 151–157. [CrossRef]

71. de Toro-Martín, J.; Arsenault, B.J.; Després, J.P.; Vohl, M.C. Precision nutrition: A review of personalized nutritional approaches for the prevention and management of metabolic syndrome. *Nutrients* **2017**, *9*, 913. [CrossRef] [PubMed]

72. Astrup, A.; Hjorth, M.F. Classification of obesity targeted personalized dietary weight loss management based on carbohydrate tolerance. *Eur. J. Clin. Nutr.* **2018**, *72*, 1300–1304. [CrossRef] [PubMed]

73. Hjorth, M.F.; Ritz, C.; Blaak, E.E.; Saris, W.H.; Langin, D.; Poulsen, S.K.; Larsen, T.M.; Sørensen, T.I.; Zohar, Y.; Astrup, A. Pretreatment fasting plasma glucose and insulin modify dietary weight loss success: Results from 3 randomized clinical trials. *Am. J. Clin. Nutr.* **2017**, *106*, 499–505. [CrossRef] [PubMed]

74. Hjorth, M.F.; Due, A.; Larsen, T.M.; Astrup, A. Pretreatment Fasting Plasma Glucose Modifies Dietary Weight Loss Maintenance Success: Results from a Stratified RCT. *Obesity* **2017**, *25*, 2045–2048. [CrossRef] [PubMed]

75. Gow, M.L.; Garnett, S.P.; Baur, L.A.; Lister, N.B. The effectiveness of different diet strategies to reduce type 2 diabetes risk in youth. *Nutrients* **2016**, *8*, 486. [CrossRef] [PubMed]

76. Larsen, T.M.; Dalskov, S.; van Baak, M.; Jebb, S.A.; Papadaki, A.; Pfeiffer, A.F.H.; Martinez, J.A.; Handjieva-Darlenska, T.; Kunešová, M.; Pihlsgård, M.; et al. Diets with high or low protein content and glycemic index for weight-loss maintenance. *N. Engl. J. Med.* **2010**, *363*, 2102–2113. [CrossRef]

77. Papadaki, A.; Linardakis, M.; Larsen, T.M.; van Baak, M.A.; Lindroos, A.K.; Pfeiffer, A.F.H.; Martinez, J.A.; Handjieva-Darlenska, T.; Kunesová, M.; Holst, C.; et al. The effect of protein and glycemic index on children's body composition: The DiOGenes randomized study. *Pediatrics* **2010**, *126*, 1143. [CrossRef]

78. Farooqi, S.; O'Rahilly, S. Genetics of obesity in humans. *Endocr. Rev.* **2006**, *27*, 710–718. [CrossRef]

79. Holzapfel, C.; Drabsch, T. A scientific perspective of personalised gene-based dietary recommendations for weight management. *Nutrients* **2019**, *11*, 617.

80. Loos, R.J. The genetics of adiposity. *Curr. Opin. Genet. Dev.* **2018**, *50*, 86–95. [CrossRef]

81. Livingstone, K.M.; Celis-Morales, C.; Papandonatos, G.D.; Erar, B.; Florez, J.C.; Jablonski, K.A.; Razquin, C.; Marti, A.; Heianza, Y.; Huang, T.; et al. FTO genotype and weight loss: Systematic review and meta-analysis of 9563 individual participant data from eight randomised controlled trials. *BMJ* **2016**, *354*, i4707. [CrossRef] [PubMed]

82. Gardner, C.D.; Trepanowski, J.F.; Del Gobbo, L.C.; Hauser, M.E.; Rigdon, J.; Ioannidis, J.P.A.; Desai, M.; King, A.C. Effect of Low-Fat vs Low-Carbohydrate Diet on 12-Month Weight Loss in Overweight Adults and the Association With Genotype Pattern or Insulin Secretion: The DIETFITS Randomized Clinical Trial. *JAMA* **2018**, *319*, 667–679. [CrossRef] [PubMed]

83. Zlatohlavek, L.; Maratka, V.; Tumova, E.; Ceska, R.; Lanska, V.; Vrablik, M.; Hubacek, J.A. Body adiposity changes after lifestyle interventions in children/adolescents and the NYD-SP18 and TMEM18 variants. *Med. Sci. Monit.* **2018**, *24*, 7493–7498. [CrossRef] [PubMed]

84. Hinney, A.; Wolters, B.; Pütter, C.; Grallert, H.; Illig, T.; Hebebrand, J.; Reinehr, T. No impact of obesity susceptibility loci on weight regain after a lifestyle intervention in overweight children. *J. Pediatric Endocrinol. Metab.* **2013**, *26*, 1209–1213. [CrossRef] [PubMed]

85. Scherag, A.; Kleber, M.; Boes, T.; Kolbe, A.; Ruth, A.; Grallert, H.; Illig, T.; Heid, I.M.; Toschke, A.M.; Grau, K.; et al. SDCCAG8 obesity alleles and reduced weight loss after a lifestyle intervention in overweight children and adolescents. *Obesity* **2012**, *20*, 466–470. [CrossRef] [PubMed]

86. Moleres, A.; Rendo-Urteaga, T.; Zulet, M.A.; Marcos, A.; Campoy, C.; Garagorri, J.M.; Martínez, J.A.; Azcona-Sanjulián, M.C.; Marti, A. Obesity susceptibility loci on body mass index and weight loss in spanish adolescents after a lifestyle intervention. *J. Pediatr.* **2012**, *161*. [CrossRef]

87. Hollensted, M.; Fogh, M.; Schnurr, T.M.; Kloppenborg, J.T.; Have, C.T.; Ruest Haarmark Nielsen, T.; Rask, J.; Lund, M.A.V.; Frithioff-Bøjsøe, C.; Johansen, M.Ø.; et al. Genetic Susceptibility for Childhood BMI has no Impact on Weight Loss Following Lifestyle Intervention in Danish Children. *Obesity* **2018**, *26*, 1915–1922. [CrossRef]

88. Trabado, S.; Al-Salameh, A.; Croixmarie, V.; Masson, P.; Corruble, E.; Fève, B.; Colle, R.; Ripoll, L.; Walther, B.; Boursier-Neyret, C.; et al. The human plasma-metabolome: Reference values in 800 French healthy volunteers; impact of cholesterol, gender and age. *PLoS ONE* **2017**, *12*, e0173615. [CrossRef]

89. Playdon, M.C.; Moore, S.C.; Derkach, A.; Reedy, J.; Subar, A.F.; Sampson, J.N.; Albanes, D.; Gu, F.; Kontto, J.; Lassale, C.; et al. Identifying biomarkers of dietary patterns by using metabolomics. *Am. J. Clin. Nutr.* **2017**, *105*, 450–465. [CrossRef]

90. Garcia-Perez, I.; Posma, J.M.; Gibson, R.; Chambers, E.S.; Hansen, T.H.; Vestergaard, H.; Hansen, T.; Beckmann, M.; Pedersen, O.; Elliott, P.; et al. Objective assessment of dietary patterns by use of metabolic phenotyping: A randomised, controlled, crossover trial. *Lancet Diabetes Endocrinol.* **2017**, *5*, 184–195. [CrossRef]

91. FiHGV. Obesity and abnormal eating behavior across the lifespan. A crosssectional and longitudinal approach of environmental and neurobiological factors (Eat4healthylife). CiberObn, Madrid, Spain. Available online: https://fihgu.general-valencia.san.gva.es/-/investigadores-de-la-fihguv-valencia-participa-en-el-estudio-eae4healthylife-para-la-identificacion-precoz-de-factores-de-riesgo-asociados-a-la-obesid (accessed on 12 November 2020).

92. Mills, S.; Stanton, C.; Lane, J.A.; Smith, G.J.; Ross, R.P. Precision Nutrition and the Microbiome, Part I: Current State of the Science. *Nutrients* **2019**, *11*, 923. [CrossRef]

93. Greenwood, D.C.; Threapleton, D.E.; Evans, C.E.; Cleghorn, C.L.; Nykjaer, C.; Woodhead, C.; Burley, V.J. Glycemic index, glycemic load, carbohydrates, and type 2 diabetes: Systematic review and dose-response meta-analysis of prospective studies. *Diabetes Care* **2013**, *36*, 4166–4171. [CrossRef] [PubMed]

94. Bashiardes, S.; Godneva, A.; Elinav, E.; Segal, E. Towards utilization of the human genome and microbiome for personalized nutrition. *Curr. Opin. Biotechnol.* **2018**, *51*, 57–63. [CrossRef] [PubMed]

95. David, L.A.; Maurice, C.F.; Carmody, R.N.; Gootenberg, D.B.; Button, J.E.; Wolfe, B.E.; Ling, A.V.; Devlin, A.S.; Varma, Y.; Fischbach, M.A.; et al. Diet rapidly and reproducibly alters the human gut microbiome. *Nature* **2014**, *505*, 559–563. [CrossRef] [PubMed]

96. Salonen, A.; Lahti, L.; Salojarvi, J.; Holtrop, G.; Korpela, K.; Duncan, S.H.; Date, P.; Farquharson, F.; Johnstone, A.M.; Lobley, G.E.; et al. Impact of diet and individual variation on intestinal microbiota composition and fermentation products in obese men. *ISME J.* **2014**, *8*, 2218–2230. [CrossRef] [PubMed]

97. Biesiekierski, J.R.; Jalanka, J.; Staudacher, H.M. Can Gut Microbiota Composition Predict Response to Dietary Treatments? *Nutrients* **2019**, *11*, 1134. [CrossRef] [PubMed]

98. De Filippis, F.; Pellegrini, N.; Laghi, L.; Gobbetti, M.; Ercolini, D. Unusual sub-genus associations of faecal Prevotella and Bacteroides with specific dietary patterns. *Microbiome* **2016**, *4*, 1–6. [CrossRef]

99. Healey, G.; Murphy, R.; Butts, C.; Brough, L.; Whelan, K.; Coad, J. Habitual dietary fibre intake influences gut microbiota response to an inulin-type fructan prebiotic: A randomised, double-blind, placebo-controlled, cross-over, human intervention study. *Br. J. Nutr.* **2018**, *119*, 176–189. [CrossRef]

100. Mills, S.; Lane, J.A.; Smith, G.J.; Grimaldi, K.A.; Ross, R.P.; Stanton, C. Precision Nutrition and the Microbiome Part II: Potential Opportunities and Pathways to Commercialisation. *Nutrients* **2019**, *11*, 1468. [CrossRef]

101. Nicolucci, A.C.; Hume, M.P.; Martínez, I.; Mayengbam, S.; Walter, J.; Reimer, R.A. Prebiotics Reduce Body Fat and Alter Intestinal Microbiota in Children Who Are Overweight or With Obesity. *Gastroenterology* **2017**, *153*, 711–722. [CrossRef]

102. Zhang, C.; Yin, A.; Li, H.; Wang, R.; Wu, G.; Shen, J.; Zhang, M.; Wang, L.; Hou, Y.; Ouyang, H.; et al. Dietary Modulation of Gut Microbiota Contributes to Alleviation of Both Genetic and Simple Obesity in Children. *EBioMedicine* **2015**, *2*, 968–984. [CrossRef] [PubMed]

103. Venkataraman, A.; Sieber, J.R.; Schmidt, A.W.; Waldron, C.; Theis, K.R.; Schmidt, T.M. Variable responses of human microbiomes to dietary supplementation with resistant starch. *Microbiome* **2016**, *4*, 1–9. [CrossRef] [PubMed]

104. Pasolli, E.; Truong, D.T.; Malik, F.; Waldron, L.; Segata, N. Machine Learning Meta-analysis of Large Metagenomic Datasets: Tools and Biological Insights. *PLoS Comput. Biol.* **2016**, *12*, e1004977. [CrossRef] [PubMed]

105. Zeevi, D.; Korem, T.; Zmora, N.; Israeli, D.; Rothschild, D.; Weinberger, A.; Ben-Yacov, O.; Lador, D.; Avnit-Sagi, T.; Lotan-Pompan, M.; et al. Personalized Nutrition by Prediction of Glycemic Responses. *Cell* **2015**, *163*, 1079–1094. [CrossRef]

106. Korem, T.; Zeevi, D.; Zmora, N.; Weissbrod, O.; Bar, N.; Lotan-Pompan, M.; Avnit-Sagi, T.; Kosower, N.; Malka, G.; Rein, M.; et al. Bread Affects Clinical Parameters and Induces Gut Microbiome-Associated Personal Glycemic Responses. *Cell. Metab.* **2017**, *25*, 1243–1253.e5. [CrossRef]

Publisher's Note: MDPI stays neutral with regard to jurisdictional claims in published maps and institutional affiliations.

 *nutrients*

Article

Interaction Effect of the Mediterranean Diet and an Obesity Genetic Risk Score on Adiposity and Metabolic Syndrome in Adolescents: The HELENA Study

Miguel Seral-Cortes [1], Sergio Sabroso-Lasa [2], Pilar De Miguel-Etayo [1,3,*],
Marcela Gonzalez-Gross [3,4,5], Eva Gesteiro [4], Cristina Molina-Hidalgo [6], Stefaan De Henauw [7],
Éva Erhardt [8], Laura Censi [9], Yannis Manios [10], Eva Karaglani [10], Kurt Widhalm [11],
Anthony Kafatos [12], Laurent Beghin [13], Aline Meirhaeghe [14], Diego Salazar-Tortosa [15],
Jonatan R. Ruiz [16], Luis A. Moreno [1,3], Luis Mariano Esteban [17] and Idoia Labayen [18]

1   Growth, Exercise, NUtrition and Development (GENUD) Research Group, Faculty of Health Sciences,
     Instituto Agroalimentario de Aragón (IA2), Instituto de Investigación Sanitaria Aragón (IIS Aragón),
     Universidad de Zaragoza, 50009 Zaragoza, Spain; mseral@unizar.es (M.S.-C.); lmoreno@unizar.es (L.A.M.)
2   Spanish National Cancer Research Centre (CNIO), 28029 Madrid, Spain; ssabroso@cnio.es
3   CIBER Fisiopatología de la Obesidad y Nutrición (CIBERobn), Instituto de Salud Carlos III,
     28029 Madrid, Spain; marcela.gonzalez.gross@upm.es
4   ImFine Research Group, Department of Health and Human Performance,
     Facultad de Ciencias de la Actividad Física y del Deporte-INEF, Universidad Politécnica de Madrid,
     28040 Madrid, Spain; eva.gesteiro@upm.es
5   Institute of Nutritional and Food Sciences, Nutritional Physiology, University of Bonn, 53113 Bonn, Germany
6   EFFECTS 262 Department of Medical Physiology, School of Medicine, University of Granada,
     18071 Granada, Spain; cristinamolinapsico@gmail.com
7   Department of Public Health and Primary Care, Faculty of Medicine and Health Sciences, Ghent University,
     9000 Ghent, Belgium; Stefaan.DeHenauw@UGent.be
8   Department of Pediatrics, Medical School, University of Pécs, 7623 Pécs, Hungary; erhardt.eva@pte.hu
9   Council for Agricultural Research and Economics-Research Center for Food and Nutrition,
     00198 Rome, Italy; laura.censi@crea.gov.it
10   Department of Nutrition and Dietetics, School of Health Science & Education, Harokopio University,
     176 71 Athens, Greece; manios@hua.gr (Y.M.); ekaragl@hua.gr (E.K.)
11   Division of Gastroenterology and Hepatology, Department of Internal Med III,
     Austria and Austrian Academic Institute for Clinical Nutrition, 1090 Vienna, Austria;
     kurt.widhalm@meduniwien.ac.at
12   Faculty of Medicine, University of Crete, 715 00 Crete, Greece; kafatos@med.uoc.gr
13   CIC-1403-Inserm-CHU, Clinical Investigation Center, LIRIC UMR 995 Inserm, CHU Lille,
     Université de Lille, 59000 Lille, France; Laurent.BEGHIN@CHRU-LILLE.FR
14   UMR1167, RID-AGE, Risk Factors and Molecular Determinants of Aging-Related Diseases, Centre Hosp,
     Institut Pasteur de Lille, Université de Lille, 59019 Lille, France; aline.meirhaeghe@pasteur-lille.fr
15   Department of Ecology and Evolutionary Biology, University of Arizona, Tucson, AZ 85719, USA;
     dftortosa@email.arizona.edu
16   PROmoting FITness and Health through Physical Activity Research Group (PROFITH),
     Department of Physical and Sports Education, School of Sports Science,
     Sport and Health University Research Institute (iMUDS), University of Granada, 18016 Granada, Spain;
     ruizj@ugr.es
17   Escuela Politécnica de La Almunia, Universidad de Zaragoza, 50100 Zaragoza, Spain; lmeste@unizar.es
18   Department of Health Sciences, Public University of Navarra, 31006 Pamplona, Spain;
     idoia.labayen@unavarra.es
*   Correspondence: pilardm@unizar.es; Tel.: +34-876-553-756

Received: 5 November 2020; Accepted: 14 December 2020; Published: 16 December 2020

Nutrients **2020**, *12*, 3841; doi:10.3390/nu12123841

Abstract: Obesity and metabolic syndrome (MetS) are worldwide major health challenges. The Mediterranean diet (MD) is associated with a better cardiometabolic profile, but these beneficial effects may be influenced by genetic variations, modulating the predisposition to obesity or MetS. The aim was to assess whether interaction effects occur between an obesity genetic risk score (obesity-GRS) and the MD on adiposity and MetS in European adolescents. Multiple linear regression models were used to assess the interaction effects of an obesity-GRS and the MD on adiposity and MetS and its components. Interaction effects between the MD on adiposity and MetS were observed in both sex groups ($p < 0.05$). However, those interaction effects were only expressed in a certain number of adolescents, when a limited number of risk alleles were present. Regarding adiposity, a total of 51.1% males and 98.7% females had lower body mass index (BMI) as a result of higher MD adherence. Concerning MetS, only 9.9% of males with higher MD adherence had lower MetS scores. However, the same effect was observed in 95.2% of females. In conclusion, obesity-related genotypes could modulate the relationship between MD adherence and adiposity and MetS in European adolescents; the interaction effect was higher in females than in males.

Keywords: metabolic syndrome; Mediterranean diet; genetic risk score; HELENA; adolescents; sex

1. Introduction

Metabolic syndrome (MetS) is known to be a major world health challenge, with increasing prevalence together with obesity and cardiovascular diseases [1]. The prevalence of overweight and obesity worldwide has drastically increased among youth in recent years, with similar numbers in males and females [2]. The prevalence of obesity and metabolic syndrome in European children and adolescents continue in the same increasing line despite the efforts of prevention programs in recent years [3,4]. The definition of MetS features a number of cardiometabolic risk factors, including total and/or central adiposity, dyslipidemia, hypertension, and insulin resistance [5]. Clustering of cardiometabolic risk factors is increasingly considered in children's and adolescents' health rather than single risk factors [6,7]. In European children, an inverse association between the Mediterranean diet (MD) and childhood obesity has been observed [8] and showed that high MD adherence at early age is associated with a lower risk of developing overweight and obesity during childhood [9]. Moreover, in children and adolescents, MD was associated with lower body mass index (BMI) and improved glucose and lipid profiles [10]. The beneficial effects associated with a high MD adherence may be influenced by the interaction with other factors, such as genetic variations, which could modulate the predisposition/risk to obesity and MetS [11]. In adults, a systematic review [12] showed that the interaction between the melanocortin 4 receptor (*MC4R*) gene (a protein-coding gene previously associated to BMI [13]) and MD modulates the development of obesity and type 2 diabetes mellitus (T2DM) phenotypes. In Chinese children and adolescents, interactions between genetic variants and dietary behaviors in relation to obesity have been observed [14].

Previous studies have shown the potential of genetic approaches that identify individuals at high risk of developing a disease. That is the case of genetic risk scores (GRS), that combine a number of single nucleotide polymorphisms (SNPs) by summing the number of risk alleles [15]. In order to try to prevent the development of obesity and MetS in European adolescents it is crucial to improve our understanding of the predisposing genetic factors [16,17].

To our knowledge, no studies have examined the interaction effect between MD adherence and an obesity-related GRS on adiposity and MetS in European adolescents. Therefore, the aim of our study is to assess whether interaction effects occur between the MD adherence and obesity-GRS on adiposity and MetS in European adolescents. We hypothesize that higher predisposition to obesity risk may attenuate the protective effect of MD adherence on adiposity and MetS in European adolescents.

2. Materials and Methods

2.1. Study Design and Population

The Healthy Lifestyle in Europe by Nutrition in Adolescence (HELENA) multicentric and cross-sectional study included a total sample of 4356 adolescents (51.6% females), aged 11–19 years [18]. Data were obtained from 10 European cities located in different geographical points within Europe, during 2006–2007. The HELENA study was designed to obtain reliable and comparable data on nutrition and health-related parameters, applying standardized procedures [19]. The HELENA study was approved by the Research Ethics Committees of each study site and followed the ethical guidelines of the Declaration of Helsinki 1964 (revision of 2000), good clinical practice, and the legislation about clinical research in humans in each one of the countries involved in the study [20]. Written informed consent was obtained by the parents/legal guardians of all participants. Blood sampling was performed in one third of the individuals randomly selected from the total sample (N = 1172) [19]. Inclusion of specific parameters to develop the present study (genomic, adiposity, cardiometabolic risk factors, and dietary data) provided a final number of 605 adolescents (51.6% female) meeting the selection criteria. Information of selection procedure is displayed in a flow chart (Figure 1).

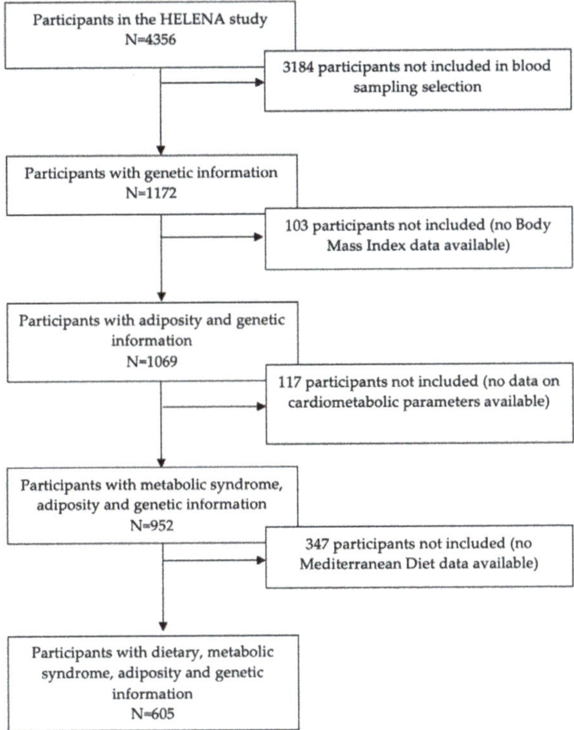

Figure 1. Flow chart of the sample selection process.

2.2. Physical Examination, Adiposity Measurements and Cardiometabolic Risk Score

Anthropometric measurements were performed by trained researchers following standard protocols [21]. Height was measured barefoot in the Frankfort plane with a telescopic height measuring instrument (Type SECA 225) to the nearest 0.1 cm, and weight was measured in underwear and without shoes with an electronic scale (Type SECA 861) to the nearest 0.1 kg. BMI was calculated from weight and height (kg/m^2) [22]. Waist circumference (WC) was measured in triplicate with a nonelastic tape (SECA 200) to the nearest 0.1 cm as the mid-point between the lowest rib and the iliac crest [21], and the average of the three measures was used. Systolic blood pressure (SBP) and diastolic blood pressure (DBP) were measured twice in sitting position separated in a 10- minute interval with a blood pressure oscillometric monitor device OMRON HEALTHCARE® (M6-HEM7001; OMRON HEALTHCARE®, Kyoto, Japan) The lowest blood pressure (BP) reading was used. Thereafter, the mean of arterial pressures (MAP) of all participants was obtained from the DBP + [(SBP − DBP)/3)] formula. Total cholesterol (TC), HDL-cholesterol (HDL-c), triglycerides (TG), and glucose were measured using enzymatic methods (Dade Behring, Schwalbach, Germany). Insulin levels were obtained from frozen serum using an Immulite 2000 analyzer (DPC Bierman GmbH, Bad Nauheim, Germany). As measurement of insulin resistance, the homeostatic model assessment (HOMA) index was calculated from glucose and insulin measurements [23]. Moreover, a clustered cardiometabolic risk score was computed from the sum of the standardized z-scores of TC/HDL ratio, WC, HOMA index, and MAP [7]. The standardized z-scores of intended variables were calculated from the age and gender specific cut-off points [24]. Lower values in the score indicate better cardiometabolic risk profile. As sensitivity analyses, a second MetS risk score, comprising HDL-c, WC, Glucose, SBP and TG, was obtained from the International Diabetes Federation (IDF) guidelines [5]. HDL-c was multiplied by −1 as characterized by lower metabolic risk with increasing values. Results obtained with both cardiometabolic risk scores in adolescents have been included in the present analyses.

2.3. Dietary Intake Assessment and Mediterranean Diet Score

Dietary habits were determined from a self-administered, computerized, validated 24-h dietary recall called the HELENA Dietary Assessment Tool (HELENA-DIAT) [25,26], a tool validated first in Flemish adolescents [26] and then adapted to be used in the 10 cities [27]. Participants completed the HELENA-DIAT on two non-consecutive days within a time span of two weeks. This method has been used and recommended to assess dietary intake in European children and adolescents [28]. In order to calculate individual usual dietary intake, the multiple source method (MSM) was used [28]. This method allows correction of dietary data for between and within individuals' variability.

We used a Mediterranean diet score (MDS) based on nine single components: vegetables, fruits and nuts, cereals and roots, pulses, fish, monounsaturated/saturated fatty acids ratio, dairy products, meat, and alcohol. A scale indicating the degree of adhesion/adherence to the traditional MD was developed [29]. The description of MD food subgroup components is described elsewhere [30]. Vegetables, fruits and nuts, cereals, legumes, fish, dairy products, and unsaturated to saturated fat ratio positively contributed to the MD adherence, whereas meat (including processed meat) and alcohol consumption were inversely considered. Of note is that dairy products are positively considered as they are recommended during growth and development periods, such as adolescence [31]. Alcohol intake was regarded as an unhealthy habit among adolescents. Therefore, in a no-alcohol consumption situation, the value 1 was assigned, while any alcohol intake was computed as value 0. The MD adherence was constructed by a 0 to 9 points scale, with higher scores indicating greater adherence [32]. The sex-specific median intake (g/day) of all subgroups forming the MDS is shown in Supplementary Table S1.

2.4. Genomic Information and Genetic Risk Score

Standard methods for blood collection, transport, and analysis was performed by a certified laboratory [33]. Blood sampling (EDTA K3 tubes) for DNA extraction, collection, and storage was performed at the Institute of Nutritional and Food Sciences (IEL) of the University of Bonn, and sent to the Laboratoire d'Analyse Genomique Centre de Ressources Biologiques (LAG-CRB) BB- 0033-00071 Institut Pasteur de Lille, F-59000 Lille, France. DNA was obtained from white blood cells with the Puregene kit (QIAGEN, Courtaboeuf, France) and stored at −20 °C. The genotyping was done by an Illumina system (Illumina, Inc., San Diego, CA, USA) using the Golden Gate technology (sampling procedure scheme, GoldenGate; Software, Inc, San Francisco, CA, USA).

To analyze the influence of genetic information on the association between MD and adiposity and cardiometabolic biomarkers, we used the obesity-GRS developed from HELENA adolescents (submitted but not yet accepted) [34] using 21 SNPs significantly associated with overweight/obesity, defined as the equivalent to BMI > 25 kg/m^2. The main characteristics of SNPs forming the obesity-GRS are displayed in Supplementary Table S2. Also, a comparative analysis by sex is shown in Supplementary Table S3.

2.5. Statistical Analysis

To test the variables´ normality, the Shapiro–Wilk test was performed. As not all variables follow a normal distribution, the descriptive sex-specific characteristics are displayed as median and interquartile range (IQR) for continuous variables; absolute and relative frequencies are shown for categorical variables. In order to compare differences by sex, the statistical Pearson's chi-square test was used for categorical variables and the Mann–Whitney–Wilcoxon test for continuous variables.

Sex-specific multiple linear regression models were used to assess the association between adiposity and cardiometabolic parameters and the interaction effect between the obesity-GRS and MD, adding to the effect of MD alone in the same model.

RStudio Version 1.2.5001 (RStudio Team (2015). RStudio: Integrated Development for R. RStudio, Inc., Boston, MA, USA, URL http://www.rstudio.com/) was used to perform all statistical analyses and $p < 0.05$ was the significance level set in the present analysis.

3. Results

3.1. Descriptive Characteristics of the Study Sample

The main characteristics of participants are shown in Table 1. In summary, there were significant differences between males and females for weight and height ($p \leq 0.001$) although no significant differences were found for BMI. Regarding cardiometabolic risk factors, girls had higher WC ($p \leq 0.001$) and HOMA ($p = 0.044$), whereas boys showed higher SBP ($p \leq 0.001$), MAP ($p \leq 0.001$), and MetS ($p \leq 0.001$) than girls. There were no significant differences for the remaining cardiometabolic parameters. The distribution of obesity-GRS among participants is displayed by sex in Figure 2. Focusing in the SNPs included in the obesity-GRS, no statistically significant differences between sex were found in Supplementary Table S3.

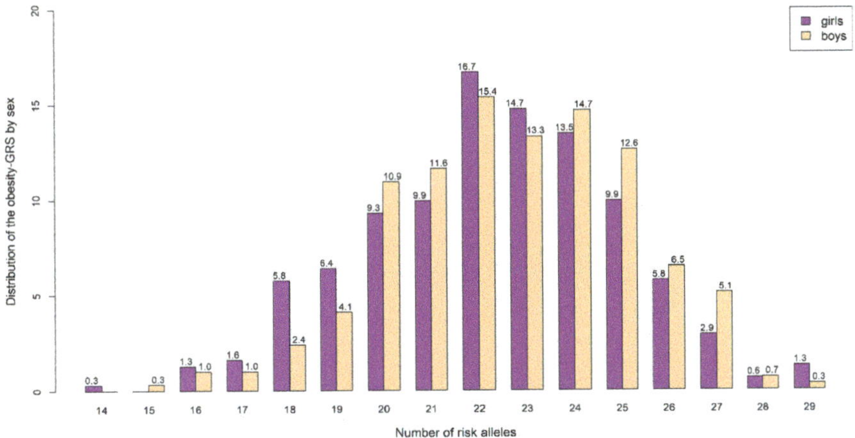

Figure 2. Distribution of the obesity-GRS (obesity-related genetic risk score) by sex (% displayed) in the HELENA cohort according to the number of risk alleles.

Table 1. Demographics and cardiometabolic characteristics of the Healthy Lifestyle in Europe by Nutrition in Adolescence (HELENA) participants displayed by sex.

	Total	Male	Female	*p*-Value
	n = 605	*n* =293	*n* =312	
Age (years)	14.7 (13.8–15.6)	14.8 (13.8–15.6)	14.8 (13.7–15.7)	0.948
Height (cm)	166.0 (159.5–172.2)	170.0 (163.9–177.0)	162.2 (157.0-167.2)	**<0.001**
Weight (kg)	58.4 (49.9–64.5)	61.1 (51.1–68.8)	55.8 (49.2–61.6)	**<0.001**
BMI (kg/m^2)	21.1 (18.6–22.9)	21.0 (18.5–22.7)	21.2 (18.7–23.0)	0.194
WC (cm)	72.1 (66.7–75.8)	65.75 (46.0–79.0)	71.5 (48.7–83)	**<0.001**
HOMA index	2.2 (1.3–2.6)	2.2 (1.3–2.5)	2.3 (1.4–2.7)	**0.044**
SBP (mmHg)	116 (108–124)	120 (112–129)	112 (105–120)	**<0.001**
DBP (mmHg)	64.4 (59.0–70.0)	64.0 (59.0-69.0)	64.8 (60.0–70.0)	0.331
MAP	0.6 (−0.01–1.1)	1.1 (0.5–1.6)	0.2 (−0.4–0.7)	**<0.001**
HDL-c (mmol/L)	55.7 (49–63)	53.3 (47–59)	57.9 (50–65)	**<0.001**
TG (mmol/L)	68.7 (47.0–80.0)	65.7 (46.0–79.0)	71.5 (48.7–83.0)	0.056
TC:HDL ratio	2.3 (2.5–1.1)	2.3 (2.5–3.2)	2.9 (2.5–3.3)	0.839
PA (mins/day)	54.8 (40.7–71.5)	65.4 (51.8–82.4)	47.3 (35.2–59.8)	**<0.001**
MetS Score *	0.02 (-1.2–1.0)	0.3 (−0.7–1.3)	−0.3 (−1.5–0.8)	**<0.001**
MDS **	4 (0–8)	4 (0–8)	4 (0–8)	0.495
Obesity-GRS ***	23 (21–24)	23 (21–25)	22 (21–24)	0.087

Abbreviations: BMI (body mass index); WC (waist circumference); HOMA index (homeostatic model assessment index); SPB (systolic blood pressure); DBP (diastolic blood pressure); MAP (mean arterial pressure); HDL-c (high density lipoprotein cholesterol); TG (triglycerides); TC:HDL ratio (total cholesterol/HDL cholesterol ratio); PA (physical activity); (MetS Score (Metabolic Syndrome score); MDS (Mediterranean diet score); and obesity-GRS (obesity-related genetic risk score). Median values (p. 25–p. 75) expressed. * Metabolic Syndrome score resulting from the mean of WC, HOMA, MAP, and TC-HDL variables combined. ** Mediterranean diet score resulting from the sum of nine food subgroups compliance. *** Genetic risk score resulting from the sum of risk alleles of HELENA participants. Boldface values indicate sig *p*-value Sig *p*-value < 0.05.

3.2. Interaction between MD and Obesity-GRS on Adiposity/Cardiometabolic Variables

Table 2 shows the association between the cardiometabolic parameters in relation to MD and the interaction between MD and the obesity-GRS in the additive model, by sex groups.

Table 2. Multiple linear regression models of obesity-related genetic risk score (obesity-GRS) and Mediterranean diet (MD) interaction, and the MD effect alone on adiposity and cardiometabolic parameters by sex.

p-Values	Male		Female	
	Obesity-GRSxMD	**MD**	**Obesity-GRSxMD**	**MD**
BMI (kg/m^2)	**0.003**	**0.008**	**<0.001**	**<0.001**
WC (cm)	**0.009**	**0.030**	**<0.001**	**<0.001**
HOMA	0.495	0.836	**0.027**	**0.013**
SBP (mmHg)	0.994	0.739	0.310	**0.047**
DBP (mmHg)	**0.005**	**0.014**	0.795	0.626
MAP	**0.031**	**0.045**	0.872	0.325
TG (mmol/L)	0.421	0.413	0.587	0.689
TC:HDL	0.465	0.530	0.184	0.118
MetS Score	**0.014**	**0.047**	**0.006**	**0.002**

Boldface values indicate sig *p*-value Sig *p*-value < 0.05.

Considered within the additive model (Table 2), MD adherence had a protective role over BMI (male $p < 0.01$ vs. female $p \leq 0.001$), WC (male $p < 0.05$ vs. female $p \leq 0.001$), and MetS (male $p < 0.05$ vs. female $p \leq 0.01$) in both sex groups. However, the association of the MD with HOMA ($p < 0.05$) and SBP ($p < 0.05$) was only observed in females, and the association of the MD with MAP ($p \leq 0.05$) and DBP ($p < 0.05$) only in males.

Furthermore, both the MD and the interaction between MD and obesity-GRS were significant in the case of BMI for both sex groups; the inverse association between the MD and BMI was higher in females than in males (females $p \leq 0.001$ vs. males $p < 0.01$). When studying the cardiometabolic variables, we observed significant interactions between obesity-GRS and MD in both sex groups for WC ($p < 0.05$) and MetS ($p < 0.05$). For both adiposity and cardiometabolic variables, females showed stronger interactions than males. Moreover, the obesity-GRS and MD showed a significant interaction on MAP ($p < 0.05$) and DBP ($p < 0.01$) for males only, whereas the interaction on HOMA was only significant for females ($p < 0.05$).

Figure 3 shows the interaction effects of the obesity-GRS and the MD on BMI, WC, and MetS for male and female participants. The relations between the MD and BMI, WC, and MetS were modulated by the obesity-GRS values. In order to interpret the abovementioned variables, different sub-figures have been displayed in a matrix panel according to each sex group for BMI, WC, and MetS. Different lines were drawn to relate MD and adiposity and cardiovascular biomarkers modulated by the distribution of the genetic predisposition to obesity in our population. It must be remarked that the majority of the adolescents were concentrated in the central parts of the graph, within 20-26 risk alleles (82.3% of the total population); therefore, extreme values should be interpreted cautiously. Thus, for those participants represented with a negative slope, a higher MD adherence could act as a protective factor in relation to cardiometabolic factors despite their genetic predisposition to obesity (high or low).

Regarding the adiposity parameters, a total of 46.8% of males (those with 22 risk alleles or below) with higher MD adherence had lower BMI; in 98.1% of females (in those individuals scoring 27 risk alleles or below), a higher MD adherence was associated with lower BMI, attenuating the genetic risk to obesity. For WC, we also observed that 19.8% males (≤20 risk alleles) with higher MD adherence, had lower WC; however, 95.2% of females (≤26 risk alleles), had also lower WC levels.

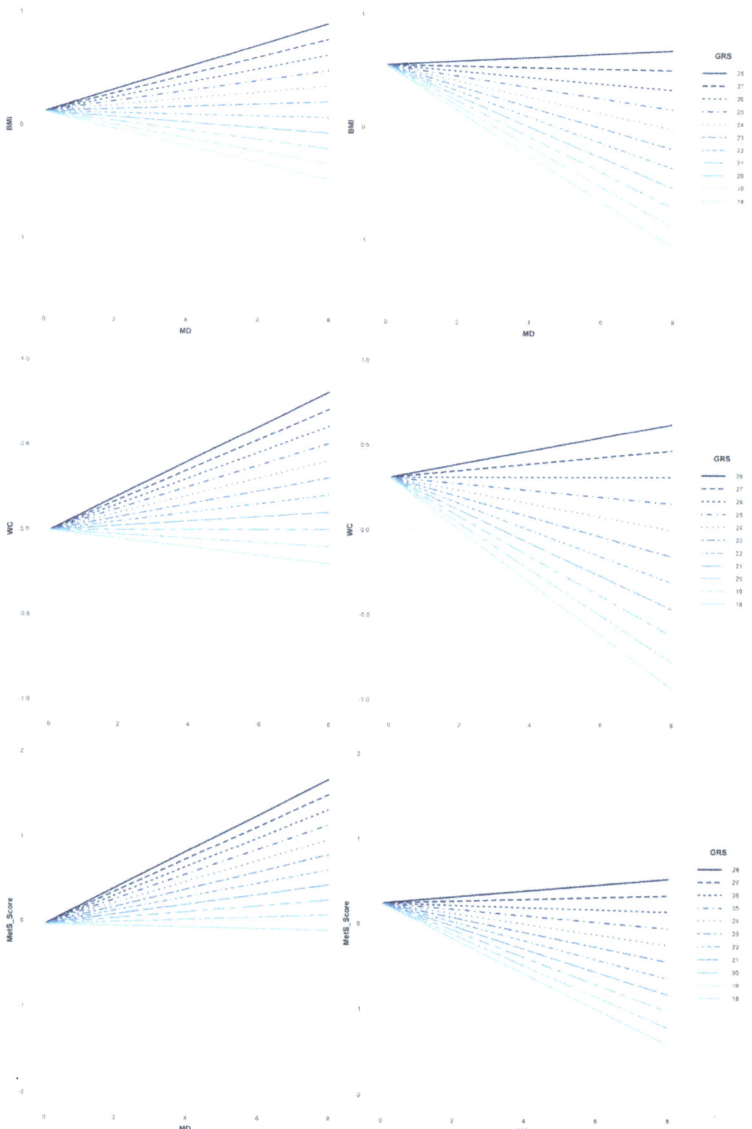

Figure 3. Matrix panel of interaction models between body mass index (BMI), waist circumference (WC), and metabolic syndrome score (MetS Score); and the Mediterranean diet (MD) according to the obesity genetic risk score (Obesity-GRS) modulation compared by sex (males left panel, females right panel). Obesity-GRS values (18–28) displayed according to our population distribution. Legend: when designing the population distribution representation, different lines were drawn as reference points to observe the trend of the studied population according to the genetic predisposition to obesity. When analyzing the results represented in these figures, a positive gradient shows the MD acting as risk factor, whereas a negative gradient indicates the protective role of the MD.

Concerning MetS, 4.8% of males with higher MD adherence had lower cardiometabolic risk score if the risk score was ≤18. More so, 95.2% of females having higher MD adherence showed higher MetS scores if the risk score was ≤26. In terms of sensitivity analyses, the MetS analyses were repeated using the MetS score following the IDF recommendations. In males, the obesity-GRS and MD showed a significant interaction on MetS ($p = 0.001$) but not on the IDF MetS score ($p = 0.092$). However, the interaction was significant for both MetS scores in females ($p = 0.005$ vs. $p = 0.003$).

There were sex-related differences in the interaction between genes and diet on other cardiometabolic parameters, such as HOMA, DBP, and MAP. Specifically, in females with higher MD adherence, those having ≤25 risk alleles had lower HOMA levels (89.4% of the adolescents). Male participants with higher MD adherence with an obesity-GRS ≤21 (31.4% of the adolescents) had lower DBP. Likewise, males with higher MD adherence and an obesity-GRS ≤22 (46.8% of the adolescents) had a lower MAP. The interaction effects of the obesity-GRS and MD on HOMA, DBP, and MAP for male and female participants are displayed in Supplementary Figure S1.

4. Discussion

The main findings of the present study indicate that the influence of high MD adherence on adiposity and MetS was only expressed when a limited number of risk alleles were present. As a result, the gene–diet interaction effect was higher in females than in males.

The MD has previously been shown to provide numerous health benefits [35], such as the reduction of risk factors for non-communicable diseases [36,37]. However, little is known about how the genetic variations among individuals determine the response to MD adherence [38]. To our knowledge, no previous studies have assessed their interaction effects using an obesity-related GRS. Instead, isolated SNPs from candidate genes previously associated in the literature with adiposity or MetS were examined. More so, no gene–diet interaction studies on adiposity and cardiometabolic parameters considering the MD were found in adolescents, as the majority of studies have been conducted in adult populations.

Concerning adiposity, similar findings were observed in a study where the *FTOrs9939609* and *MC4Rrs17782313* polymorphisms showed an interaction with the MD adherence, which reduced the risk of obesity and T2DM [39]. In line with our findings, one study also showed low adiposity levels in relation with the interaction effect with different allele combinations, and considering other dietary approaches different to the MD: low polyunsaturated fatty acids (PUFA) intake showed an inverse association with obesity risk (BMI ≥ 30 kg/m^2) when the *ADAM17i33708A* polymorphism was present [40]. Another study, considering high saturated fat intake, showed a significantly higher BMI in the GG carriers of the *THRArs1568400* than in the A carriers [41], suggesting counter-productive effects in comparison to our study. In the current study, not only BMI, but also WC, a surrogate marker of abdominal adiposity, showed lower levels in both male and female adolescents, as a result of the obesity-GRS–MD interaction; to our knowledge, no similar findings have been reported in other studies with similar characteristics.

Regarding MetS, we found no studies assessing a gene–diet interaction effect on a MetS risk score. We have also assessed cardiometabolic parameters individually. Our results showed low HOMA levels in the majority of females modulated by the obesity-GRS when adhering to the MD. One study assessing the effect of increasing the ratio of saturated fat to carbohydrate intake, showed higher HOMA levels in carriers of the minor allele (*PLIN11482G>A*) [42].

Concerning DBP, the male adolescents with higher MD adherence had lower levels of DBP when having 21 or fewer risk alleles. Our findings are coincident to one intervention study promoting the Dietary Approaches to Stop Hypertension (DASH) diet, where AA carriers of the Angiotensinogen genotype (G-6A ANG polymorphism) showed the greatest reduction in DBP [43]. No interaction studies considering the MAP levels were reported in the literature.

Despite the fact that we did not observe an effect on HDL-c, we included the TC:HDL ratio variable in our MetS score, showing a significant association with the gene–diet interaction. In this sense,

Ordovás et al. observed that women carriers of the A allele of the *APOA1* gene (G-A polymorphism) responded with higher HDL-c concentrations to a high PUFA intake, whereas the opposite effect was observed in the G carriers [44]. Other authors have also pointed novel genes–nutrients interactions with high-carbohydrate diets in GG carriers of *KCTD10i5642G>C* and *MMAB3U3527G>C* and C allele carriers of *KCTD10V206VT>C*, contributing to lower HDL-c concentrations [45].

When adhering to the MD, no associations were found in terms of TG levels. However, another study showed that, after 12 months of a MD-based intervention, higher levels of HDL-c and TG were seen in those individuals carrying the T allele of the *CETPrs3764261* than in those with the GG genotype [46]. Another study showed that the *TNF-alphars1800629* GG subjects had higher levels of TG than A carriers in MetS patients after another MD intervention [47]. Finally, other authors assessed the interaction between *PPARAL162V* and PUFA intake, noticing lower TG levels with higher intake in the V carriers [48].

In order to compare the present results to other cardiometabolic definitions, the development of our MetS score was compared to another MetS score, following the IDF recommendations. The IDF score showed no association with the GRS–MD interaction in males. This fact could be due to the different age and gender specific criteria selected to define the cut-off points between authors. Nevertheless, positive associations of the GRS–MD interaction were seen in the female group for both MetS scores. The consensus to use unified criteria to identify adolescents at risk of MetS remains under development.

The present study has some limitations. As the HELENA project is a cross-sectional study, cause–effect relationships cannot be established. Moreover, only selected risk loci are available in the HELENA study. When constructing the obesity-GRS, it has been calculated that BMI changes can be explained by a small proportion of genetic variants discovered so far [49], so potential rarer variants yet to be found might emerge when Genome-Wide Association Studies (GWAS) are carried out.

On the other hand, the study presents several strengths. Most analyses from previous studies were focused on specific single SNPs interactions, whereas the present study used an obesity-GRS, considered as a useful genetic tool [50], to predict adiposity and MetS in European adolescents. More so, the development of the cardiometabolic risk score was considered appropriate for the present study as it provides higher sensitivity and low susceptibility to errors compared to other approaches [51]. Furthermore, we included the whole MD pattern developing a cluster of different food groups in an adherence scale rather than considering single macronutrients or individual specific food groups´ intake. Additionally, different effects were identified in males and females. There is different behavior dependent on sex, not attributable to genetic predisposition, maybe associated with physical activity. Due to the multicentric design of the HELENA study, congregating participants from 10 European cities, the researchers have been provided with large datasets from diversely distributed populations across Europe. Finally, the study is focused on adolescents, a population age that is understudied and where early detection plays a key role in the development of obesity and metabolic syndrome. Little has been found in the literature using a similar approach, where most similar studies are available on adult populations.

As MetS and excess of adiposity may occur at any stage from childhood to adulthood, early detection and diagnosis is fundamental to elaborate on health prevention programs among youth to effectively reduce the risk of cardiovascular diseases and T2DM [52,53]. At the same time, it has been previously shown that greater adherence to MD was associated with a significant improvement in overall health status among youth, suggesting the implementation of a MD dietary pattern for primary prevention of major chronic diseases [54]. The effect of the interventions would be heterogeneous depending on the genetic background of the individuals and should be considered in the efficacy analyses.

5. Conclusions

Obesity-related genotypes had a modulation effect in the relationship between MD adherence and obesity and MetS risk in European adolescents. The genes–diet interaction effect on MetS was stronger in females than in males. These observations strengthen the idea of applying genomic information to promote targeted dietary advice.

Supplementary Materials: The following are available online at http://www.mdpi.com/2072-6643/12/12/3841/s1, Table S1: Mediterranean diet score items specified in g/day displayed in sex-specific median intake in HELENA participants with dietary information available, Table S2: Main characteristics of the 21 single nucleotide polymorphisms (SNPs) included in the obesity genetic risk score (obesity-GRS), Table S3: Comparative analysis of the 21 single nucleotide polymorphisms (SNPs) included in the obesity genetic risk score (obesity-GRS) by sex, Figure S1: Interaction models between HOMA, diastolic blood pressure (DBP) and mean arterial pressure (MAP) and Mediterranean Diet (MD) according to the obesity genetic risk score (obesity-GRS) modulation in female. Obesity-GRS values (18–28) are displayed according to our population distribution.

Author Contributions: L.A.M. is M.S.-C.'s supervisor. P.D.M.-E. and I.L. are M.S.-C.'s co-supervisors. M.S.-C., S.S.-L., P.D.M.-E., L.M.E., L.A.M., and I.L. contributed to the conceptualization, methodology, formal analysis and writing—original draft preparation. I.L. supervised the genomic data. M.G.-G. was responsible for the blood sampling and analysis procedure. M.G.-G., E.G., C.M.-H., S.D.H., É.E., L.C., Y.M., E.K., K.W., A.K., L.B., A.M., D.S.-T., and J.R.R. participated in the writing—review and editing. L.A.M. was coordinator of the HELENA project, and together with M.G.-G. and S.D.H. formed the Core Management Group of the project. All authors have read and agreed to the published version of the manuscript.

Funding: Miguel Seral-Cortes, the corresponding author, has received funding from the Iberus Talent Pre-doctoral fellowships 2018, under the European Union's H2020 research and innovation programme under Marie Sklodowska-Curie grant agreement No 801586. Pilar De Miguel-Etayo was supported by ISCIII–CB15/00043 (Centro de Investigación Biomédica en Red de Fisiopatología de la Obesidad y Nutrición, CIBERObn). This work was part of the HELENA Study (http://www.helenastudy.com/). We gratefully acknowledge financial support of the European Community Sixth RTD Framework Programme (contract FOOD-CT-2005-007034).The data for this study were gathered under the auspices of the HELENA project, and further analysis was additionally supported by the Spanish Ministry of Economy and Competitiveness (Grants RYC-2010-05957 and RYC-2011-09011), the Instituto de Salud Carlos III, Centro de Investigación Biomédica en Red de Fisiopatología de la Obesidad y Nutrición (CIBERObn).

Acknowledgments: The autors acknowledge the Laboratoire d'Analyse Génomique Centre de Ressources Biologiques (LAG-CRB) BB-0033-00071 Institut Pasteur de Lille, F-59000 Lille, France for managing human samples. We are grateful for the support provided by school boards, headmasters, teachers, school staff and communities, and for the effort of all study nurses and our data managers.

Conflicts of Interest: The authors declare no conflict of interest. The funding sources had no role in the design, conduct, and analysis of the study or in the decision to submit the manuscript for publication.

References

1. Eckel, R.H.; Grundy, S.M.; Zimmet, P.Z. The metabolic syndrome. *Lancet* **2005**, *365*, S0140–S6736. [CrossRef]

2. Abarca-Gómez, L.; Abdeen, Z.A.; Hamid, Z.A.; Abu-Rmeileh, N.M.; Acosta-Cazares, B.; Acuin, C.; Adams, R.J.; Aekplakorn, W.; Afsana, K.; Aguilar-Salinas, C.A.; et al. Worldwide trends in body-mass index, underweight, overweight, and obesity from 1975 to 2016: A pooled analysis of 2416 population-based measurement studies in 128·9 million children, adolescents, and adults. *Lancet* **2017**, *390*, 2627–2642. [CrossRef]

3. Spinelli, A.; Buoncristiano, M.; Kovacs, V.; Yngve, A.; Spiroski, I.; Obreja, G.; Starc, G.; Pérez, N.; Rito, A.; Kunešová, M.; et al. Prevalence of Severe Obesity among Primary School Children in 21 European Countries. *Obes. Facts* **2019**, *12*. [CrossRef]

4. Steene-Johannessen, J.; Kolle, E.; Anderssen, S.; Andersen, L. Cardiovascular disease risk factors in a population-based sample of Norwegian children and adolescents. *Scand. J. Clin. Lab. Investig.* **2009**, *69*. [CrossRef]

5. Zimmet, P.A.; Alberti, K.G.M.; Kaufman, F.; Tajima, N.; Silink, M.; Arslanian, S.; Wong, G.; Bennett, P.; Shaw, J.; Caprio, S. The metabolic syndrome in children and adolescents—An IDF consensus report. *Pediatric Diabetes* **2007**, *8*. [CrossRef]

6. Rendo-Urteaga, T.; De Moraes, A.C.F.; Collese, T.S.; Manios, Y.; Hagströmer, M.; Sjöström, M.; Kafatos, A.; Widhalm, K.; Vanhelst, J.; Marcos, A.; et al. The combined effect of physical activity and sedentary behaviors on a clustered cardio-metabolic risk score: The Helena study. *Int. J. Cardiol.* **2015**, *186*. [CrossRef]

7. Cristi-Montero, C.; Chillón, P.; Labayen, I.; Casajus, J.A.; Gonzalez-Gross, M.; Vanhelst, J.; Manios, Y.; Moreno, L.A.; Ortega, F.B.; Ruiz, J.R. Cardiometabolic risk through an integrative classification combining physical activity and sedentary behavior in European adolescents: HELENA study. *J. Sport Health Sci.* **2019**, *8*. [CrossRef]

8.   Tognon, G.; Hebestreit, A.; Lanfer, A.; Moreno, L.; Pala, V.; Siani, A.; Tornaritis, M.; De Henauw, S.; Veidebaum, T.; Molnár, D.; et al. Mediterranean diet, overweight and body composition in children from eight European countries: Cross-sectional and prospective results from the IDEFICS study. *NMCD* **2014**, *24*. [CrossRef] [PubMed]

9.   Notario-Barandiaran, L.; Valera-Gran, D.; Gonzalez-Palacios, S.; Garcia-de-la-Hera, M.; Fernández-Barrés, S.; Pereda-Pereda, E.; Fernández-Somoano, A.; Guxens, M.; Iñiguez, C.; Romaguera, D.; et al. High adherence to a mediterranean diet at age 4 reduces overweight, obesity and abdominal obesity incidence in children at the age of 8. *Int. J. Obes.* **2020**, *44*. [CrossRef] [PubMed]

10.   Velázquez-López, L.; Santiago-Díaz, G.; Nava-Hernández, J.; Muñoz-Torres, A.V.; Medina-Bravo, P.; Torres-Tamayo, M. Mediterranean-style diet reduces metabolic syndrome components in obese children and adolescents with obesity. *BMC Pediatrics* **2014**, *14*. [CrossRef] [PubMed]

11.   Zhang, D.; Li, Z.; Wang, H.; Yang, M.; Liang, L.; Fu, J.-F.; Wang, C.-L.; Ling, J.; Zhang, Y.; Zhang, S.; et al. Interactions between obesity-related copy number variants and dietary behaviors in childhood obesity. *Nutrients* **2015**, *7*, 3054–3066. [CrossRef] [PubMed]

12.   Koochakpoor, G.; Hosseini-Esfahani, F.; Daneshpour, M.S.; Hosseini, S.A.; Mirmiran, P. Effect of interactions of polymorphisms in the Melanocortin-4 receptor gene with dietary factors on the risk of obesity and Type 2 diabetes: A systematic review. *Diabet. Med.* **2016**, *33*. [CrossRef] [PubMed]

13.   Lotta, L.; Mokrosiński, J.; Mendes de Oliveira, E.; Li, C.; Sharp, S.; Luan, J.; Brouwers, B.; Ayinampudi, V.; Bowker, N.; Kerrison, N.; et al. Human Gain-of-Function MC4R Variants Show Signaling Bias and Protect against Obesity. *Cell* **2019**, *177*. [CrossRef] [PubMed]

14.   Lv, D.; Zhang, D.-D.; Wang, H.; Zhang, Y.; Liang, L.; Fu, J.-F.; Xiong, F.; Liu, G.-L.; Gong, C.-X.; Luo, F.-H.; et al. Genetic variations in SEC16B, MC4R, MAP2K5 and KCTD15 were associated with childhood obesity and interacted with dietary behaviors in Chinese school-age population. *Gene* **2015**, *560*. [CrossRef] [PubMed]

15.   Janssens, A.; Aulchenko, Y.; Elefante, S.; Borsboom, G.; Steyerberg, E.; Van Duijn, C. Predictive testing for complex diseases using multiple genes: Fact or fiction? *Genet. Med.* **2006**, *8*. [CrossRef] [PubMed]

16.   Seyednasrollah, F.; Mäkelä, J.; Pitkänen, N.; Juonala, M.; Hutri-Kähönen, N.; Lehtimäki, T.; Viikari, J.; Kelly, T.; Li, C.; Bazzano, L.; et al. Prediction of Adulthood Obesity Using Genetic and Childhood Clinical Risk Factors in the Cardiovascular Risk in Young Finns Study. *Circ. Cardiovasc. Genet.* **2017**, *10*. [CrossRef] [PubMed]

17.   Viljakainen, H.; Dahlström, E.; Figueiredo, R.; Sandholm, N.; Rounge, T.; Weiderpass, E. Genetic risk score predicts risk for overweight and obesity in Finnish preadolescents. *Clin. Obes.* **2019**, *9*. [CrossRef]

18.   Moreno, L.A.; Gottrand, F.; Huybrechts, I.; Ruiz, J.R.; Gonzalez-Gross, M.; DeHenauw, S. Nutrition and lifestyle in european adolescents: The HELENA (Healthy Lifestyle in Europe by Nutrition in Adolescence) study. *Adv. Nutr.* **2014**, *5*, 615s–623s. [CrossRef]

19.   Moreno, L.A.; De Henauw, S.; Gonzalez-Gross, M.; Kersting, M.; Molnar, D.; Gottrand, F.; Barrios, L.; Sjostrom, M.; Manios, Y.; Gilbert, C.C.; et al. Design and implementation of the Healthy Lifestyle in Europe by Nutrition in Adolescence Cross-Sectional Study. *Int. J. Obes.* **2008**, *32*, S4–S11. [CrossRef]

20.   Beghin, L.; Castera, M.; Manios, Y.; Gilbert, C.C.; Kersting, M.; De Henauw, S.; Kafatos, A.; Gottrand, F.; Molnar, D.; Sjostrom, M.; et al. Quality assurance of ethical issues and regulatory aspects relating to good clinical practices in the HELENA Cross-Sectional Study. *Int. J. Obes.* **2008**, *32*, S12–S18. [CrossRef]

21.   Nagy, E.; Vicente-Rodriguez, G.; Manios, Y.; Beghin, L.; Iliescu, C.; Censi, L.; Dietrich, S.; Ortega, F.B.; De Vriendt, T.; Plada, M.; et al. Harmonization process and reliability assessment of anthropometric measurements in a multicenter study in adolescents. *Int. J. Obes.* **2008**, *32*, S58–S65. [CrossRef] [PubMed]

22.   Cole, T.J.; Bellizzi, M.C.; Flegal, K.M.; Dietz, W.H. Establishing a standard definition for child overweight and obesity worldwide: International survey. *BMJ* **2000**, *320*, 1240–1243. [CrossRef]

23.   Matthews, D.R.; Hosker, J.P.; Rudenski, A.S.; Naylor, B.A.; Treacher, D.F.; Turner, R.C. Homeostasis model assessment: Insulin resistance and beta-cell function from fasting plasma glucose and insulin concentrations in man. *Diabetologia* **1985**, *28*. [CrossRef]

24.   Stavnsbo, M.; Resaland, G.K.; Anderssen, S.A.; Steene-Johannessen, J.; Domazet, S.L.; Skrede, T.; Sardinha, L.B.; Kriemler, S.; Ekelund, U.; Andersen, L.B.; et al. Reference values for cardiometabolic risk scores in children and adolescents: Suggesting a common standard. *Atherosclerosis* **2018**, *278*. [CrossRef]

25. Diethelm, K.; Huybrechts, I.; Moreno, L.; De Henauw, S.; Manios, Y.; Beghin, L.; Gonzalez-Gross, M.; Le Donne, C.; Cuenca-Garcia, M.; Castillo, M.J.; et al. Nutrient intake of European adolescents: Results of the HELENA (Healthy Lifestyle in Europe by Nutrition in Adolescence) Study. *Public Health Nutr.* **2014**, *17*, 486–497. [CrossRef]

26. Vereecken, C.A.; Covents, M.; Matthys, C.; Maes, L. Young adolescents' nutrition assessment on computer (YANA-C). *Eur. J. Clin. Nutr.* **2005**, *59*, 658–667. [CrossRef]

27. Vereecken, C.; On behalf of the HELENA Study Group; Covents, M.; Sichert-Hellert, W.; Alvira, J.M.F.; Le Donne, C.; De Henauw, S.; De Vriendt, T.; Phillipp, M.K.; Béghin, L.; et al. Development and evaluation of a self-administered computerized 24-h dietary recall method for adolescents in Europe. *Int. J. Obes.* **2008**, *32*. [CrossRef]

28. Andersen, L.F.; Lioret, S.; Brants, H.; Kaic-Rak, A.; De Boer, E.J.; Amiano, P.; Trolle, E. Recommendations for a trans-European dietary assessment method in children between 4 and 14 years. *Eur. J. Clin. Nutr.* **2011**, *65*, S58–S64. [CrossRef]

29. Trichopoulou, A.; Costacou, T.; Bamia, C.; Trichopoulos, D. Adherence to a Mediterranean diet and survival in a Greek population. *N. Engl. J. Med.* **2003**, *348*, 2599–2608. [CrossRef]

30. Arenaza, L.; Huybrechts, I.; Ortega, F.B.; Ruiz, J.R.; De Henauw, S.; Manios, Y.; Marcos, A.; Julián, C.; Widhalm, K.; Bueno, G.; et al. Adherence to the Mediterranean diet in metabolically healthy and unhealthy overweight and obese European adolescents: The Helena study. *Eur. J. Nutr.* **2019**, *58*. [CrossRef]

31. Moreno, L.A.; Bel-Serrat, S.; Santaliestra-Pasias, A.; Bueno, G. Dairy products, yogurt consumption, and cardiometabolic risk in children and adolescents. *Nutr. Rev.* **2015**, *73*, 8–14. [CrossRef]

32. Trichopoulou, A. Traditional Mediterranean diet and longevity in the elderly: A review. *Public Health Nutr.* **2004**, *7*, 943–947. [CrossRef]

33. González-Gross, M.; On Behalf of the HELENA Study Group; Breidenassel, C.; Martínez, S.G.; Ferrari, M.; Beghin, L.; Spinneker, A.; Díaz, L.E.; Maiani, G.; Demailly, A.; et al. Sampling and processing of fresh blood samples within a European multicenter nutritional study: Evaluation of biomarker stability during transport and storage. *Int. J. Obes.* **2008**, *32*. [CrossRef]

34. Seral-Cortes, M.; Sabroso-Lasa, S.; De Miguel-Etayo, P.; Gonzalez-Gross, M.; Gesteiro, E.; Molina-Hidalgo, C.; De Henauw, S.; Gottrand, F.; Mavrogianni, C.; Manios, Y.; et al. Development of a Genetic Risk Score to predict the risk of overweight and obesity in European adolescents: The Helena study. *Under Rev.* **2020**.

35. Serra-Majem, L.; Román-Viñas, B.; Sanchez-Villegas, A.; Guasch-Ferré, M.; Corella, D.; La Vecchia, C. Benefits of the Mediterranean diet: Epidemiological and molecular aspects. *Mol. Asp. Med.* **2019**, *67*. [CrossRef]

36. Trichopoulou, A.; Orfanos, P.; Norat, T.; Bueno-de-Mesquita, B.; Ocké, M.; Peeters, P.; van der Schouw, Y.; Boeing, H.; Hoffmann, K.; Boffetta, P.; et al. Modified Mediterranean diet and survival: EPIC-elderly prospective cohort study. *BMJ (Clin. Res. Ed.)* **2005**, *330*. [CrossRef]

37. Esposito, K.; Maiorino, M.I.; Ceriello, A.; Giugliano, D. Prevention and control of type 2 diabetes by Mediterranean diet: A systematic review. *Diabetes Res. Clin. Pract.* **2010**, *89*. [CrossRef] [PubMed]

38. Corella, D.; Ordovas, J.M. How does the Mediterranean diet promote cardiovascular health? Current progress toward molecular mechanisms: Gene-diet interactions at the genomic, transcriptomic, and epigenomic levels provide novel insights into new mechanisms. *BioEssays* **2014**, *36*. [CrossRef]

39. Ortega-Azorín, C.; Sorlí, J.V.; Asensio, E.M.; Coltell, O.; Martínez-González, M.Á.; Salas-Salvadó, J.; Covas, M.I.; Arós, F.; Lapetra, J.; Serra-Majem, L.; et al. Associations of the FTO rs9939609 and the MC4R rs17782313 polymorphisms with type 2 diabetes are modulated by diet, being higher when adherence to the Mediterranean diet pattern is low. *Cardiovasc. Diabetol.* **2012**, *11*. [CrossRef]

40. Junyent, M.; Parnell, L.D.; Lai, C.-Q.; Arnett, D.K.; Tsai, M.Y.; Kabagambe, E.K.; Straka, R.J.; Province, M.; An, P.; Smith, C.E.; et al. ADAM17_i33708A>G polymorphism interacts with dietary n-6 polyunsaturated fatty acids to modulate obesity risk in the Genetics of Lipid Lowering Drugs and Diet Network study. *NMCD* **2010**, *20*. [CrossRef]

41. Real, J.M.F.; Corella, D.; Goumidi, L.; Mercader, J.M.; Valdés, S.; Martínez, G.R.; Ortega, F.; Martínez-Larrad, M.T.; Gómez-Zumaquero, J.M.; Salas-Salvado, J.; et al. Thyroid hormone receptor alpha gene variants increase the risk of developing obesity and show gene-diet interactions. *Int. J. Obes.* **2013**, *37*. [CrossRef]

42. Smith, C.E.; Arnett, D.K.; Corella, D.; Tsai, M.Y.; Lai, C.-Q.; Parnell, L.D.; Lee, Y.-C.; Ordovas, J.M. Perilipin polymorphism interacts with saturated fat and carbohydrates to modulate insulin resistance. *NMCD* **2012**, *22*. [CrossRef] [PubMed]

43. Svetkey, L.P.; Moore, T.J.; Simons-Morton, D.G.; Appel, L.J.; Bray, G.A.; Sacks, F.M.; Ard, J.D.; Mortensen, R.M.; Mitchell, S.R.; Conlin, P.R.; et al. Angiotensinogen genotype and blood pressure response in the Dietary Approaches to Stop Hypertension (DASH) study. *J. Hypertens.* **2001**, *19*. [CrossRef] [PubMed]

44. Ordovas, J.M.; Corella, D.; Cupples, L.A.; Demissie, S.; Kelleher, A.; Coltell, O.; Wilson, P.W.F.; Schaefer, E.J.; Tucker, K. Polyunsaturated fatty acids modulate the effects of the APOA1 G-A polymorphism on HDL-cholesterol concentrations in a sex-specific manner: The Framingham Study. *Am. J. Clin. Nutr.* **2002**, *75*. [CrossRef] [PubMed]

45. Junyent, M.; Parnell, L.D.; Lai, C.-Q.; Lee, Y.-C.; Smith, E.C.; Arnett, N.K.; Tsai, M.Y.; Kabagambe, E.K.; Straka, R.J.; Province, M.; et al. Novel variants at KCTD10, MVK, and MMAB genes interact with dietary carbohydrates to modulate HDL-cholesterol concentrations in the Genetics of Lipid Lowering Drugs and Diet Network Study. *Am. J. Clin. Nutr.* **2009**, *90*. [CrossRef] [PubMed]

46. Garcia-Rios, A.; Alcalá-Diaz, J.F.; Delgado-Lista, J.; Delgado-Lista, J.; Marin, C.; León-Acuña, A.; Camargo, A.; Rodriguez-Cantalejo, F.; Blanco-Rojo, R.; Quintana-Navarro, G.; et al. Beneficial effect of CETP gene polymorphism in combination with a Mediterranean diet influencing lipid metabolism in metabolic syndrome patients: CORDIOPREV study. *Clin. Nutr.* **2018**, *37*. [CrossRef]

47. Gomez-Delgado, F.; Alcala-Diaz, J.F.; Garcia-Rios, A.; Delgado-Lista, J.; Ortiz-Morales, A.; Rangel-Zuñiga, O.; Tinahones, F.J.; Gonzalez-Guardia, L.; Malagon, M.M.; Bellido-Muñoz, E.; et al. Polymorphism at the TNF-alpha gene interacts with Mediterranean diet to influence triglyceride metabolism and inflammation status in metabolic syndrome patients: From the CORDIOPREV clinical trial. *Mol. Nutr. Food Res.* **2014**, *58*. [CrossRef]

48. Tai, E.S.; Corella, D.; Demissie, S.; Cupples, L.A.; Coltell, O.; Schaefer, E.J.; Tucker, K.L.; Ordovas, J.M. Polyunsaturated fatty acids interact with the PPARA-L162V polymorphism to affect plasma triglyceride and apolipoprotein C-III concentrations in the Framingham Heart Study. *J. Nutr.* **2005**, *135*. [CrossRef]

49. Locke, A.E.; Kahali, B.; Berndt, S.I.; Justice, A.E.; Pers, T.H.; Day, F.R.; Powell, C.; Vedantam, S.; Buchkovich, M.L.; Yang, J.; et al. Genetic studies of body mass index yield new insights for obesity biology. *Nature* **2015**, *518*, 197–206. [CrossRef]

50. Morrison, A.C.; Bare, L.A.; Chambless, L.E.; Ellis, S.G.; Malloy, M.; Kane, J.P.; Pankow, J.S.; Devlin, J.J.; Willerson, J.T.; Boerwinkle, E. Prediction of coronary heart disease risk using a genetic risk score: The Atherosclerosis Risk in Communities Study. *Am. J. Epidemiol.* **2007**, *166*. [CrossRef]

51. Steele, R.M.; Brage, S.; Corder, K.; Wareham, N.J.; Ekelund, U. Physical activity, cardiorespiratory fitness, and the metabolic syndrome in youth. *J. Appl. Physiol.* **2008**, *105*. [CrossRef] [PubMed]

52. Magnussen, C.G.; Koskinen, J.; Chen, W.; Thomson, R.; Schmidt, M.D.; Srinivasan, S.R.; Kivimäki, M.; Mattsson, N.; Kähönen, M.; Laitinen, T.; et al. Pediatric metabolic syndrome predicts adulthood metabolic syndrome, subclinical atherosclerosis, and type 2 diabetes mellitus but is no better than body mass index alone: The Bogalusa Heart Study and the Cardiovascular Risk in Young Finns Study. *Circulation* **2010**, *122*. [CrossRef] [PubMed]

53. Morrison, J.A.; Friedman, L.A.; Wang, P.; Glueck, C.J. Metabolic syndrome in childhood predicts adult metabolic syndrome and type 2 diabetes mellitus 25 to 30 years later. *J. Pediatrics* **2008**, *152*. [CrossRef] [PubMed]

54. Sofi, F.; Cesari, F.; Abbate, R.; Gensini, G.F.; Casini, A. Adherence to Mediterranean diet and health status: Meta-analysis. *BMJ* **2008**, *337*. [CrossRef]

Publisher's Note: MDPI stays neutral with regard to jurisdictional claims in published maps and institutional affiliations.

MDPI

St. Alban-Anlage 66

4052 Basel

Switzerland

Tel. +41 61 683 77 34

Fax +41 61 302 89 18

www.mdpi.com

Nutrients Editorial Office

E-mail: nutrients@mdpi.com

www.mdpi.com/journal/nutrients